Ultrasonic Scattering in Biological Tissues

Edited by
K. Kirk Shung
Gary A. Thieme

CRC Press

Boca Raton   Ann Arbor   London   Tokyo

Library of Congress Cataloging-in-Publication Data

Ultrasonic scattering in biological tissues / edited by K. Shung, Gary
  A. Thieme.
     p. cm.
  Includes bibliographical references and index.
  ISBN 0-8493-6568-6 (alk. paper)
    1. Ultrasonic waves--Physiological effect. 2. Ultrasonic in
medicine. 3. Diagnosis, Ultrasonic. I. Shung, K. Kirk.
II. Thieme, Gary A.
    [DNLM: 1. Scattering, Radiation. 2. Ultrasonics.
3. Ultrasonography. WB 289 U4475]
QP82.2.U37U475 1992
616.07'543--dc20
DNLM/DLC
for Library of Congress   92-25433
                                            CIP

PREFACE

Although ultrasonic imaging has been in existence for more than 30 years, it is still advancing at a very rapid rate and new diagnostic applications are constantly emerging. It is second only to conventional X-ray in terms of number of clinical procedures performed. The immense popularity of ultrasound lies in its many advantages over other imaging modalities. Ultrasonic imaging is noninvasive, capable of providing both anatomical and blood flow information in real time, relatively inexpensive, and portable. Despite the fact that it is widely used and much diagnostic information can be retrieved from ultrasonic images, the genesis of ultrasonic speckle or texture exhibited by a tissue is by no means well-understood. Since the ultrasonic image is formed from the echoes backscattered from tissues, a systemic and thorough understanding of the phenomenon of ultrasonic scattering in biological tissues is crucial for the proper and accurate interpretation of the images and for the further development of the modality. It is for this reason that ultrasonic scattering processes have been under intensive investigation for many years. This book is a compilation of the most important accomplishments that have been made in these years.

This book is intended to give an overview of the field and consists of 14 chapters written by authorities detailing theoretical and experimental aspects of the ultrasonic scattering phenomenon in biological tissues. Introductory materials are presented in the first three chapters. Theoretical treatments are discussed in Chapters 4 and 5. Experimental approaches are described in Chapters 6 to 8. *In vitro* results on selective tissues are given in Chapter 9, whereas *in vivo* results on various tissues are discussed in Chapters 10 to 13. The last chapter, Chapter 14, reports current status of quantitative backscatter imaging.

It is hoped that this book will serve as a handy reference for the workers in the field and will be able to provide readers who are interested in learning more about the field a clear perspective.

THE EDITORS

K. Kirk Shung, Ph.D., is Professor of Bioengineering in the Bioengineering Program, Pennsylvania State University, University Park, Pennsylvania.

Dr. Shung received his B.S. degree in Electrical Engineering from National Cheng-Kung University in Taiwan in 1968. He obtained his M.S. and Ph.D. in Electrical Engineering from the University of Missouri in 1970 and University of Washington in 1975. Following one year of postdoctoral work at Providence Medical Center in Seattle, Washington, he was appointed as a research engineer at the same institution while also holding an appointment as a research scientist at the Institute of Applied Physiology and Medicine. In 1979, he moved to the Bioengineering Program at Pennsylvania State University as an Assistant Professor. He became an Associate Professor of Bioengineering in 1985 and a Professor in 1989.

Dr. Shung is a fellow of the American Institute of Ultrasound in Medicine and the Acoustical Society of America and a senior member of Engineering in Medicine and Biology Society and the Ultrasonics, Ferroelectrics, and Frequency Control Society of the Institute of Electrical and Electronic Engineers. He is a founding fellow of the recently formed American Institute of Medical and Biological Engineering. He served as a member on the National Institutes of Health Diagnostic Radiology Study Section from 1985 to 1989. He was the recipient of the IEEE Engineering in Medicine and Biology Society early career achievement award in 1985.

Dr. Shung has been the principal investigator of research grants from the National Institutes of Health, National Science Foundation, American Heart Association, and private industry. He has published more than 60 papers and is the author of a textbook, *Principles of Medical Imaging,* published by Academic Press. His research interest is in ultrasonic imaging and tissue characterization, and ultrasonic contrast blood flow measurements.

Gary A. Thieme, M.D., is Chief of the Diagnostic Ultrasound Division of the Department of Radiology, The Milton S. Hershey Medical Center, Pennsylvania State University, Hershey, Pennsylvania.

Dr. Thieme graduated from Purdue University in 1972 with a B.S. degree in Physics. He obtained his M.D. degree in 1977 from the University of Colorado School of Medicine and served his internship in surgery at the University of Utah Health Sciences Center in Salt Lake City, Utah. By 1982 he had completed resident training in Diagnostic Radiology and subspecialty fellow training in clinical and research ultrasound at the University of Colorado Health Sciences Center.

Dr. Thieme is a member of the American Institute of Ultrasound in Medicine, Radiologic Society of North America, and American Roentgen Ray Society. In addition to his clinical diagnostic responsibilities, he teaches ultrasound physics, instrumentation, and technique to residents, sonographers, and bioengineering students.

CONTRIBUTORS

Jeffrey P. Astheimer, Ph.D.
66 Sibley Road
Honeoye Falls, New York 14472

E. J. Boote, Ph.D.
Assistant Professor
Department of Radiology
University of Missouri
Columbia, Missouri

Richard A. Bowerman, M.D.
Associate Professor
Department of Radiology
University of Michigan Hospitals
Ann Arbor, Michigan

David G. Brown, Ph.D.
Chief Scientist
Division of Electronics and
  Computer Science
Center for Devices and Radiological
  Health
United States Food and Drug
  Administration
Rockville, Maryland

Paul L. Carson, Ph.D.
Professor
Department of Radiology
University of Michigan Medical
  Center
Ann Arbor, Michigan

Richard S. C. Cobbold, Ph.D.
Professor
Institute of Biomedical Engineering
University of Toronto
Toronto, Ontario, Canada

Ernest J. Feleppa, Ph.D.
Manager
Biomedical Engineering Laboratory
Riverside Research Institute
New York, New York

Brian S. Garra, M.D.
Assistant Professor
Director of Ultrasound
Department of Radiology
Georgetown University Medical
  Center
Washington, D.C.

Michael F. Insana, Ph.D.
Associate Professor
Department of Diagnostic
  Radiology
University of Kansas Medical
  Center
Kansas City, Kansas

Frederic L. Lizzi, Eng.Sc.D.
Research Director
Biomedical Engineering
Riverside Research Institute
New York, New York

Z. F. Lu, M.S.
Department of Medical Physics
University of Wisconsin
Madison, Wisconsin

Ernest L. Madsen, Ph.D.
Associate Professor
Department of Medical Physics
University of Wisconsin
Madison, Wisconsin

Charles E. Meyer, Ph.D.
Associate Professor
Department of Radiology
University of Michigan Hospitals
Ann Arbor, Michigan

James G. Miller, Ph.D.
Professor
Department of Physics
Washington University
St. Louis, Missouri

Larry Y. L. Mo, Ph.D.
Senior Physicist
Applied Science Laboratory
General Electric Medical Systems
Milwaukee, Wisconsin

Julio E. Pérez, M.D.
Associate Professor
Department of Medicine
Washington University School of
  Medicine
St. Louis, Missouri

John M. Reid, Ph.D.
Calhoun Chair Professor
Biomedical Engineering and
  Science Institute
Department of Electrical and
  Computer Engineering
Drexel University
Philadelphia, Pennsylvania

K. Kirk Shung, Ph.D.
Professor
Department of Bioengineering
The Pennsylvania State University
University Park, Pennsylvania

Gary A. Thieme, M.D.
Assistant Professor
Department of Radiology
The Milton S. Hershey Medical
  Center
The Pennsylvania State University
Hershey, Pennsylvania

Robert C. Waag, Ph.D.
Professor
Departments of Electrical
  Engineering and Radiology
University of Rochester
Rochester, New York

Samuel A. Wickline, M.D.
Associate Professor of Medicine
Adjunct Assistant Professor of
  Physics
Department of Cardiology
Jewish Hospital at Washington
  University
St. Louis, Missouri

L. X. Yao, Ph.D.
Siemens Quantum, Inc.
Issaquah, WA

J. A. Zagzebski, Ph.D.
Professor
Department of Medical Physics
University of Wisconsin
Madison, Wisconsin

TABLE OF CONTENTS

Chapter 1

INTRODUCTION

K. Kirk Shung

TABLE OF CONTENTS

0-8493-6568-6/93/$0.00 + $.50
© 1993 by CRC Press, Inc.

1

I. HISTORY

The potential of ultrasound as a diagnostic tool was recognized as early as the late 1940s although it was used for industrial material evaluation much earlier.[1] Among the pioneers in using ultrasound for medical diagnosis are Drs. K. Tanaka, T. Wagai, Y. Kikuchi, S. Satomura, and Y. Nimura in Japan; Dr. K. T. Dussik in Europe, and Drs. G. D. Ludwig, R. H. Bolt, T. Heuter, J. J. Wild, J. M. Reid, D. Howry, and W. J. Fry in the U.S. Among the most notable in these earlier developments related to the subject addressed by this book are Satomura and Nimura's work on Doppler ultrasound[2] as well as Wild and Reid's work which demonstrated that tissue pathology might be characterized by echoes reflected or scattered back from internal structures of the tissue.[3] When using a prototype Doppler instrument to detect heart valvular motion, Satomura found that there was a noisy signal present in addition to the echoes from the heart valves and walls and postulated that the noise was due to turbulence of blood flow in the cardiac chambers. Although later findings by Kato et al.,[4] which were corroborated by other investigations,[5-7] showed that this noisy signal was related to red blood cells, more recent studies,[8] the details of which are discussed in Chapters 5 and 9 of this book, indicated that the contribution to ultrasonic signals backscattered from blood in the heart chambers from flow turbulence may still be important after all. Wild and Reid, by demonstrating that cancerous stomach wall could be differentiated from normal tissue based upon the echo patterns from these tissues, laid the foundation for much of the tissue characterization research that blossomed in the 1970s and is still today an area of active research of paramount interest to many scientists in the world.[9-15]

Ultrasound imaging did not become a well-accepted diagnostic imaging modality until the early 1970s when gray-scale ultrasonography was introduced.[16] The primary reason is that, in gray-scale ultrasonography, the returned echoes from tissues are nonlinearly compressed to reduce the dynamic range of the echo amplitude so that both specularly reflected echoes from tissue boundaries and diffusely scattered echoes from tissue parenchyma can be included in an image since a typical display device like a video monitor cannot display information with a dynamic range much greater than 20 dB. Prior to that, only specularly reflected echoes were used to form an ultrasound image whereas the smaller (typically 40 to 50 dB lower than specular echoes) diffusely scattered echoes were completely discarded. Clearly, the fact that the smaller scattered echoes from tissue internal structures, which carried much diagnostically valuable information about the state of the tissue, were utilized in the formation of an image might have contributed to the success of gray-scale ultrasound. It is therefore of no wonder that a flurry of activities were launched in the 1970s to study physical mechanisms of ultrasound and tissue interaction including ultrasonic scattering in biological tissues, hoping to better understand the genesis of ultrasonic images.

Since echoes from tissue parenchyma are used in gray-scale ultrasound, despite the fact that the returned echoes are distorted by the intervening tissues

The acoustic velocity and impedance for a few common materials and biological tissues are listed in Table 1. For a more comprehensive listing, see References 57 and 58.

The intensity of a wave is defined as the average power carried by the wave per unit area normal to the direction of propagation. For ultrasonic propagation, the instantaneous intensity, $i(t)$, is related to the instantaneous medium velocity and pressure by the following relationship:

$$i(t) = p_z(t)u_z(t) \tag{5}$$

For sinusoidal propagation, the average intensity, I, can be found by averaging i(t) over a cycle.

$$I = \tfrac{1}{2} p_0 u_0 \tag{6}$$

where p_0 and u_0 represent peak values. However, most current ultrasound imaging techniques use pulsed ultrasound (i.e., only a few cycles of the oscillation are used), and the spatial intensity of the beam produced by an ultrasonic transducer generally is not uniform.

Spatial average intensity is defined as the average intensity over the ultrasound beam. The temporal average intensity is defined as the average intensity over a pulse repetition period, which is the temporal peak intensity multiplied by the duty factor (the ratio of pulse duration to pulse repetition period). Whenever biological effects of ultrasound are considered, it is essential to state or understand which definition of intensity is being used. Spatial average temporal average intensity I_{SATA} or spatial peak temporal average I_{SPTA} intensity are frequently used in the literature.[59]

C. REFLECTION AND REFRACTION

When a wave impinges on an interface between two media, it is reflected and refracted as shown in Figure 2 where i, r, and t refer to incident, reflected, and transmitted waves, respectively. Using the boundary conditions that the pressure and particle velocity have to be continuous across the boundary, it can be shown that as in optics, Snell's law applies:

$$\theta_i = \theta_r \qquad \text{and} \qquad \sin \theta_i / \sin \theta_t = c_1/c_2$$

$$p_r/p_i = \frac{Z_2 \cos \theta_i - Z_1 \cos \theta_t}{Z_2 \cos \theta_i + Z_1 \cos \theta_t} \tag{7}$$

and

$$p_t/p_i = \frac{2Z_2 \cos \theta_i}{Z_2 \cos \theta_i + Z_1 \cos \theta_t} \tag{8}$$

TABLE 1
Acoustic Properties of Biological Tissues and Pertinent Materials

Material	Acoustic velocity[a] (10^5 cm/s)	Impedance (10^6 kg/m^2/s)	Attenuation (neper/ cm at 1 MHz)	Backscatter (cm^{-1} steradian^{-1} at 5 MHz)
Water	1.4839	1.48	2.5×10^{-4}	—
Aluminum	6.42	17.0	2.1×10^{-3}	—
Air (stp)	0.343	4.5×10^{-4}	1.38	—
Plexiglass	2.67	3.2	0.23	—
Blood	1.55	1.61	0.02	1.4×10^{-5}
Myocardium (perpendicular to fibers)	1.55	1.62	0.10	7×10^{-4}
Liver	1.57	1.65	0.08	2×10^{-3}
Kidney	1.56	1.62	0.11	1.1×10^{-3}
Aorta	1.57	1.69	—	—
Vena cava	1.56	1.65	—	—
Bone[b]	3.36[c]	6.0	1.3	—

a   Room temperature (20–25°C).
b   Skull bone.
c   Longitudinal velocity.

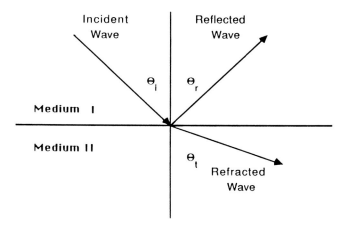

FIGURE 2. Reflection and refraction of plane wave at boundary between medium I with acoustic impedance Z_1 and medium II with acoustic impedance Z_2.

where p_r/p_i and p_t/p_i are, respectively, the pressure reflectivity and the pressure transmittivity of the interface.

As an acoustic wave propagates through inhomogeneous media such as biological tissue, part of its energy is lost due to absorption and scattering which will be discussed below and is known to be a function of the dimension and elastic properties of the inhomogeneities and frequency of the wave, and part of its energy is lost due to specular reflection at the boundary of two adjacent tissue layers. Gray-scale ultrasonic images are generated from the specularly reflected echoes as well as the diffusely scattered echoes. Therefore, changes in elastic properties of the tissue may be detectable in the acoustic image. This feature has been the principal rationale behind conventional ultrasonic imaging techniques.[16] Since both scattering and reflection depend on the elastic properties of the tissues, and the properties of connective tissues such as collagen and elastin are known to be very different from other tissues, it is reasonable to speculate that echographic visuability of tissues is affected by their connective tissue content.[28]

D. ATTENUATION, SCATTERING, AND ABSORPTION

When a wave propagates through a medium, its energy is reduced as a function of distance. The energy may be diverted by reflection or scattering, or absorbed by the medium and converted to heat. The pressure of a plane wave propagating in the z direction can be expressed as:

$$p_z = p_{z0}e^{-\beta z} \tag{9}$$

where p_{z0} is the pressure at $z = 0$ and β is the pressure attenuation coefficient in nepers per centimeter. Attenuation coefficients of a number of biological tissues and relevant materials can also be found in Table 1.

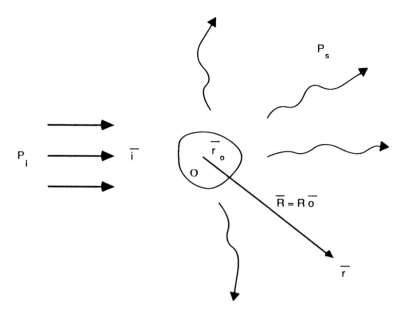

FIGURE 3.   Scattering of plane wave incident upon object, O, located at $\bar{r}_o$.

Scattering of a wave by a single scatterer is frequently described in the literature by the term "scattering cross-section". This term is defined as the total power scattered by the particle per unit incident intensity. Volumetric scattering cross-section, or scattering cross-section per unit volume, is used to describe the bulk scattering properties of a medium.

If a plane incident wave is assumed, the incident pressure p_i at a position $\bar{r}_0$ shown in Figure 3 can be expressed as

$$p_i(\bar{r}_0) = p_0 e^{-jk \cdot \bar{r}_0} \tag{10}$$

where $\bar{k} = k\bar{i}$ is the wave number vector. The scattered pressure at $\bar{r}$ is then:

$$p_s(\bar{r}) = f(\bar{o}, \bar{i}) \frac{e^{jkR}}{R} p_i(\bar{r}_0) \tag{11}$$

where $\bar{i}$ and $\bar{o}$ are directions of incidence and observation assuming that the observation point is in the far field of the scatterer and $R = |\bar{r} - \bar{r}_0|$. In Equation 11, $f(\bar{o}, \bar{i})$ is called the scattering amplitude function, which describes the scattering properties of the object. Since the incident intensity is given by:

$$I_i = \frac{1}{2} \frac{|p_i|^2}{Z} \tag{12}$$

we have from the scattered intensity:

$$I_s = \frac{1}{2} \frac{|p_s|^2}{Z} \tag{13}$$

$$I_s = (|f(\bar{o}, \bar{i})|^2/R^2)I_i \tag{14}$$

The differential scattering cross-section, $\sigma_d(\bar{i}, \bar{o})$, which is defined as the power scattered in $\bar{o}$ direction per solid angle per unit incident intensity can be written:

$$\boldsymbol{\sigma}_d(\bar{o}, \bar{i}) = |f(\bar{o}, \bar{i})|^2 \tag{15}$$

The scattering cross-section, σ_s, is then given by:

$$\boldsymbol{\sigma}_s = \int_{4\pi} \boldsymbol{\sigma}_d \, d\Omega \int_{4\pi} |f(\bar{o}, \bar{i})|^2 \, d\Omega \tag{16}$$

where $d\Omega$ is the differential solid angle.

Similarly, we can define the absorption cross-section of an object, σ_a, as the total power per unit incident intensity absorbed by that object. The energy loss due to the presence of the object is

$$2\beta = \boldsymbol{\sigma}_a + \boldsymbol{\sigma}_s$$

where 2β is the intensity attenuation coefficient. If a number of objects are present, then the intensity attenuation coefficient is given by:

$$2\beta = n(\boldsymbol{\sigma}_a + \boldsymbol{\sigma}_s) \tag{17}$$

where n is the particle concentration. This relation is valid only if n is small. As n increases, multiple scattering occurs and there may be particle-particle interactions. Under these circumstances, Equation 17 is no longer valid. More detail on this subject can be found in Chapter 5.

Solving exactly for the scattering cross-section of an object of arbitrary shape is not an easy task. However, a number of approximations exist that can simplify the problem considerably. One is the Born approximation,[60] which assumes that the wave inside the object is the same as the incident wave. This is generally a good assumption if the dimension of the scattering

object is much smaller than the wavelength or if the acoustic properties of
the scatterer are similar to the surrounding medium. By applying this approximation and using the wave equation, the differential scattering cross-
section at an angle θ (angle between $\bar{i}$ and $\bar{o}$) and scattering cross-section for
an arbitrary-shaped object whose dimension is much smaller than the wave-
length can be shown to be[24]

$$\sigma_s = \frac{4\pi k^4 a^6}{9} \left[\left| \frac{G_e - G}{G} \right|^2 + \frac{1}{3} \left| \frac{3\rho_e - 3\rho}{2\rho_e + \rho} \right|^2 \right] \tag{18}$$

$$\sigma_d(\theta) = \frac{k^4 a^6}{9} \left| \left(\frac{G_e - G}{G} + \frac{3\rho_e - 3\rho}{2\rho_e + \rho} \cos \theta \right) \right|^2 \tag{19}$$

where a is the dimension of the particle, G_e and G are the compressibilities
of the particle and the surrounding medium, and ρ_e and ρ are the corresponding
densities. Equation 18 can be applied to calculate the ultrasonic scattering
properties of red blood cells, since their diameter (~3 μm) is much smaller
than the wavelength of ultrasound in the frequency range of 1 to 20 MHz
(~1500 μm). Using these values[5,7,8] $G_e = 34.1 \times 10^{-12}$ cm^2/dyne and ρ_e
$= 1.092$ g/cm^3 for the cells, and $G = 40.9 \times 10^{-12}$ cm^2/dyne and $\rho =$
1.021 g/cm^3 for the plasma; then $\sigma_s = 1.1 \times 10^{-12}$ cm^2 at 10 MHz, a
relatively small scattering cross-section. To obtain backscattering cross-section, set θ to be 180°. Volumetric backscattering cross-section or backscattering cross-section per unit volume of scatterers is sometimes called backscattering coefficient.[17]

For small or tenuous cylindrical scatterers, it has been found that the
scattering intensity is proportional to the 3rd power of frequency and 4th
power of radius of the cylinder.[24] For larger and nontenuous scatterers, solutions can only be found for scatterers with simple geometry.[60,61]

The absorption mechanisms in biological tissues are quite complex and
have been assumed to arise from (1) classical absorption due to viscosity,
and (2) a relaxation phenomenon. Both mechanisms depend on the wave
frequency. Classical absorption describes the frictional loss associated with
a viscous medium. It has been shown that in media such as water or air,
where classical absorption dominates, the absorption is proportional to f^2.
However, in most biological materials this is not the case. It has been postulated that absorption in biological tissues can be best described by a relaxation mechanism. When a molecule is forced to a new position and the force
is then released, a finite time is required for the molecule to return to its

neutral position. This time is called the relaxation time of the molecule. For a medium composed of only one type of molecule, the relaxation time of the molecule is also the relaxation time of the medium. If the relaxation time is short compared to the period of the wave, the relaxation effect will be small. However, if the relaxation time is comparable to the wave period, the molecule may not be able to return completely before a second compression wave arrives. When this phenomenon occurs, the compression wave is moving in one direction while the molecule is moving in the opposite direction. More energy is thus required to reverse the direction of the molecule. On the other hand, if the frequency is increased high enough so that the molecules simply cannot follow the wave motion, the relaxation effect does not occur. Maximum absorption occurs when the relaxation motion of the molecules is completely out of synchronization with the wave motion. Therefore, a relaxation process is characterized by an absorption peak at one frequency and negligible absorption elsewhere. Mathematically, ultrasonic absorption in biological tissues, β_a, can be represented by:

$$\frac{\beta_a}{f^2} = A + \sum_i \frac{B_i}{(f/f_{ri})^2} \tag{20}$$

where A and B_is are constants representing the magnitudes of classical absorption and absorption due to relaxation, respectively, f_{ri} = relaxation frequency of the ith molecule = $1/T_i$, and T_i = relaxation time of the ith molecule. Since a biological tissue has many different types of molecules, the many relaxation processes overlap, resulting in an approximately linear function of frequency in the ultrasound range which was experimentally observed for many soft tissues.[57,58] In summary, the attenuation of ultrasound in biological tissue can be attributed to two major mechanisms: (1) scattering and (2) absorption. The relative importance of these mechanisms in determining tissue attenuation is a controversial issue and remains to be exploited, although several recent reports indicate that the contribution to attenuation due to scattering is small, a few percent at most.[62,63]

III. SPECKLE

A. ORIGIN OF SPECKLE
Ultrasonic textural pattern or speckle exhibited by biological tissues in B-mode images which much resembles the laser speckle has been used routinely for clinical diagnosis. The speckle patterns for different tissues are different. Tissue pathology which causes tissue anatomical structures to change also may result in a change in its speckle appearance in a B-mode image. Although the origin of these patterns is still not well understood, it is believed to be the result of a wave interference phenomenon of the echoes arriving at the transducer scattered by structures in the tissue and to be related to both

tissue properties and the system characteristics of the imager.[64-71] On the other end, various schemes have been implemented to smooth out the pattern in attempting to improve the ability of the scanner to resolve small objects. Present knowledge appears to indicate that tissue speckle pattern is determined both by the microstructure of the tissue and by the characteristics of the imaging device such as frequency and bandwidth of the ultrasonic pulse, frequency response of the transducer and electronic system, and acoustic properties of the tissues.[66] An analysis of the texture characteristics both from computer simulation and experiments clearly demonstrates that the echoes seem to resemble laser speckle resulting from a random interference phenomenon; but embedded in the random signal there is a coherent component.[67,68] Moreover, there is a fundamental difference between laser speckle and ultrasonic speckle in that ultrasonic signals used in imaging systems are only partially coherent unlike coherent laser speckle so that the ultrasonic speckle patterns do contain information related to tissue structure.[67-69]

B. SPECKLE REDUCTION

Since the presence of speckle may obscure smaller structures, thus degrading the spatial resolution of an ultrasonic imaging device, various schemes have been used to reduce the speckle appearance of an image. Most notable are frequency compounding techniques and spatial compounding techniques, both of which are based on the principle of incoherent averaging of images with different speckle patterns.[72-75] In frequency compounding, images collected at different frequencies are averaged whereas in spatial compounding, images collected at slightly different spatial locations are averaged. Understandably, these speckle reduction algorithms would lengthen the image acquisition time and also slightly affect the spatial resolution. Consequently, they have not been widely implemented on commercial scanners.

REFERENCES

1. **Goldberg, B. B. and Kimmelman, B. A.,** Medical Diagnostic Ultrasound: A Retrospective on its 40th Anniversary, Eastman Kodak, Rochester, NY, 1988.
2. **Satomura, S.,** A study on examining the heart with ultrasonics. I. Principle. II. Instrument, *Jpn. Circ. J.,* 20, 227, 1956.
3. **Wild, J. J. and Reid, J. M.,** Further pilot echographic studies on the histologic structure of tumors of the living intact human breast, *Am. J. Pathol.,* 28, 893, 1952.
4. **Kato, K., Kido, Y., Motomiya, M., Kaneko, Z., and Kotani, H.,** On the mechanisms of generation of detected sound in ultrasonic flowmeter, *Mem. Inst. Sci. Ind. Res. Osaka Univ.,* 19, 51, 1962.
5. **Shung, K. K., Sigelmann, R. A., and Reid, J. M.,** Scattering of ultrasound by blood, *IEEE Trans. Biomed. Eng.,* BME-23, 460, 1976.
6. **Borders, S. H., Fronek, A., Kemper, W. S., and Franklin, D.,** Ultrasonic energy backscattered from blood: an experimental determination of the variation of sound energy with hematocrit, *Ann. Biomed. Eng.,* 6, 83, 1978.

7. **Yuan, Y. W. and Shung, K. K.**, Ultrasonic backscatter from flowing whole blood. I. Dependence on shear rate and hematocrit, *J. Acoust. Soc. Am.*, 84, 52, 1988.

8. **Shung, K. K., Yuan, Y. W., Fei, D. Y., and Tarbell, J. M.**, Effect of flow disturbance on ultrasonic backscatter from blood, *J. Acoust. Soc. Am.*, 75, 1265, 1984.

9. **Linzer, M.**, Ultrasonic Tissue Characterization I, Spec. Publ. 453, National Bureau of Standards, Gaithersburg, MD, 1976.

10. **Linzer, M.**, Ultrasonic Tissue Characterization II, Spec. Publ. 525, National Bureau of Standards, Gaithersburg, MD, 1979.

11. **Waag, R. C.**, A review of tissue characterization from ultrasonic scattering, *IEEE Trans. Biomed. Eng.*, BME-31, 884, 1984.

12. **Shung, K. K.**, Ultrasonic characterization of biological tissues, *ASME J. Biomech. Eng.*, 107, 309, 1985.

13. **Greenleaf, J. F.**, *Tissue Characterization with Ultrasound*, CRC Press, Boca Raton, FL, 1986.

14. **Taylor, K. J. W. and Wells, P. N. T.**, Tissue characterization, *Ultras. Med. Biol.*, 15, 421, 1989.

15. **Shung, K. K.**, Basic principles of ultrasound tissue characterization, in *Noninvasive Techniques in Biology and Medicine*, Freeman, S. E., Fukushima, E., and Greene, E. R., Eds., San Francisco Press, San Francisco, 1990.

16. **Kossoff, G.**, Display technique in ultrasonic pulse echo investigations: a review, *J. Clin. Ultras.*, 2, 61, 1974.

17. **Reid, J. M., Sigelmann, R. A., Nasser, M., and Baker, D.**, The scattering of ultrasound by human blood, *Proc. 8th Int. Conf. Med. Biol. Eng.*, 10–7, 1969.

18. **Sigelmann, R. A. and Reid, J. M.**, Ultrasound scattering from biological tissues, in Interaction of Ultrasound and Biological Tissues, Reid, J. M. and Sikov, M. R., Eds., U.S. Dept. Health, Education and Welfare Publ. (FDA) 73-8008, 1972.

19. **Sigelmann, R. A. and Reid, J. M.**, Analysis and measurement of ultrasound backscattering from an ensemble of scatterers excited by sine wave bursts, *J. Acoust. Soc. Am.*, 53, 1351, 1973.

20. **Senapati, N., Lele, P. P., and Woodin, A.**, A study of scattering of sub-millimeter ultrasound from tissues and organs, 1972 IEEE Ultrasonics Symp. Proc. (IEEE cat. # 72 CHO 708-8 SU), 59, 1972.

21. **Lele, P. P., Mansfield, A. B., Murphy, A. I., Namery, J., and Senapati, N.**, Tissue characterization by ultrasonic frequency-dependent attenuation and scattering, in Ultrasonic Tissue Characterization I, Linzer, M., Ed., National Bureau of Standards Spec. Publ. 453, Gaithersburg, MD, 1976.

22. **Nicholas, D. and Hill, C. R.**, Tissue characterization by an acoustic Bragg scattering, *Nature*, 257, 305, 1975.

23. **Waag, R. C. and Lerner, R. M.**, Tissue microstructure determination with swept-frequency ultrasound, 1973 IEEE Ultrasonics Symp. Proc. (IEEE cat. # 73 CHO 807-8 SU) 63, 1973.

24. **Morse, P. M. and Ingard, K. U.**, *Theoretical Acoustics*, McGraw-Hill, New York, 1968.

25. **Lax, P. D. and Philips, R. S.**, *Scattering Theory*, Academic Press, New York, 1967.

26. **Twersky, V.**, On scattering of waves by random distributions. I. Free-space scatterer formalism, *J. Math. Phys.*, 3, 700, 1962.

27. **Twersky, V.**, On scattering of waves by random distributions. II. Two-space scatterer formalism, *J. Math. Phys.*, 3, 724, 1962.

28. **Fields, S. and Dunn, F.**, Correlation of echographic visualization of tissue with biological composition and physiologic state, *J. Acoust. Soc. Am.*, 54, 809, 1973.

29. **O'Donnell, M. and Miller, J. G.**, Broad band integrated backscatter: an approach to spatially localized tissue characterization, 1979 IEEE Ultrasonics Symp. Proc. (IEEE cat. # 79 CH1482-9), 175, 1979.

30. O'Donnell, M. and Miller, J. G., Quantitative broad band backscatter: an approach to nondestructive evaluation in acoustically inhomogeneous materials, *J. Appl. Phys.*, 52, 1056, 1981.

31. Madsen, E. L., Insana, M. F., and Zagzebski, J. A., Method of data reduction for accurate determination of acoustic backscatter coefficients, *J. Acoust. Soc. Am.*, 76, 913, 1984.

32. Insana, M. L., Madsen, E. L., Hall, T. J., and Zagzebski, J. A., Tests of the accuracy of a data reduction method for determination of acoustic backscatter coefficients, *J. Acoust. Soc. Am.*, 79, 1230, 1986.

33. Hall, T. J., Madsen, E. L., Zagzebski, J. A., and Boote, E. J., Accurate depth-independent determination of acoustic backscatter coefficients with focused transducers, *J. Acoust. Soc. Am.*, 85, 2410, 1989.

34. Campbell, J. A. and Waag, R. C., Measurements of calf liver ultrasonic differential and total scattering cross sections, *J. Acoust. Soc. Am.*, 75, 603, 1984.

35. Waag, R. C. and Astheimer, J. P., Characterization of measurement system effects in ultrasonic scattering measurements, *J. Acoust. Soc. Am.*, 88, 2418, 1990.

36. Bamber, J. C. and Hill, C. R., Acoustic properties of normal and cancerous human liver. I. Dependence on pathological condition, *Ultras. Med. Biol.*, 7, 121, 1979.

37. Bamber, J. C., King, J. A., and Hill, C. R., Acoustic properties of normal and cancerous liver. II. Dependence on tissue structure, *Ultras. Med. Biol.*, 7, 135, 1979.

38. Nassiri, D. K. and Hill, C. R., The differential and total bulk scattering cross sections of some human and animal tissues, *J. Acoust. Soc. Am.*, 79, 2034, 1986.

39. O'Donnell, M., Mimbs, J. W., and Miller, J. G., Relationships between collagen and ultrasonic backscatter in myocardial tissue, *J. Acoust. Soc. Am.*, 69, 580, 1981.

40. Fei, D. Y., Shung, K. K., Wilson, T. M., Ultrasonic backscatter from bovine tissues: variation with pathology, *J. Acoust. Soc. Am.*, 81, 166, 1987.

41. Fei, D. Y. and Shung, K. K., Ultrasonic backscatter from mammalian tissues, *J. Acoust. Soc. Am.*, 78, 871, 1985.

42. Haberkorn, U., Zuna, I., Lorenz, A., Zerban, H., Layer, G., Van Kaick, G., and Rath, U., Echographic tissue characterization in diffuse parenchymal disease: correlation of image structure with histology, *Ultras. Imaging*, 12, 155, 1990.

43. Cohen, R. D., Mottley, J. G., Miller, J. G., Kurnik, P. B., and Sobel, B. E., Detection of ischemic myocardium in vivo through the chest wall by quantitative ultrasonic tissue characterization, *Am. J. Cardiol.*, 50, 838, 1982.

44. Wickline, S. A., Thomas, L. J., Miller, J. G., Sobel, B. E., and Perez, J. E., Sensitive detection of the effects of reperfusion on myocardium by ultrasonic tissue characterization with integrated backscatter, *Circulation*, 74, 389, 1986.

45. Fellepa, E. J., Lizzi, F. L., Coleman, D. J., and Yaremko, M. M., Diagnostic spectrum analysis in ophthalmology: a physical perspective, *Ultras. Med. Biol.*, 12, 633, 1986.

46. Insana, M. F., Wagner, R. F., Garra, B. S., Momenan, R., and Shawker, T. A., Pattern recognition methods for optimizing multivariate tissue signatures in diagnostic ultrasound, *Ultras. Imag.*, 8, 165, 1986.

47. Miller, J. G., Barzilai, B., Milunski, M. R., Mohr, G. A., Perez, J. E., Thomas, L. J., Wear, K. A., Wickline, S. A., Vered, Z., and Sobel, B. E., Myocardial tissue characterization: clinical confirmation of laboratory results, 1989 IEEE Ultrasonics Symp. Proc. (IEEE cat. # 89 CH2791-2), 1029, 1989.

48. King, D., Lizzi, F. L., Feleppa, E. J., Wai, P., Yaremko, M., Rorke, M. C., and Herbst, J., Focal and diffuse liver disease studies by quantitative microstructural sonography, *Radiology*, 155, 457, 1985.

49. Chivers, R. C. and Hill, C. R., A spectral approach to ultrasonic scattering from human tissue: methods, objectives, and backscattering measurements, *Phys. Med. Biol.*, 20, 799, 1975.

50. Chivers, R. C., The scattering of ultrasound by human tissues — some theoretical models, *Ultras. Med. Biol.*, 3, 1, 1977.

51. **Atkinson, P. and Berry, M. V.**, Random noise in ultrasonic echoes diffracted by blood, *J. Phys. A: Math. Nucl. Gen.*, 7, 1293, 1974.
52. **Angelsen, B. J.**, A theoretical study of the scattering of ultrasound from blood, *IEEE Trans. Biomed. Eng.*, BME-27, 61, 1980.
53. **Lizzi, F. L., Greenebaum, M., Feleppa, E. J., and Elbaum, M.**, Theoretical framework for spectrum analysis in ultrasonic tissue characterization, *J. Acoust. Soc. Am.*, 73, 1366, 1983.
54. **Shung, K. K.**, On the ultrasound scattering from blood as a function of hematocrit, *IEEE Trans. Ultras. Sonics*, SU-29, 327, 1982.
55. **Lucas, R. J. and Twersky, V.**, Inversion of ultrasonic scattering data for red cell suspensions under different flow conditions, *J. Acoust. Soc. Am.*, 82, 794, 1987.
56. **Mo, L. Y. L. and Cobbold, R. S. C.**, A stochastic model of the backscattered Doppler ultrasound from blood, *IEEE Trans. Biomed. Eng.*, BME-33, 20, 1986.
57. **Goss, S. A., Johnston, R. L., and Dunn, F.**, Comprehensive compilation of empirical ultrasonic properties of mammalian tissues, *J. Acoust. Soc. Am.*, 64, 423, 1978.
58. **Goss, S. A., Johnston, R. L., and Dunn, F.**, Comprehensive compilation of empirical ultrasonic properties of mammalian tissues. II, *J. Acoust. Soc. Am.*, 68, 93, 1980.
59. National Council on Radiation Protection and Measurements Rep. No. 74, Biological Effects of Ultrasound: Mechanisms and Clinical Implications, Natl. Coun. Radiat. Protect. Meas., Bethesda, MD, 1983.
60. **Ishimaru, A.**, *Wave Propagation and Scattering in Random Media*, Academic Press, San Diego, CA, 1978.
61. **Tsang, L., Kong, J. A., and Shin, R. T.**, *Theory of Microwave Remote Sensing*, John Wiley & Sons, New York, 1985.
62. **Pauly, H. and Schwan, H. P.**, Mechanism of absorption of ultrasound in liver tissue, *J. Acoust. Soc. Am.*, 50, 692, 1970.
63. **Parker, K. J.**, Ultrasonic attenuation and absorption in liver tissue, *Ultras. Med. Biol.*, 9, 363, 1983.
64. **Burckhard, C. R.**, Speckle in ultrasonic B-mode scans, *IEEE Trans. Sonics Ultras.*, SU-25, 1, 1978.
65. **Abbott, J. G. and Thurstone, F. L.**, Acoustic speckle: theory and experimental analysis, *Ultras. Imag.*, 1, 303, 1979.
66. **Flax, S. W., Glover, G. H., and Pelc, N. J.**, Textural variations in B-mode ultrasonography, *Ultras. Imag.*, 3, 235, 1981.
67. **Wagner, R. F., Smith, S. W., Sandrick, J. M., and Lopez, H.**, Statistics of speckle in ultrasound B-scans, *IEEE Trans. Sonics Ultras.*, SU-30, 156, 1983.
68. **Smith, S. W. and Wagner, R. F.**, Ultrasound speckle size and lesion signal to noise ratio: verification of theory, *Ultras. Imag.*, 6, 174, 1984.
69. **Dainty, J. C.**, *Laser Speckle and Related Phenomena*, 2nd ed., Springer-Verlag, Berlin, 1984.
70. **Bamber, J. C. and Dickinson, R. J.**, Ultrasonic B-scanning: a computer simulation, *Phys. Med. Biol.*, 25, 463, 1980.
71. **Tuthill, T. A., Sperry, R. H., and Parker, K. J.**, Deviations from Rayleigh statistics in ultrasonic speckle, *Ultras. Imag.*, 10, 81, 1988.
72. **Magnin, P. A., von Ramm, O. T., and Thurstone, F. L.**, Frequency compounding for speckle contrast reduction in phase array images, *Ultras. Imag.*, 4, 267, 1982.
73. **Trahey, G. E., Smith, S. W., and von Ramm, O. T.**, Speckle pattern correlation with lateral aperture translation: experimental results and implication for spatial compounding, IEEE Trans. Ultras. Ferroelectr. Frequency Cont. UFFC-33, 257, 1986.

Chapter 2

CLINICAL RELEVANCE OF SCATTERING

Gary A. Thieme

TABLE OF CONTENTS

0-8493-6568-6/93/$0.00 + $.50

I. INTRODUCTION

The importance of diffuse scattering to modern gray-scale ultrasound imaging cannot be understated. This physical phenomenon is the basic sound interaction responsible for the soft tissue detail seen in clinical images. The counterpart of diffuse scattering is specular reflection, which is the physical interaction responsible for bistable images of the 1960s and early 1970s and for the bright interfaces seen in modern images. Only gray-scale imaging by the pulse-echo mode of operation will be considered here. The goals of this chapter are (1) to demonstrate the effects of operator-controlled system variables on the image generated from diffuse scattering processes and (2) to illustrate the clinical aspects of diffuse scattering through the use of normal and pathologic imaging examples.

The diagrams in Figure 1 schematically represent diffuse scattering (Figure 1A) and specular reflection (Figure 1B) situations. Both are important to clinical diagnosis. Diffuse scattering is responsible for the gray shade texture observed in solid tissue or semisolid structures. Specular reflections are responsible for the bright boundary interfaces between structures. The specific aspects of sound interactions in tissues are presented in several texts, journal references, and in Chapters 1, 3, 4, and 5 of this book.[1-10]

Diffuse scattering consists of sound interactions with rough tissue interfaces where the structural features are usually smaller than the wavelength of the interrogating sound energy. Most of the sound energy is directed away from the path of the beam. Only a small fraction of the original energy is actually reflected back to the transducer along the path of the beam. Small changes in acoustic impedance produce these diffuse scattering interactions. Transforming these low level energy reflections into gray shades in an image is a complex process. The resultant image features are determined both by the manufacturer through design specifications and by the user via adjustments of the ultrasound system controls.

An important component of diffuse scattering is Rayleigh scattering. Rayleigh scattering results from acoustic interactions with structures much smaller than the wavelength of sound. As Sample[2] points out, any tissue can be thought of as a combination of resolvable structures (major vessels and connective tissue interfaces) and of nonresolvable structures (small vessels and microscopic anatomy interfaces). Most tissues consist of a dominant combination of these nonsolvable scatterers and a smaller component of resolvable scatterers. Rayleigh scattering has an approximate fourth power relationship with sound frequency. Both diffuse scattering and Rayleigh scattering are important contributors to ultrasound image texture features and must be considered jointly in the discussions that follow.

Specular reflections consist of sound interactions with smooth tissue interfaces where the surface features are much larger than the wavelength of the interrogating sound energy. For these reflections to be detected, the sound

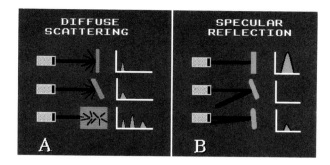

FIGURE 1. Sound interaction diagram. (A) Diffuse scattering is usually independent of the incident angle and consists of low amplitude reflections in many directions. Only a small amount of the incident energy is reflected back along the beam path to the transducer. Sample A-mode tracings. (B) Specular reflection is highly angle dependent. When the beam is perpendicular to the interface, a large amount of the incident energy is received by the transducer. Otherwise, it is reflected elsewhere, and no energy is received. Sample A-mode tracings.

beam must strike the interface at a perpendicular angle. These specular echoes generally originate from high acoustic impedance difference boundaries such as fluid-tissue or bone-tissue interfaces. A large fraction of the incident energy is then reflected directly back to the transducer to produce a bright interface in the image. Transforming these high level reflections into gray shades is much less dependent upon user adjustment of system controls, since these echo levels are usually at or near the saturation level.

Clinical examples of specular reflections and diffuse scattering are seen in the image of the placenta, amniotic fluid, and membrane separating the twin gestational sacs (Figure 2). The amniotic fluid contains no scatterers and does not reflect any sound; it is black. The placenta contains a network of connective tissue and cells which diffusely scatter sound energy resulting in a spectrum of intermediate gray levels. Only where the membrane is perpendicular to the sound beam can it be seen as a bright line interface, a specular reflection. Elsewhere, the membrane is represented by a low level gray line where the ultrasound beam is not perpendicular to the membrane and is deflected into directions other than the backward direction. For these faintly visualized segments, the sound interaction mechanism is diffuse scattering. Other examples of specular reflectors are the bones of the fetal skeleton.

It is important to recognize that the features observed in ultrasound images are both a reflection of the actual tissue structure and a product of the imaging system characteristics. In Chapter 3 the relationships between the spatial and contrast resolution of ultrasound systems and the macro- and microstructural features of different organ tissues will be presented. Experienced sonographers (technologists) and sonologists (physicians) recognize that the same structural features of tissues will vary in appearance from system to system, from changes in processing technique, and from differences in body habitus and

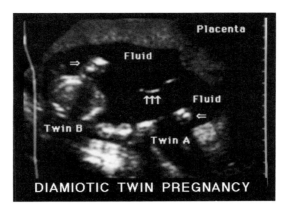

FIGURE 2. Clinical example of sound interactions. A membrane (arrows) separates the fetuses in this image of a twin pregnancy. Where the membrane is perpendicular to the sound beam, a thin brightly reflective surface is seen. Elsewhere, it acts like a diffuse scatterer and is only faintly seen. The placenta is a diffuse scatterer and has an inhomogeneous gray texture. Amniotic fluid contains no significant scatterers and is black. Fetal bones act like specular reflectors (open arrows).

scanning conditions. At times, it may be difficult to determine what features are due to structural alterations of tissues and to ultrasound system parameters. In this chapter, both phantoms and clinical material will be used to illustrate how system variables and scanning conditions influence our interpretation of diffuse scattering in ultrasound images.

II. SYSTEM FACTORS

When evaluating images and imaging system performance, it is important to attempt to separate spatial resolution, contrast resolution, texture, and speckle features. These imaging parameters are described in detail in journal and textbook references.[7-13] Thijssen[10] provides perhaps the most comprehensive and recent examination of this complex subject. These four parameters govern our ability to detect focal lesions and diffuse pathologic conditions in the clinical setting. Their representation in ultrasound images is significantly dependent upon the design and construction of the ultrasound system. Spatial resolution measures the smallest distance between two structures before they can no longer be seen as separate. It is a three-dimensional parameter. The volume or voxel is determined by the axial resolution, lateral resolution, and slice thickness. All sound interactions take place within this voxel and are incorporated into the two-dimensional image representation. Contrast resolution measures the smallest mean echo amplitude difference necessary to distinguish the presence of two different scattering media. The backscatter amplitude is a function of the density, size, and spacing of the scatterers as well as the homogeneity or heterogeneity of the scatterer types. The backscatter level is also dependent upon acoustic impedance differences and upon sound frequency. Generally, spatial resolution and contrast resolution are inversely related and are a function of the pulse length, frequency, and focusing properties of the transducer.

Texture is the pattern formed by an ensemble of scatterers whose volume is generally many times the physical limits of the spatial and contrast resolution limits. The pattern is composed of many small dots of varying echo levels, where the actual dot size is smaller than the spatial resolution as determined by the voxel parameters. Speckle is the term that describes this fine dot pattern seen in ultrasound images. The axial speckle dimension is proportional to the pulse length. The lateral speckle dimension changes with depth and is proportional to the lateral resolution or beam/width. Some frequently used terms to describe texture patterns are homogeneous, heterogeneous, mixed, coarse, fine, smooth, and rough. While macroscopic structures several millimeters or greater in size (e.g., blood vessels) may be seen in texture patterns, there is no direct correspondence between the fine dot pattern and histologic features.

After the transducer has transformed the returning sound energy into an electrical signal, the perception of diffuse scattering in a clinical image depends upon the signal processing chain of events occurring in the ultrasound system and upon the adjustment of controls by the sonographer or sonologist. Briefly, we can list these influences as

1. Transducer frequency and focusing characteristics
2. System gain
3. Time gain compensation
4. System dynamic range
5. Preprocessing functions
6. Postprocessing functions
7. System power
8. Frame averaging

General information can be found in texts and journals.[10-31]

A. TRANSDUCER FACTORS

The process begins with the selection of a transducer appropriate for the clinical imaging task. Each transducer has frequency and focusing characteristics which significantly influence our perception of diffuse scattering features, as described previously. Higher frequency transducers provide better spatial resolution than lower frequency transducers because of shorter pulse length and narrower beam focus. At higher frequencies finer structural detail can be seen because the diffuse scattering events occur in a smaller volume of tissue. At a lower frequency, scattering events are averaged over a larger

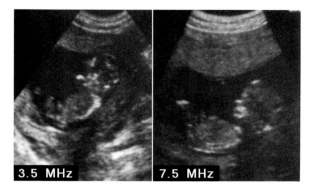

FIGURE 3. Sound frequency and spatial resolution. At 3.5 MHz features of the fetal head and body are coarse. At 7.5 MHz the fetal head and body are seen in much finer detail. Now the forehead, nose, and mouth can be resolved.

volume of tissue, in effect blurring detail. This point is graphically illustrated in Figure 3 where the same 12-week fetus has been imaged with both 3.5 and 7.5 MHz transducers. Finer structural detail is clearly evident in the higher frequency image.

The contrast (brightness difference) observed between two tissues in an ultrasound image depends upon the difference in the backscatter properties of the tissues, as described previously. Ultrasonic backscatter strength determines echo amplitude. The echo amplitude relationship between two adjacent tissues could be independent, linearly dependent, or non-linearly dependent upon the sound frequency. For example, a mass in the liver could be more apparent or less apparent or unchanged in echo amplitude when different transducer frequencies are used. Although ultrasonic backscatter is a function of frequency for most tissues, changing the transducer frequency sometimes may not affect echo amplitude relationships (contrast) significantly. In the following clinical example (Figure 4), diffuse scattering from a hemangioma in the right lobe of the liver is much greater than that from surrounding normal liver. However, the frequency dependence of the backscatter coefficients is not apparent in these images since the tissues have approximately the same echo amplitude relationship at 5 MHz as at 3 MHz.

As described previously, the texture patterns observed in ultrasound images are an ensemble of scattering interactions which are represented by many small dots, termed speckle. The dot pattern is dependent upon both the frequency and the focusing properties of the transducer. This heterogeneous pattern, as a function of the distance from the transducer face, is best illustrated by imaging a tissue-mimicking phantom with known uniform scatterer distribution throughout its volume. In Figure 5, three different patterns are observed in the near zone, focal zone, and far zone. The differences are best appreciated when sample sections are placed in a horizontal row. The fine-

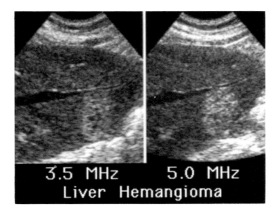

FIGURE 4. Frequency-dependent scattering. A liver hemangioma appears as a discrete hyperechoic mass and is imaged with both 3.5 and 5.0 MHz annular array transducers. Visually, the overall echo amplitude of the hemangioma when compared to normal adjacent liver may be slightly greater at 5 MHz than at 3.5 MHz. However, using a higher frequency transducer does not seem to have any clinically significant benefit with respect to lesion detection.

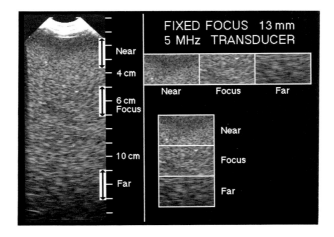

FIGURE 5. Speckle patterns. The depth-dependent speckle pattern of a fixed-focus 5 MHz transducer shows three distinct patterns in the near, focus, and far zones. Centimeter scale provides size and depth reference. Small regions placed adjacent to one another illustrate the fine dot, coarser dot, and elongated blob patterns characterizing the near, focus, and far zones. The scatterers are known to be uniformly distributed in this tissue mimicking phantom.

grained appearance in the near zone is actually a false representation of the known poor spatial resolution which is present at this superficial depth for this fixed-focus 13 mm diameter 5 MHz transducer. A coarser but still uniform dot pattern is observed in the focal zone. In the far zone where beam focusing is again poor, the pattern changes to very coarse laterally elongated blobs.

Flax[8] and Thijssen[10] used similar images from a phantom and from computer-generated wavefront modeling techniques to show that the "fineness" of the texture pattern has a complex dependence upon the transducer's properties. By comparing images from focused, nonfocused, and divergent beam transducers, they showed that the fine texture provides an illusion of high resolution even when the actual spatial resolution of the transducer is poor and that the texture pattern varies with depth as well. Consequently, the conventional echo amplitude ultrasound image does not accurately portray the uniform scatterer distribution known to exist in the phantom and does not accurately represent its physical architecture.

These conclusions have obvious serious adverse implications for those investigators wishing to do tissue characterization by direct visual analysis. Fortunately, tissue texture analysis can be accomplished if the original radio frequency (RF) data are corrected for transducer performance characteristics, diffraction effects, and attenuation. These sophisticated techniques are described in subsequent chapters.

A reasonable solution to the above problem associated with fixed-focus transducers can be accomplished by using electronically focused arrays in a special mode. During dynamic sequential focusing, a single A-line is synthesized from a series of pulse-echo sequences where the transmit focus is successively placed at greater depths. Each depth segment has optimized focusing so that lateral resolution does not vary significantly throughout the depth of field of view. This process is enhanced by utilizing dynamic apodization where the width of the array and number of elements increases with depth of focus. If focus on receive is also implemented, then a further improvement of lateral resolution is attained. The penalty is a slow frame rate since each A-line in the image requires many pulse-echo sequences. However, the speckle pattern and texture are virtually uniform throughout the depth of field of view. Under this special circumstance, direct visual analysis or computer analysis of tissue texture features may be possible in conventional commercial B-scan gray-scale images.

B. SYSTEM GAIN

Perception and interpretation of scattering interactions of sound in the body are greatly influenced by operator control of the system gain. This stage of signal processing raises or lowers the overall amplification applied to the returning echo waveforms. In Figure 6, echo amplitude histograms from a single region of interest of a tissue-mimicking phantom are compared for various system gain settings. The mean echo amplitude increases, but the echo amplitude histogram width, height, and shape remain virtually unchanged as system gain is increased. Other variables, such as dynamic range and post-processing, have been held constant.

The effect of changing system gain can be compared to a shifting window with constant width. Increasing the system gain amplifies lower level echoes

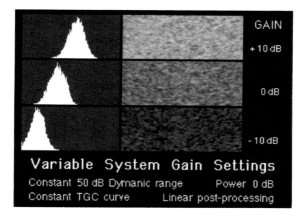

FIGURE 6. System gain effects: phantom. A uniform region of a tissue mimicking phantom is compared at system gain steps of 10 dB. Other parameters are held constant. Echo amplitude histograms show a simple translation of the distribution which has nearly constant width and shape. System gain shifts the window and changes overall image brightness but does not alter contrast.

to make them more visible, but higher level echoes may become saturated. Decreasing the system gain results in loss of lower level echoes and less saturation of higher level echoes. The clinical example (Figure 7) of hemorrhagic debris in an ovarian cyst dramatically illustrates the effects of system gain changes. As the system gain is raised, the debris becomes visible within the cyst, but other surrounding features are less evident because of saturation. For both the phantom and the clinical examples, image brightness features change, but contrast is not altered. Note that the texture pattern is not significantly altered for these modest changes in gain.

C. TIME GAIN COMPENSATION

Closely related to system gain is time gain compensation. This variable determines the amount of signal amplification as a function of depth. The controls must be set so that the natural attenuation of sound energy as a function of depth is properly compensated. Otherwise, it is impossible to compare the echo amplitude levels at a shallow depth with the echo amplitude levels from a deeper depth. In Figure 8B, the TGC is properly set so that echo amplitude throughout the depth of field of view is the same. In Figure 8C, the TGC is set so that the slope is too high; echo amplitude increases and then decreases when the TGC curve saturates. In Figure 8A, the TGC is set so that the slope is too low; echo amplitude decreases throughout the depth of field of view. Diffuse scattering is known to be the same at all depths in this uniform tissue-mimicking phantom. Thus, proper TGC setting is important for proper interpretation of strength of scattering of sound from tissues.

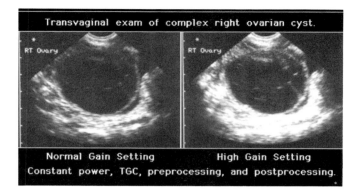

FIGURE 7. System gain effects: clinical. In this example, only the system gain has been altered. The debris within the hemorrhagic ovarian cyst cannot be appreciated at the lower gain setting but is readily seen at the higher gain setting. However, surrounding structures become saturated due to the upward shift of all echoes at higher gain.

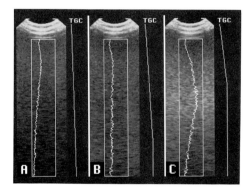

FIGURE 8. Time gain compensation: phantom. The tissue mimicking phantom is imaged with a 5 MHz fixed-focus transducer at low, medium, and high slope settings. (A) The slope setting does not adequately compensate for attenuation of sound energy. The average echo amplitude declines with depth. (B) The slope setting provides the correct compensation for attenuation. The average echo amplitude is steady. (C) The slope setting overcompensates for attenuation. The average echo amplitude increases with depth.

The images in Figure 9 represent the clinical equivalent of the phantom example. The liver represents a relatively uniform scattering medium that should be isoechoic when the slope is set to compensate for the natural attenuation of sound from diffuse scattering and absorption. In Figure 9A the TGC has been set to provide uniform echo amplitude throughout the depth of field of view. In Figure 9B the slope is set too high, and echo amplitude increases with depth until the TGC saturates. A bright band artifact is gen-

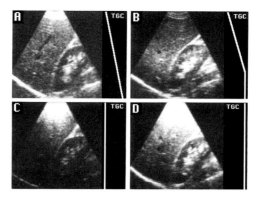

FIGURE 9. Time gain compensation: clinical. The liver provides a uniform body standard for setting the slope to compensate for the natural attenuation of sound energy. (A) The echo amplitudes of structures at all depths can be compared when the slope is set to compensate for the natural attenuation of sound in liver. This setting is unique. (B) Slope is set too high, creating a bright band artifact at the depth of the kidney. Echo amplitudes from diffuse scattering cannot be compared. (C) and (D) Slope is set too low. Gain adjustment only shifts the narrow band of intermediate grays deeper. Comparison of echo amplitudes at different depths is not possible.

erated across the liver and kidney. In Figures 9C and 9D the slope is set too low (flat). There is no compensation for the natural attenuation of sound. Echo amplitude decreases throughout the field-of-view. Increasing the system gain only results in saturation of echoes near the transducer. The TGC setting in Figure 9A is the only one where diffuse scattering from the liver and kidney can be evaluated and compared in a clinically meaningful way.

D. DYNAMIC RANGE

The dynamic range of signals processed through the system also affects the display of structures which diffusely scatter sound. At the transducer level, the typical signal dynamic range is 100 to 120 dB. After application of system gain, time gain compensation, and detection and enveloping stages, the signal dynamic range has been compressed into the 30 to 60 dB range prior to display on the viewing monitor. Many systems allow the user to adjust the dynamic range variable.

In Figure 10 echo amplitude histograms from a single region of interest are compared as the dynamic range variable is changed. At 60 dB, the band of echo amplitudes from diffuse scattering interactions in the tissue mimicking phantom is rather narrow and occupies a small segment of the available digital range. Contrast is low. At 30 dB the same band of echo amplitudes from the same diffuse scattering interaction nearly spans the available digital range. Contrast is high. The same linear gray scale is applied to the range of digital values from each dynamic range setting. Thus, the post-processing function does not affect the results. For graphic illustration purposes the system gain

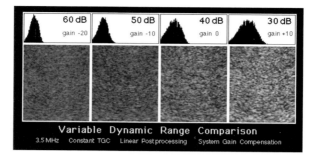

FIGURE 10.   Dynamic range effects: phantom. A uniform region of a tissue mimicking phantom is compared at dynamic range values from 30 to 60 dB. System gain has been adjusted to keep the mean echo amplitude approximately constant. Other variables are constant. Echo amplitude histograms show how the gray-shade distribution broadens with decreasing dynamic range. Contrast increases with decreasing dynamic range.

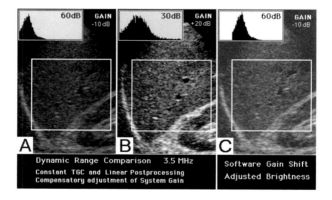

FIGURE 11.   Dynamic range effect: clinical. Normal liver and kidney are compared at 30 and 60 dB dynamic range settings. The lower dynamic range provides greater contrast as seen in the echo amplitude histogram comparison and in visual inspection of the images. To show that system gain has no effect on contrast, a constant value has been added to the 60 dB image to make the mean amplitudes of the 30 and 60 dB images similar.

has been adjusted to compensate for the shift in the mean echo amplitude and to keep image brightness constant. As was illustrated in the system gain section, this control only affects the level of the mean amplitude and does not alter the distribution. Thus, system gain only changes image brightness. Dynamic range clearly alters the distribution of echo amplitudes and changes image contrast. Note that the texture pattern dramatically changes from fine to coarse as the dynamic range decreases and the gray scale distribution is changed.

In the clinical imaging example (Figure 11), for normal liver and kidney, the differences between structures are subtle when the spectrum of echo amplitudes is distributed over the 60 dB dynamic range in Figure 11A. By comparison, the spectrum of echo amplitudes occupies most of the 30 dB dynamic range in Figure 11B. Contrast is higher for tissues in the 30 dB image. However, the subtle texture features represented by the low level echoes in the 60 dB image are not as evident in the 30 dB image. In Figure 11C a software gain shift has been used to equalize the mean amplitudes of the echo amplitude histograms. As was demonstrated in the system gain section, this gain change merely alters the brightness and does not change contrast. However, it does make comparison of the images easier. Thus, signal processing clearly affects our clinical perception of diffuse scattering from tissues by altering tissue texture and contrast relationships.

E. PRE-PROCESSING AND POST-PROCESSING

Two other system variables can also affect the perception and interpretation of tissue features due to diffuse scattering interactions. *Pre-processing* defines the mapping of the analog signal to the digital level, which represents the echo amplitude in the digital memory. The dynamic range control is a pre-processing function. *Post-processing* defines the mapping of digital levels to gray shades on the display monitor. The critical factor is the number of digital levels. In early 4-bit ultrasound systems with only 16 digital levels and 16 gray levels, representation of low level echoes was limited and very dependent upon pre-processing.

In modern 8-bit ultrasound systems with 256 digital levels and 256 gray levels, this dynamic range of 48 dB is sufficient to adequately display with simple linear pre-processing a range from subtle low level echoes to high level echoes. In fact, some manufacturers have eliminated the traditional pre-processing function and have replaced it with the dynamic range control. In modern ultrasound system designs, post-processing is used to alter the visual emphasis of low, middle, and high level echoes. This function is necessary because of the limited number of gray shades that the viewing monitor can display and the human visual system can perceive. Generally, this is accepted as 32 shades (5 bits) for clinical images and 90 shades for a gray bar under carefully controlled conditions.

Since the observer can perceive only about 32 (1/8th) of the potential 256 digital and gray levels available, selection of the post-processing curve can significantly influence the tissue contrast perceived in gray-scale images. Subtle changes in diffuse scattering relationships between different tissue types (e.g., normal and neoplastic tissues) may be overlooked if the post-processing assignment of digital values to gray shades is not optimal.

In the following clinical example, changing the pre-processing curve shape alters the assignment of the analog signal voltages to digital levels. This alters image contrast and makes the fibroids (small masses) within the uterus easier to see when the curve shape is changed from *S-curve* to *linear* curve to *light* curve (Figure 12). In another clinical example, changing the

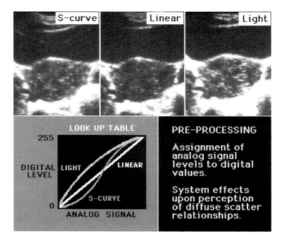

FIGURE 12. Pre-processing effects. Pre-processing assigns analog signals representing echo amplitudes to digital levels. Three different curves illustrate the shifts in tissue contrast for the fibroids (hypoechoic small masses) of the uterus. Surrounding structures have become saturated during image preparation.

post-processing curve shape alters the contrast and hence the perception of kidney features (Figure 13). Note that the changes in histogram peak, width, and shape are subtle. Yet, they intuitively reflect the visual effect of changing the gray-scale assignment to the digital levels.

F. SYSTEM POWER

The user-controlled power variable can have a significant but more subtle effect on the imaging of diffuse scattering interactions, especially for low level scatterers. Power is the amount of energy that is transmitted into the phantom (Figure 14). As the power is reduced, the amount of energy reflected by diffuse scatterers will decrease to a level where increases in the system gain can no longer effectively amplify the small signals above the system noise level. At this point, the lower level echoes "disappear" in the system noise. This can be seen as filling in of the cyst with low level echoes. This effect may be so subtle that it is difficult to perceive the texture change that has taken place in the adjacent tissue-mimicking gel. Obviously, lesions could be masked by the noise and overlooked in a clinical situation where system gain and power variables have not been properly adjusted.

G. FRAME AVERAGING

Most modern ultrasound systems allow the user to control the number of image sweeps (or frames) that are displayed on the viewing monitor. That is, the viewer can observe the real time study as a series of single frame images or as a series of time-integrated images (temporal averaging) where two or more time-adjacent frames are averaged and presented as a single

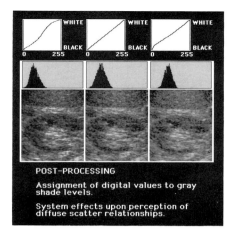

FIGURE 13. Post-processing effects. Post-processing assigns digital levels to gray shades. Three different curves illustrate the shifts in tissue contrast for this kidney cross-section with liver above it. The echo amplitude histogram distributions reflect the processing curve changes.

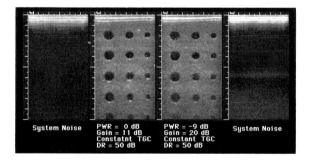

FIGURE 14. Power setting effects. As the power is reduced, the system gain and/or time gain compensation must be correspondingly increased to keep echo amplitude levels constant. However, as gain is increased to amplify weaker signals returning as a result of reduced power, a level of gain is finally reached where noise becomes visible as low level echoes falsely filling in cystic structures and reducing tissue contrast. The different levels of system noise are readily appreciated when the transducer is not in contact with the phantom.

image (Figure 15). Averaging the echo amplitude of each pixel in an image over time improves the signal-to-noise ratio and reduces speckle because there is usually some minor movement of the tissue scatterers relative to the transducer, due to motion from vascular pulsations, patient movement, or transducer movement. Both reduced noise and reduced speckle from the averaging of these slightly decorrelated images result in an image which appears significantly smoother than a single frame image. By comparison, the single-frame image appears ''grainy'' and not as aesthetically pleasing to the son-

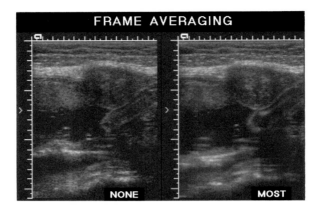

FIGURE 15. Temporal averaging. Obstetrical scan shows placenta, fibroid, fluid, and fetal parts. The left image is a captured single frame with no time averaging. It has a grainy appearance when compared to the smoother right image where the maximum amount of averaging of a series of time-adjacent image frames is employed. Temporal averaging reduces speckle and increases signal-to-noise ratio.

ographer and sonologist as is the temporally averaged image. Which is the better representation of the diffuse scattering interactions remains controversial. However, the principle is similar to the compounding effect obtained by using a static (articulated arm) scanner where highly decorrelated images from different viewing angles are averaged. Compound scanning reduces speckle and increases signal-to-noise ratio so that boundary definition and lesion detection are improved. Thus, clinical interpretation of diffuse scattering is influenced in a minor way by the temporal averaging function as long as blurring (loss of spatial resolution) from tissue or transducer motion is minimal.

H. SUMMARY OF SYSTEM FACTORS

In summary, changing a system variable (dynamic range, system gain, time gain compensation, power, pre-processing, post-processing, and frame averaging) alters the signal processing in ways which can change the echo amplitude relationships among structures seen in an ultrasound image. It is important to remember that diffuse scattering interactions of sound in body tissues (or phantoms) are *not* altered by changes in these system variables. Only the visual representation of the diffuse scattering interactions is affected. On the other hand, the choice of transducer frequency and focusing properties will fundamentally alter diffuse scattering interactions of sound in body tissues.

Both the choice of system variables and the selection of transducer properties will affect spatial resolution, contrast resolution, texture, and speckle features. If used with knowledge and skill, these parameters can be adjusted by the operator to enhance the diagnostic content of ultrasound images. Conversely, there are far more combinations of these variables which reduce or mask diagnostic information. Clearly, this operator dependence can critically influence the clinical interpretation of ultrasound studies.

III. QUALITATIVE ASSESSMENT OF SCATTERING IN THE CLINICAL SETTING

In daily clinical imaging, differences in the degree of scattering are observed as changes in brightness, contrast, and texture. Unlike Hounsfield units in computed tomography where each tissue type has a specific normal numeric range of values, brightness levels in ultrasound images are all relative and can be easily altered with a simple increase or decrease of the system gain, time gain compensation, or power controls. Furthermore, the echo amplitude relationships can be altered in non-linear ways by changing the dynamic range, pre-processing, and post-processing controls. Thus, quantitative analysis of echo levels representing the diffuse scattering properties of tissues in conventional B-scan images has significant limitations. Fortunately, qualitative interpretation is sufficient for day-to-day clinical work.

A. GENERAL TERMINOLOGY AND EXAMPLES

In clinical imaging the strength and pattern from diffuse scattering in body tissues are used to identify normal anatomy as well as to recognize pathology. The following qualitative terms are commonly used to compare the scattering properties of one tissue to another.

Anechoic: No internal echoes indicating no diffuse scatterers
Hypoechoic: The tissue of interest has lower echo levels than the comparison tissue
Isoechoic: The tissue of interest has the same echo level as the comparison tissue
Hyperechoic: The tissue of interest has higher echo levels than the comparison tissue
Homogeneous: Uniform overall echo level and pattern
Heterogeneous: Mixture of regions of higher and lower echo levels
Complex: Mixture of echogenic and anechoic components

Anechoic is defined as no visible echoes. Usually this means no internal scatterers. In Figure 16A two simple kidney cysts are examples of *anechoic* structures. The breast cyst in Figure 16B has a similar appearance except that the tissues deep to the cyst are brighter than the adjacent structures. Both are fluid-filled sacs and contain no internal reflectors of sound. The artificially high brightness level (posterior acoustic enhancement) seen deep to the breast cyst is a TGC (time-gain-compensation) artifact due to the low attenuation

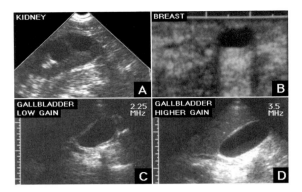

FIGURE 16.   Anechoic structures. The kidney cysts (A), breast cyst (B), and gallbladder (C) are all anechoic structures. No internal echoes from diffuse scatterers are seen. The gallbladder (D) illustrates the importance of transducer frequency and proper adjustment of system gain. Sludge not seen in (C) is now visible.

of sound by the cyst fluid. The TGC has been set to compensate for the higher attenuation of the surrounding soft tissues. Not all cysts have associated posterior acoustic enhancement. The renal cyst fluid most likely contains a sufficiently high protein content to absorb enough sound energy to attenuate the beam, similar to the surrounding solid tissues. The breast cyst fluid protein content must be low since there is little absorption of sound energy. Recognizing that fluid-filled structures can modify the sound beam by absorption of sound even though there are no diffuse scatterers is often important with respect to correct interpretation of images.[32]

In Figure 16C the lumen of the gallbladder appears *anechoic* at low gain with the 2.25 MHz transducer. By using a higher gain setting and a higher frequency transducer, the lumen no longer appears completely *anechoic* in Figure 16D. A layer of sludge (bile crystal precipitant) is now visible as low-level diffuse scatterers, and a layer of anechoic bile is seen above the sludge. Thus, system gain and transducer selection also influence clinical interpretation of *anechoic* structures.

When the echo levels of two tissues are compared, the conditions under which the observations are made determine whether the comparison is valid. Basically, the two tissues should be directly adjacent and should be at the same distance from the transducer so that the TGC effect is the same. In addition, the intervening structures should have similar properties. Also, the TGC must be properly set to compensate for the natural attenuation of sound.

A classic example is the comparison of the right kidney to the liver. In Figure 17 a uniform thickness of the abdominal wall and liver is interposed between the kidney and the adjacent liver. Conditions are ideal and a comparison of relative scattering of renal cortex to liver is valid. Diffuse scattering from kidney cortex is less than liver. Thus, kidney cortex is *hypoechoic* with

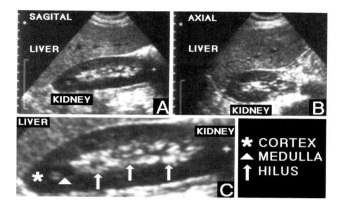

FIGURE 17. Scattering relationships. This normal kidney and liver illustrates the use of the terms hyperechoic, hypoechoic, and isoechoic. The terms homogeneous and heterogeneous apply to portions of the images as well.

respect to liver. Conversely, the liver is *hyperechoic* with respect to kidney cortex.

In Figure 17C, other valid comparisons can also be made in smaller regions of interest where technical conditions are known to be uniform. One such region is the kidney itself where the central sinus (hilus) scattering is greater than the surrounding cortex and medulla. The hilus is *hyperechoic* with respect to kidney parenchyma, but it is *isoechoic* with respect to the adjacent retroperitoneal structures. The renal pyramids (medulla) are *hypoechoic* with respect to the renal cortex. Thus, these tissues represent a broad spectrum of diffuse scattering interactions and illustrate the use of these terms.

Among the images already presented are other examples of the terms used to qualitatively compare scattering properties of tissues. In Figure 2 the amniotic fluid is *anechoic*. The bones (open arrows) and membrane (simple arrows) are *hyperechoic*. The liver hemangioma in Figure 4 is a *hyperechoic* mass whose diffuse scattering features are characteristic of this benign tumor. The small dark fibroids of the uterus in Figure 12 are examples of *hypoechoic* masses. The hemorrhagic ovarian cyst in Figure 7 is classified as *complex* because there is a mixture of solid tissues and fluid containing debris.

B. SPECIFIC CLINICAL EXAMPLES

Alterations in the diffuse scattering properties of tissues can reflect maturational changes as well as pathologic responses to benign and malignant diseases. The organ response may be immediate, delayed, or progressive. The changes may be visible in the mild early form of the disease or may only become evident when the disease process is in an advanced severe stage. Correlation of the ultrasound images with gross specimen and histologic preparations may provide insight into common associations between the disease process and diffuse scattering properties of the affected tissues.

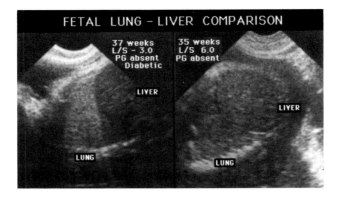

FIGURE 18. Diffuse scattering and tissue maturation. The fetal lung is hyperechoic with respect to fetal liver for this fetus with immature L/S ratio at 37 weeks. Fetal lung and liver echo amplitudes are equal for another fetus at 35 weeks with mature L/S ratio. These cases imply that diffuse scattering from fetal lung changes as it matures.

Changes in diffuse scattering from fetal lung have been investigated with respect to prediction of lung maturity (ability to support respiration *ex utero*). The fact that the lung development progresses through three stages (pseudoglandular, canalicular, and terminal sac) during the second and third trimesters suggests that diffuse scattering properties should change. Liver development is one stage and uniform throughout the same time period. While echogenicity differences between lung and liver have been shown to change with gestational age (Figure 18), a high correlation between lung maturity and diffuse scattering properties has never been established and remains an elusive research goal. Details will be presented in Chapter 13.

Disease processes which affect organs in a global manner generally can be classified into two categories. One process infiltrates the organ leaving the underlying architecture intact. The other alters or destroys the underlying architecture, often replacing it with fibrous connective tissue. The process may or may not be reversible depending upon the disease characteristics.

Because of the liver's large size and the common occurrence of diseases within it, the body of knowledge is extensive.[33-40] One of the most common benign entities affecting the liver is fatty infiltration where there is a generalized increased amount of fat in the hepatocytes. Normally, the liver is slightly hyperechoic with respect to kidney cortex. Fatty infiltration causes increased backscatter by the liver which is probably due to increased scatterer density. The liver becomes very bright with respect to the normal right kidney (Figure 19).

A second but less common disease process globally affecting the liver is cirrhosis. The hepatocytes are destroyed and replaced by fibrous connective tissue. Intuitively, one might expect an increase in liver echogenicity; however, this has not been observed when it is a pure process (no accompanying

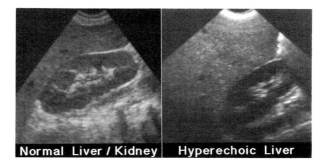

FIGURE 19. Normal and fatty liver. The echo amplitude relationship of normal liver and kidney is shown on the left. Fatty infiltration of the liver produces a much higher level of diffuse scattering and liver becomes hyperechoic with respect to kidney.

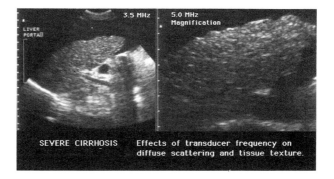

FIGURE 20. Cirrhosis of the liver. Diffuse scattering from nodular cirrhosis is seen as a heterogeneous echo pattern. The ascites (fluid) allows the nodular surface of the liver to be seen. Compare the architecture seen here to the homogeneous appearance of normal liver in Figure 19.

fatty infiltration). The fibrosis causes architectural changes best described as nodules in the advanced stage. In earlier stages, the appearance may be indistinguishable from normal liver. In Figure 20 the nodular appearance of advanced cirrhosis is accentuated by the ascites which outlines the rippled liver surface (normally smooth). At 5 MHz and magnified, the resolution is improved (compared to 3.5 MHz) and the texture pattern of cirrhosis is better appreciated.

Most of the disease processes which globally alter the kidney are either inflammatory conditions, vascular conditions (renal artery stenosis), or infiltrative processes (leukemia or lymphoma).[41-48] The body of knowledge is most extensive for the inflammatory conditions (e.g., glomerulonephritis). Changes in the interstitial, glomerular, and tubular architecture often result in a dramatic increase in ultrasonic backscatter with respect to the liver (normal

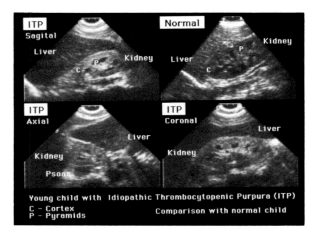

FIGURE 21. Systemic kidney disease. Different views and acoustic windows of the right kidney and liver of this child with increased kidney cortical echogenicity are compared to the normal child where the normal kidney cortex is slightly hypoechoic with respect to the liver. The ITP disease state has increased the diffuse scattering level from the kidney cortex but has not affected the hypoechoic pyramids.

and unchanged). Hricak et al. studied renal parenchymal disease which produces varying degrees of globally increased echogenicity. They found that the specific type of renal disease causing the hyperechoic change could not be named. However, there was a positive correlation between cortical echogenicity and both renal function and histopathologic findings. While these changes are usually attributed to the long term effects of fibrous tissue replacement of the cortex and medulla, a dramatic increase in acoustic reflectivity can be seen acutely in the kidney cortex as a result of systemic diseases such as idiopathic thrombocytopenic purpura (Figure 21). The medullary pyramids are spared. Their normal mild hypoechoic character is greatly accentuated by the presence of the increased diffuse scattering in the cortex. Three different acoustic windows through liver are shown and compared to a normal newborn kidney where cortical echogenicity is similar to liver.

Global alterations in scattering properties of tissue can also be observed in fetal disease processes. One of the most dramatic is infantile polycystic kidneys[49] (Figure 22). The enlarged non-functional mildly hyperechoic kidneys contain many tiny cysts too small to be resolved by clinical ultrasound imaging. The individual cyst size is near the wavelength of the interrogating sound. It is postulated that these tiny fluid-filled spaces accentuate diffuse scattering resulting in the increased reflectivity observed clinically.

Benign and malignant neoplasms have a broad spectrum of echographic appearances. Any of the qualitative terms used to describe the scattering properties of tissues may apply. The following clinical examples are illustrative.

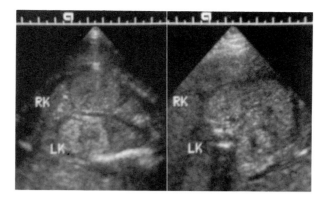

FIGURE 22. Infantile polycystic kidneys. These enlarged hyperechoic fetal kidneys do not function. The normal black space of amniotic fluid is absent. This diffuse scattering pattern is diagnostic for this fatal fetal malformation.

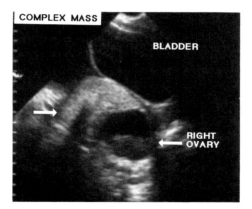

FIGURE 23. Complex mass. The anechoic bladder provides an acoustic window to visualize the complex mass called a dermoid tumor which has replaced the right ovary tissue. The hypoechoic dependent debris layer in the cystic component of this larger hyperechoic mass is a characteristic pattern for this neoplasm.

In Figure 23 a complex mass in the right ovary has features characteristic of a dermoid.[50] The cystic component (right arrow) contains a fluid-debris level. A horizontal line separates the anechoic fluid layer from the hypoechoic debris precipitated in the dependent aspect. The hyperechoic solid component (left arrow) contains a small region which highly scatters and absorbs the sound to produce a narrow acoustic shadow.

A hyperechoic mixed pattern of hydatidiform mole[51] is surrounded by the hypoechoic muscular wall of the uterus (Figure 24). The mixed echo texture

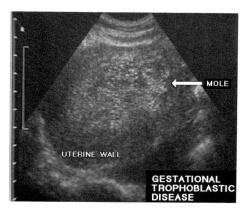

FIGURE 24. Heterogeneous echo pattern. This heterogeneous hyperechoic mass surrounded by homogeneous hypoechoic uterine muscle wall is characteristic of gestational trophoblastic disease (mole). Scattering from many tiny hypoechoic cysts within the mole contribute to the mixed echo amplitude appearance.

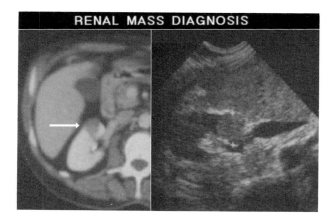

FIGURE 25. Solid homogeneous mass. Ultrasound can easily determine that the kidney mass seen on the CT scan is solid, not cystic. The homogeneous diffuse scattering from this hyperechoic mass with respect to kidney is characteristic of a renal cell carcinoma. It is isoechoic with respect to the liver and is separated by a thin hyperechoic boundary.

of gestational trophoblastic disease is the result of diffuse scattering from the tiny fluid-filled spaces of this neoplasm.

Many primary neoplasms and secondary neoplasms (metastases) appear as discrete focal masses with well-defined boundaries. The mass may be hypoechoic, isoechoic, or hyperechoic with respect to the surrounding organ tissues. The renal cell carcinoma (Figure 25) originally seen on a computed tomography (CT) study is characterized as a discrete solid homogeneous mass

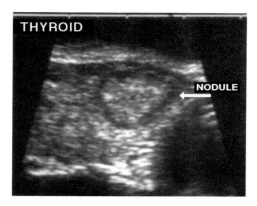

FIGURE 26. Isoechoic mass with halo. This isoechoic thyroid mass is seen only because of the hypoechoic ring boundary (halo).

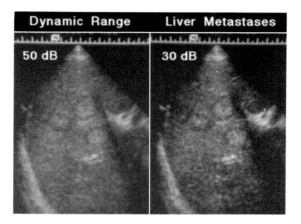

FIGURE 27. Hyperechoic liver masses with halo. The mildly increased diffuse scattering from these liver metastases is accentuated by the hypoechoic halo boundary. This image set also illustrates the contrast enhancement of the smaller dynamic range image.

(between the calipers) on the adjacent ultrasound image.[51-53] It is mildly hyperechoic with respect to kidney and isoechoic with respect to the adjacent liver. Ultrasound is often used to determine whether a mass seen on a CT exam is cystic or solid.

The isoechoic nodule in the thyroid (Figure 26) can only be identified because of the hypoechoic rim outline.[54] Histologic studies have shown that the rim consists of compressed tissues displaced by the growing neoplasm. Originally thought to be associated with benign neoplasms, this interesting diffuse scattering pattern has been demonstrated with malignant neoplasms as well. In Figure 27 mildly hyperechoic masses in the liver are metastatic

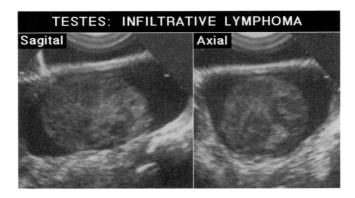

FIGURE 28.   Heterogeneous infiltrative pattern. Lymphoma cells infiltrate this testis and create a heterogeneous scattering pattern where boundaries are poorly defined. The gradual transition from normal to abnormal tissue (hypoechoic) is reflected in the gradual change in diffuse scattering at some boundaries.

adenocarcinoma. Decreasing the dynamic range from 50 to 30 dB enhances the contrast and makes the halo easier to see. The term target lesion is sometimes used to describe this scattering pattern.[55]

Malignant neoplasms are often infiltrative and multi-focal. The irregular invasion of cancer cells into surrounding normal tissue causes a gradual change in the diffuse scattering pattern. Infiltrating lymphoma of the testicle[56-57] (Figure 28) is representative of this type of texture pattern. Irregular hypoechoic regions with poorly defined margins have replaced much of the normal testicular tissue. The surrounding fluid (anechoic) is a reactive hydrocele.

Blood clots or hematomas are the result of bleeding into soft tissues or fluid spaces. Generally, the appearance will change with time. In the acute phase the hematoma usually appears hyperechoic with respect to surrounding organ tissues. As repair processes break down the blood products, the diffuse scattering level decreases. In the isoechoic phase about 1 week later, the hematoma may be difficult to identify. As the hematoma debris is removed, it liquifies and becomes hypoechoic weeks later. This rather unique process is illustrated in Figure 29 where a traumatic subcapsular hematoma of the spleen is seen in its acute and chronic phases.[58]

Inflammatory processes due to bacterial infection can have a broad spectrum of patterns. However, the most common pattern is a reduced level of diffuse scattering. This is attributed to edema of tissues in the early stage and liquifaction in the late stage. The intermediate stage may have a mixed or complex pattern.

In Figure 30 the left testis is mildly enlarged and hypoechoic with respect to the normal right testis.[56,57] This acute phase of bacterial infection (orchitis) has caused edema of the testis. The increased fluid content may separate scatterers and reduce the amount of scattering per unit volume of tissue. An

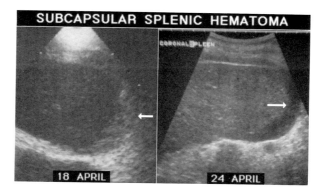

FIGURE 29. Temporal change in diffuse scattering. The hyperechoic crescent of blood clot is transformed into a hypoechoic rim as this subcapsular hematoma of the spleen undergoes lysis. Over the 6 days, the density of cellular material in the hematoma is reduced as it is gradually changed to mostly fluid. Diffuse scattering decreases with time.

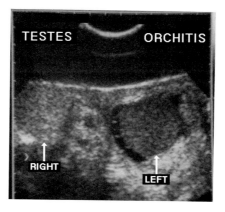

FIGURE 30. Tissue edema. The bacterial infection of the left testis produces tissue edema which probably increases scatterer spacing and reduces scatterer density. Diffuse scattering from the affected left testis is decreased, and it appears hypoechoic with respect to the normal right testis.

acoustic stand-off pad has been used to provide an equidistant path to the testes so that TGC effect will be balanced.

Bacterial inflammation can be focal. Multiple hypoechoic masses (Figure 31) in the liver of this patient with AIDS represent the necrotic stage of an infectious process.[59-60] The inflammatory debris within abscesses has a reduced level of diffuse scattering. Without clinical information, this appearance of hypoechoic masses might also be interpreted as metastatic carcinoma to the liver. This case illustrates the overlapping character of disease processes.

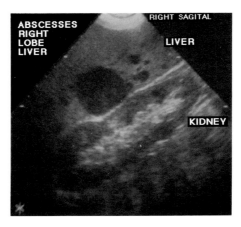

FIGURE 31. Hypoechoic liver mass. The one large and multiple smaller abscesses of the liver are hypoechoic. Apparently, the tissue edema and cellular debris do not provide as many interfaces for diffuse scattering as do adjacent liver and normal kidney.

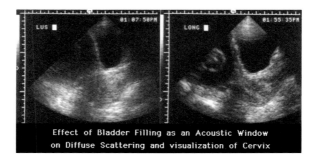

FIGURE 32. Acoustic window effects. The larger urinary bladder 48 min later provides a better acoustic window for the transmission of sound. In the earlier image the diffuse scattering pattern does not accurately portray the lower uterine segment and cervix structures, which are much better visualized in the later image. Diffuse scattering in deeper tissues is affected by sound interactions in intervening structures through their effects on the ultrasound beam.

Finally, we must recognize the effects of the acoustic window on the diffuse scattering process. Adverse acoustic conditions for sound transmission can result in significant image degradation. Obtaining the optimal acoustic window is both a function of the operator's scanning skills and a function of the patient's body habitus. Additional factors such as proper bladder filling for pelvic and obstetric exams can also dramatically improve visualization (Figure 32). The larger urinary bladder 48 min later provides a better acoustic window for the transmission of sound. In the earlier image the diffuse scat-

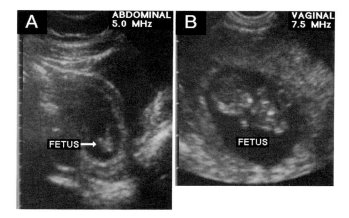

FIGURE 33. Transabdominal endovaginal comparison. (A) The thick intervening abdominal wall tissues of this obese female and the lack of a good acoustic window from the nearly empty bladder degrade the image quality so severely that the fetus and gestational sac are poorly seen. (B) Placing a probe in the vagina directly adjacent to the uterus eliminates the offending abdominal wall tissues. Diffuse scattering from the fetus and placenta is not degraded by intervening tissues, and visualization is dramatically improved.

tering pattern does not accurately portray the lower uterine segment and cervix structures which are much better visualized in the later image. Diffuse scattering in deeper tissues is affected by sound interactions in intervening structures through their effects on the ultrasound beam.

Diffuse scattering in the thick adipose tissue of the abdominal wall of obese patients can produce severe image degradation from loss of both spatial resolution and contrast resolution. Reducing the dynamic range to increase contrast, reducing the transducer frequency to increase penetration and sensitivity and decrease noise, and changing the post-processing function to enhance contrast are all methods which may improve image quality.

If the application is suitable, eliminating the abdominal wall by using an endoluminal probe (such as transvaginal, transrectal, or transesophageal) offers the best solution. This latter approach can provide a dramatic improvement in image quality (Figure 33). From a transabdominal approach, the early intrauterine gestation is poorly seen. From a transvaginal approach, a higher frequency transducer can be used, and scattering from intervening tissues is virtually eliminated. Detail is dramatically improved.

C. CONTRAST AGENTS FOR ULTRASOUND

Ultrasound contrast agents in the form of free microbubbles were first used for opacification of the right heart chambers in echocardiography. Initial preparations had microbubbles too large to pass through the pulmonary capillary bed. Later preparations of encapsulated microbubbles with diameters less than 8 μm and small enough for transpulmonary passage after peripheral venous injection allowed opacification of left heart chambers.[61]

Contrast agents in ultrasound were reviewed by Ophir and Parker.[62] The agents developed thus far have been designed for intravascular administration and consist of free gas bubbles, encapsulated microbubbles, emulsions, and colloidal suspensions. Ultrasound properties which may be changed include backscatter, attenuation, and speed of sound. Potentially observable or measurable effects include increased or decreased signal strength, altered gray-scale texture, enhanced or diminished penetration, and enhanced Doppler signal. The microbubble effects are short-lived and have only been clinically successful for imaging the right and left heart chambers, via a peripheral venous injection. Lipid emulsions have not demonstrated any enhancement effects in the liver. A colloidal suspension of micron-size particles may increase scattering by agglomeration of scatterers in Kupffer cells of the reticuloendothelial system of the liver and spleen. Perfluorochemicals, gelatin, and iodipamide ethyl ester particle suspensions have demonstrated enhancement properties in both animal and human experimental studies. However, colloidal agents require long time delays of about 48 hours before increased echogenicity effects are seen. To date, only microbubble agents for heart chamber enhancement have been approved for clinical use.

IV. CONCLUSIONS

Diffuse scattering is responsible for the gray shade appearance of tissues in modern ultrasound images. The quality of an ultrasound image depicting diffuse scattering is influenced by operator adjustment of system variables, by the scanning skills of the sonographer or sonologist, and by the body habitus of the patient. These factors affect diagnostic interpretation in the clinical setting. The effect of pathologic processes on body tissue scattering of sound is learned through clinical experience. Image analysis is both visual and qualitative. Clinical applications of quantitative analysis are limited by deterministic factors, the basic nature of pathologic processes, and the present qualitative character of commercial imaging systems.

REFERENCES

1. **McDicken, W. N.,** Ultrasound in tissue, in *Diagnostic Principles and Use of Instruments,* 3rd ed., McDicken, W. N., Ed., Churchill Livingstone, New York, 1991, chap. 4.
2. **Sample, W. F. and Erikson, K.,** Basic principles of diagnostic ultrasound, in *Diagnostic Ultrasound Text and Cases,* Sarti, D. A. and Sample, W. F., Eds., G. K. Hall Medical Publishers, Boston, 1980, chap. 1.
3. **Sarti, D. A. and Kimme-Smith, C.,** Physics of diagnostic ultrasound, in *Diagnostic Ultrasound Text and Cases,* 2nd ed., Sarti, D. A., Ed., Year Book Medical Publishers, Chicago, 1987, chap. 1.

4. **Powis, R. L. and Powis, W. T.,** Effects of tissue on ultrasound, in *A Thinker's Guide to Ultrasonic Imaging,* Powis, R. and Powis, W. T., Eds., Urban and Schwarzenberg, Baltimore, 1984, chap. 9.

5. **Chivers, R. C.,** The scattering of ultrasound by human tissue: some theoretical models, *Ultras. Med. Biol.,* 3, 1, 1977.

6. Ultrasound, in *Christensen's Introduction to the Physics of Diagnostic Radiology,* 3rd ed., Curry, T. S., III, Dowdey, J. E., and Murry, R. C., Eds., Lea and Febiger, Philadelphia, 1984, chap. 25.

7. **Wells, P. N. T. and Halliwell, M.,** Speckle in ultrasonic imaging, *Ultrasonics,* Sept. 1981, 225.

8. **Flax, S. W., Glover, G. H., and Pelc, N. J.,** Textural variations in B-mode ultrasonography: a stochastic model, *Ultras. Imag.,* 3, 235, 1981.

9. **Jones, J. P. and Kimme-Smith, C.,** An analysis of the parameters affecting texture in a B-mode ultrasonogram, *Acoust. Imag.,* 10, 295, 1982.

10. **Thijssen, J. M. and Oosterveld, B. J.,** Texture in tissue echograms: speckle or information?, *J. Ultras. Med.,* 9, 215, 1990.

11. Axial Resolution, Aero-Tech Rep., 1(1), KB-Aerotech, Lewistown, PA 17044.

12. Lateral Resolution, Aero-Tech Rep., 1(3), KB-Aerotech, Lewistown, PA 17044.

13. **McDicken, W. N.,** Performance of real-time B-scan instruments, in *Diagnostic Ultrasonic Principles and Use of Instruments,* 3rd ed., McDicken, W. N., Ed., Churchill Livingstone, New York, 1991, chap. 12.

14. **McDicken, W. N.,** Manipulation of ultrasonic waves, in *Diagnostic Ultrasonic Principles and Use of Instruments,* 3rd ed., McDicken, W. N., Ed., Churchill Livingstone, New York, 1991, chap. 5.

15. **McDicken, W. N.,** Controls of equipment for pulse-echo imaging, in *Diagnostic Ultrasonic Principles and Use of Instruments,* 3rd ed., McDicken, W. N., Ed., Churchill Livingstone, New York, 1991, chap. 7.

16. **McDicken, W. N.,** Scan converters and computers in grey-scale imaging, in *Diagnostic Ultrasonic Principles and Use of Instruments,* 3rd ed., McDicken, W. N., Ed., Churchill Livingstone, New York, 1991, chap. 10.

17. **McDicken, W. N.,** Using real-time B-scan instruments, in *Diagnostic Ultrasonic Principles and Use of Instruments,* 3rd ed., McDicken, W. N., Ed., Churchill Livingstone, New York, 1991, chap. 13.

18. **McDicken, W. N.,** Further details of ultrasonic transducers, in *Diagnostic Ultrasonic Principles and Use of Instruments,* 3rd ed., McDicken, W. N., Ed., Churchill Livingstone, New York, 1991, chap. 27.

19. **Powis, R. L. and Powis, W. T.,** Making and receiving ultrasound: transducers and how they work, in *A Thinker's Guide to Ultrasonic Imaging,* Powis, R. L. and Powis, W. T., Eds., Urban and Schwarzenberg, Baltimore, 1984, chap. 3.

20. **Powis, R. L. and Powis, W. T.,** Making use of preprocessing, in *A Thinker's Guide to Ultrasonic Imaging,* Powis, R. L. and Powis, W. T., Eds., Urban and Schwarzenberg, Baltimore, 1984, chap. 15.

21. **Powis, R. L. and Powis, W. T.,** Interrogating gray-scale images with postprocessing, in *A Thinker's Guide to Ultrasonic Imaging,* Powis, R. L. and Powis, W. T., Eds., Urban and Schwarzenberg, Baltimore, 1984, chap. 14.

22. **Powis, R. L. and Powis, W. T.,** TCG and the sonographer: building informative images, in *A Thinker's Guide to Ultrasonic Imaging,* Powis, R. L. and Powis, W. T., Eds., Urban and Schwarzenberg, Baltimore, 1984, chap. 13.

23. **Powis, R. L. and Powis, W. T.,** Gray-scale vision and the human eye, in *A Thinker's Guide to Ultrasonic Imaging,* Powis, R. L. and Powis, W. T., Eds., Urban and Schwarzenberg, Baltimore, 1984, chap. 16.

24. **Thieme, G. A., Price, R. R., and James, A. E., Jr.,** Ultrasound instrumentation and its practical applications, in *The Principles and Practice of Ultrasonography in Obstetrics and Gynecology,* 3rd ed., Sanders, R. C. and James, A. E., Eds., Appleton-Century-Crofts, Norwalk, CT, 1985, chap. 3.
25. **Kimme-Smith, C. and Jones, J. P.,** The relative effects of system parameters on texture in gray-scale ultrasonograms, *Ultras. Med. Biol.,* 10(3), 299, 1984.
26. **Alasaarela, E. and Koivukangas, J.,** Evaluation of image quality of ultrasound scanners in medical diagnostics, *J. Ultras. Med.,* 9, 23, 1990.
27. **Zagzebski, J. A., Banjavic, R. A., Madsen, E. L., et al.,** Focused transducer beams in tissue mimicking material, *J. Clin. Ultras.,* 10, 159, 1982.
28. **Zagzebski, J. A. et al.,** Focused transducer beams in tissue-mimicking material, *Semin. Ultras.,* 4(1), 44, 1983.
29. **Sommer, F. G. and Sue, J. Y.,** Imaging processing to reduce ultrasonic speckle, *J. Ultras. Med.,* 2, 413, 1983.
30. **Thickman, D. I., Ziskin, M. C., and Goldenberg, N. J.,** Effect of display format on detectability, *J. Ultras. Med.,* 2, 117, 1983.
31. **Smith, S. W.,** A contrast-detail analysis of diagnostic ultrasound imaging, *Med. Phys.,* 9(1), 4, 1982.
32. **Filly, R. A., Sommer, F. G., and Minton, M. J.,** Characterization of biological fluids by ultrasound and computed tomography, *Radiology,* 134, 167, 1980.
33. **Scatarige, J. C., Scott, W. W., Donovan, P. J., Siegelman, S. S., and Sanders, R. C.,** Fatty infiltration of the liver: ultrasonographic and computed tomographic correlation, *J. Ultras. Med.,* 3, 9, 1984.
34. **Lee, J. K. T., Dixon, W. T., Ling, D., et al.,** Fatty infiltration of the liver: demonstrations by proton spectroscopic imaging. Preliminary observations, *Radiology,* 153, 1954, 1984.
35. **King, D. L., Lizzi, F. L., Feleppa, E. J., et al.,** Focal and diffuse liver disease studied by quantitative microstructural sonography, *Radiology,* 155, 457, 1985.
36. **Sommer, F. G., Gregory, P. B., Fellingham, L. L., et al.,** Measurement of attenuation and scatterer spacing in human liver tissue, preliminary results, *J. Ultras. Med.,* 3, 557, 1984.
37. **Taylor, J. W., Riely, C. A., Hammers, L., et al.,** Quantitative US attenuation in normal liver and in patients with diffuse liver disease. Importance of fat, *Radiology,* 160, 65, 1986.
38. **DiLelio, A., Cestari, C., Lomazzi, A., et al.,** Cirrhosis: diagnosis with sonographic study of the liver surface, *Radiology,* 172, 389, 1989.
39. **Taylor, K. J. W., Gorelick, F. S., Rosenfield, A. T., et al.,** Ultrasonography of alcoholic liver disease with histological correlation, *Radiology,* 141, 157, 1981.
40. **Shawker, T. H., Moran, B., Linzer, M., et al.,** B-scan echo-amplitude measurement in patients with diffuse infiltrative liver disease, *J. Clin. Ultras.,* 9, 293, 1981.
41. **Hricak, H., Cruz, C., Romanski, R., et al.,** Renal parenchymal disease: sonographic-histologic correlation, *Radiology,* 144, 141, 1982.
42. **Rosenfield, A. T., Taylor, K. J. W., Crade, M., and DeGraaf, C. S.,** Anatomy and pathology of the kidney by gray scale ultrasound, *Radiology,* 128, 737, 1978.
43. **Hayden, C. K., Santa-Cruz, F. R., and Amparo, E. G.,** Ultrasonographic evaluation of the renal parenchyma in infancy and childhood, *Radiology,* 152, 413, 1984.
44. **Hricak, H., Slovis, T. I., Callen, C. W., et al.,** Neonatal kidneys: sonographic anatomic correlation, *Radiology,* 147, 699, 1983.
45. **Scheible, W. and Leopold, G. R.,** High-resolution real-time ultrasonography of neonatal kidneys, *J. Ultras. Med.,* 1, 133, 1982.
46. **Haller, J. O., Berdon, W. E., and Friedman, A. P.,** Increased renal cortical echogenicity: a normal finding in neonates and infants, *Radiology,* 142, 173, 1982.
47. **Choyke, P. L., Grant, E. G., Hoffer, F. A., et al.,** Cortical echogenicity in the hemolytic uremic syndrome: clinical correlation, *J. Ultras. Med.,* 7, 439, 1988.

48. **Graif, M., Shohet, I., Strauss, S., Yahav, J., and Itzchak, Y.,** Hemolytic uremic syndrome: sonographic-clinical correlation, *J. Ultras. Med.,* 3, 563, 1984.
49. **Mahony, B. S.,** The genitourinary system, in *Ultrasonography in Obstetrics and Gynecology,* 2nd ed., Callen, P. W., Ed., W. B. Saunders, Philadelphia, 1988, chap. 11.
50. **Neiman, H. L. and Mendelson, E. B.,** Ultrasound evaluation of the ovary, in *Ultrasonography in Obstetrics and Gynecology,* 2nd ed., Callen, P. W., Ed., W. B. Saunders, Philadelphia, 1988, chap. 21.
51. **Callen, P. W.,** Ultrasound evaluation of gestational trophoblastic disease, in *Ultrasonography in Obstetrics and Gynecology,* 2nd ed., Callen, P. W., Ed., W. B. Saunders, Philadelphia, 1988, chap. 20.
52. **Coleman, B. G., Arger, P. H., and Mulhern, C. B.,** Gray-scale sonographic spectrum of hypernephromas, *Radiology,* 137, 757, 1980.
53. **Charboneau, J. W., Hattery, R. R., Ernest, E. C., et al.,** Spectrum of sonographic findings in 125 renal masses other than benign simple cyst, *Am. J. Radiol.,* 140, 87, 1983.
54. **Butch, R. T., Simeone, J. F., and Mueller, P. R.,** Thyroid and parathyroid ultrasonography, *Radiol. Clin. North Am.,* 23, 57, 1985.
55. **Marchal, G. J., Pylyser, K., Tshibwabwa-Tumba, E. A., et al.,** Anechoic histologic correlation, *Radiology,* 156, 479, 1985.
56. **Carroll, B. A. and Gross, D. M.,** High-frequency scrotal sonography, *Am. J. Radiol.,* 140, 511, 1983.
57. **Rifkin, M. D., Kurtz, A. B., Pasto, M. E., and Goldberg, B. B.,** Diagnostic capabilities of high resolution scrotal ultrasonography: prospective evaluation, *J. Ultras. Med.,* 4, 13, 1985.
58. **Mittelstaedt, C. A.,** The spleen, in *Abdominal Ultrasound,* Mittlestaedt, C. A., Ed., Churchill Livingstone, New York, 1987, 582.
59. **Newlin, N., Silver, T. M., and Stuck, K. J.,** Ultrasonic features of pyogenic liver abscesses, *Radiology,* 139, 155, 1981.
60. **Kuligowska, E., Connors, S. K., and Shapiro, H. H.,** Liver abscess: sonography in diagnosis and treatment, *Am. J. Radiol.,* 138, 253, 1982.
61. **Berwing, K. and Schlepper, M.,** Echocardiographic imaging of the left ventricle by peripheral intravenous injection of echo contrast agent, *Am. Heart J.,* 115, 399, 1988.
62. **Ophir, J. and Parker, K. J.,** Contrast agents in diagnostic ultrasound, *Ultras. Med. Biol.,* 15, 319, 1989.

Chapter 3

BIOLOGICAL TISSUES AS ULTRASONIC SCATTERING MEDIA

K. Kirk Shung and Gary A. Thieme

TABLE OF CONTENTS

0-8493-6568-6/93/$0.00 + $.50

© 1993 by CRC Press, Inc.

I. INTRODUCTION

Historically biological tissues have been treated either as a continuum with varying density and compressibility or as a random distribution of scatterers whose acoustic properties differ from the surrounding medium[1-5] (detailed discussion of this topic can be found in Chapters 4 and 5. Regardless of what theoretical approach is taken, it is recognized that ultrasonic scattering process in biological tissues is primarily affected by (1) the sizes of tissue structures responsible for scattering which determine the correlation lengths of the continuum model and (2) acoustic properties of tissue structures. In this chapter, these issues which are important in assessing the relevance and validity of the theoretical models will be addressed.

Biological tissues are complex structures consisting of cells of different sizes and of different composition interspersed among which are blood vessels carrying blood to and from these cellular structures and ductal networks. This may be the reason that the fundamental ultrasonic scattering structures in a majority of the tissues are still unknown. Fields and Dunn proposed in a paper published in 1972[6] that echographic appearance of a tissue may be related to the connective tissue content of a tissue since acoustic properties of connective tissues such as collagen and elastin seem to be significantly different from those of other tissue components, e.g., fat, water, protein, etc. While several investigations[7-10] later confirmed that collagen plays a significant role in determining tissue ultrasonic properties including scattering, experimental results obtained by Bamber et al.[11] showed a poor correlation between ultrasonic properties and connective tissue content in organs like liver, spleen, and kidney. The reason for this discrepancy may be that variation in connective tissue content among these tissues is too small to be detectable by ultrasound. A hypothesis that sizes of tissue structures may be a dominant factor in the consideration of scattering was thus suggested.[12,13] Whether this model is correct remains to be proven although results from blood, whose biological composition is much simpler, explicitly show that the volume of red blood cells affects ultrasonic scattering to a great extent.[14-16] A similar relationship was also observed for myocardium.[12,13]

II. BLOOD

A. GENERAL PROPERTIES OF BLOOD[17-22]

The main physiological task of the blood is to supply oxygen to the living tissues and return carbon dioxide to the lungs. It is also responsible for collecting nutrients from the gastrointestinal tract, eliminating nongaseous metabolites in the kidney, dissipating heat through various surfaces, and neutralizing foreign biological agents which enter the system.

The blood comprises formed elements including red blood cells (erythrocytes), white blood cells (leukocytes), and platelets (thrombocytes) sus-

TABLE 1
Normal Ranges and Means for Blood Values in Humans and Some Domestic Animals

	RBC[a]	RBC[b] M.D.	PCV[c]	MCV[d]	MCHC[e]	WBC[f]	Thrombocyte[g]
Human man	4.6–6.2	7.4–9.4	40–54	70–94	30–40	4.5–11	2.7–5.5
	(5.4)	(8.4)	(47)	(87)	(33.5)	(7.4)	(4.1)
Human woman	4.2–5.4	7.4–9.4	37–47	74–98	30–40	4.5–11	2.7–5.5
	(4.8)	(8.4)	(42)	(87)	(33.5)	(7.4)	(4.1)
Dog	5.5–8.5	6.7–7.2	37–55	60–77	31–34	6–18	2.9
	(6.8)	(7)	(45)	(70)	(33)	(11)	(4.7)
Pig	5–8	4–8	32–50	50–68	30–34	11–22	3.2–7.2
	(6.5)	(6)	(42)	(63)	(32)	(16)	(5.2)
Cow	5–10	4.5–8	24–48	40–60	26–34	4–12	1–8
	(7)	(5.5)	(35)	(45–55)	(31)	(8)	(5)
Horse	7–13	4–8	32–55	37–50	31–35	7–14	1–6
	(9.8)	(5.7)	(42)	(42)	(33)	(10)	(3.3)
Sheep	8–16	3.2–6.6	24–50	23–48	29–35	4–12	2.5–7.5
	(12)	(4.5)	(38)	(32)	(32)	(9)	(4)

[a]   RBC (red blood cell) — 10^6 per cubic mm of blood.
[b]   RBC M.D. (RBC mean diameter) — microns.
[c]   PCV (packed cell volume) — volume percent.
[d]   MCV (mean corpuscular volume) — cubic microns.
[e]   MCHC (mean corpuscular hemoglobin conc.) — volume percent.
[f]   WBC (white blood cell) — 10^3 per cubic mm of blood.
[g]   Thrombocyte — 10^5 per cubic mm of blood.

pended in a saline solution of three major types of protein: fibrinogen, globulin, and albumin. This continuous suspending medium is called plasma. Cellular elements occupy approximately 45% volume of the whole blood and the remainder is plasma. The normal ranges and means for blood values in humans and some domestic animals are listed in Table 1.[17-23]

When blood comes into contact with any substance other than the wall of a healthy blood vessel, the coagulation processes are initiated and eventually the blood becomes a clot. To prevent the blood from clotting as the blood is withdrawn from the vascular system, anticoagulants such as heparin, sodium citrate, and ethylenediaminetetraacetic acid (EDTA) are typically used. The specific gravity of whole blood ranges between 1.048 and 1.066. The pH of fresh, normal human blood is approximately 7.4.

B. FORMED ELEMENTS
1. Red Cells

The red cells dominate the particular matter in blood, occupying on the average about 40 to 45% by volume of the whole blood. After a known volume of whole blood is centrifuged at a constant speed for a certain period of time, the percentage of the total volume occupied by packed red cells is

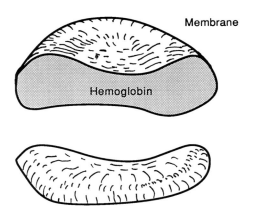

FIGURE 1.   Schematic diagram showing the morphology of a red blood cell.

called hematocrit. In normal human blood, the mean values of hematocrit for the male and female adult are 47 and 42%, respectively. The mean corpuscular volume (MCV) of an erythrocyte for human is approximately 87 μm^3, which is larger than those of pig, 63 μm^3, and cow, 50 μm^3.

The red cell is primarily composed of about 60% water and 31 to 34% hemoglobin. The hemoglobin which carries oxygen is exclusively associated with the red cells. Each cell consists of a fine flexible membrane surrounding the hemoglobin. The red cell shape is generally that of a biconcave discoid as shown in Figure 1. The membrane accounts for only 3% of the mass of the human red cell and its thickness is between 60 and 200 Å. The two concave surfaces of red cells provide an optimal surface area over which exchange of oxygen and carbon dioxide can occur.

The red cells in any animal are never of identical size. For example, the diameters of human erythrocytes range from 7.4 to 9.4 μm, and 67% of them are between 8 and 9 μm.[24] The size distribution of erythrocytes in cow blood is even wider.[21] Further, it has been observed that although the diameter of erythrocyte varies a little from mammal to mammal, as listed in Table 1, the thickness of erythrocyte is practically constant and is about 1.6 to 2 μm.[25]

Human red cells are in osmotic equilibrium with plasma, which has an osmotic pressure similar to that of a 0.9% saline solution. Therefore, when the red cell is suspended in a 0.9% saline solution, its volume would not change. However, if the red cell is suspended in a hypotonic solution, it takes on water and becomes spherical. Although the red cell membrane is flexible, when the erythrocyte swells beyond its critical volume, hemolysis takes place. The negative surface charge of red cells results mainly from the presence of ionogenic carboxyl groups of sialic acid on the cell surface.[26,27]

2. White Cells

White cells, or leukocytes, are responsible for protecting the body from diseases. Their diameter varies from 7 to 22 μm. Under conditions of normal health their volume concentration in the blood is totally insignificant compared to the red cells. There is about only one white cell to every 1000 red cells in human blood. Even in acute infection the increased number does not constitute a significant change.

3. Platelets

Platelets are an essential ingredient for blood to coagulate. They are much smaller than red or white cells, being 7 to 8 μm^3 in volume. Their shape is irregular and their volume concentration in normal blood is only about 0.3%. Because the number of platelets present is only one tenth of the red cells and each has a much smaller volume than a red cell, their direct effect on the resistance to flow of blood or on the acoustic properties of blood is probably insignificant.[24]

4. Plasma

Blood plasma consists of a solution of plasma protein in an aqueous medium in which Na^+s and Cl^-s are the major cations and anions. Other important ions such as K^+, Ca^{++}, Mg^{++}, HCO_3^-, phosphate, and sulfate are also present in small quantities. Proteins constitute in total some 7% of the total volume of the plasma. They can be divided into three main groups: (1) albumin (4%), (2) globulin (2.7%), and (3) fibrinogen (0.3%). Plasma also contains hormones, enzymes, and antibodies.

In human blood plasma the mean fibrogen concentration is about 300 mg/dl. Fibrinogen is a globulin with a molecular weight of approximately 450,000. It is one of the largest molecules among the plasma proteins and its length is approximately 20 times its width. This large asymmetry in shape results in an intrinsic viscosity of about 27.[28] Therefore, although its concentration in blood is relatively small, it is an important factor in determining the total viscosity of plasma. In addition, plasma fibrinogen is known to be responsible for red blood cell aggregation in whole blood,[29,30] which may be of great significance in affecting the blood viscosity at low shear rates.[24,31,32]

C. RED BLOOD CELL AGGREGATION

Since Fahraeus discovered the RBC aggregation phenomenon in 1921,[29] RBC aggregation and its various effects have been extensively studied. Some of the properties of RBC aggregation in whole blood relevant to ultrasonic scattering are summarized below.

It is known from experimental observations that RBC aggregation in whole blood depends on the presence of certain large asymmetric macromolecules in the plasma such as fibrinogen and some of the globulin fractions.[29,30] The process will not occur in their absence, for example, with red cells suspended in saline[33] or serum albumin at normal concentration.[29] The occurrence of RBC aggregation is also dependent on the state of blood flow and animal species.[34,35] In stationary whole blood or in whole blood at low shear rates,

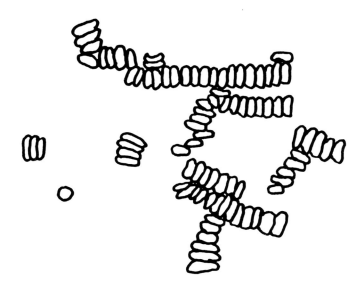

FIGURE 2.   Schematic diagram showing a network of normal human red blood cells forming an aggregate.

red blood cells aggregate to form rouleaux. Porcine and human erythrocytes have been found to have a stronger aggregation tendency than bovine red cells. However, at high shear rates they are monodispersed and aggregation is prevented. In normal human blood, it was shown that red cell disaggregation is essentially complete when the apparent shear rate is increased above 50/ s.[30,31,36]

RBC aggregation was observed to occur progressively faster with increasing fibrinogen concentration. In a steady flow of low shear rate, the degree of cell aggregation appeared to be dependent on the concentration of fibrinogen.[37] It was further observed that an increase in hematocrit up to the normal physiological value generally promotes RBC aggregation because the formation of RBC aggregates requires the presence of an adequate cell concentration to provide sufficient probability of cell encounter.[33]

Figure 2 shows the schematic drawing of a network of normal human red cells forming aggregates from a microscopic picture, where many red cells line up in chains by contacts between their discoidal surfaces (side-to-side) to give rouleaux (primary aggregation) and the secondary aggregation is due to end-to-side contact. Microscopic examination showed that the surfaces of adjacent red cells in the rouleaux follow each other in curvature, resulting in a rather uniform intercellular distance.[38] This distance of rouleaux in fibrinogen suspensions averages 25 nm, which is shorter than the molecular length of fibrinogen, 65 nm.[38]

Rouleaux can develop *in vivo* and their behavior is similar to that *in vitro*. RBC aggregation also occurs in the suspensions with such macromolecules as dextrans (plasma expanders).[37] In addition, in many diseases such as myeloma, cholangitis, and myocardial infarction, the speed and the degree of red cell aggregate formation are found to increase. In fact, erythrocyte sedimentation rate (ESR),[29] based on the fact that RBC aggregation would cause the erythrocytes to fall faster, is widely used in hospital as a quick check on the health of patients.

RBC aggregation is important not only because of clinical and diagnostic consideration but also because of its influence on the viscosity of blood at low shear rate. The flow properties of normal human whole blood are known to be approximately Newtonian at high shear rates but become non-Newtonian at low shear rates.[24,31,32] This non-Newtonian behavior, i.e., the apparent viscosity increases with decreasing shear rate, is believed to be caused by RBC aggregation at low shear rates. As will be discussed in Chapters 5 and 9, RBC aggregation plays an important role in affecting the ultrasonic backscatter from blood.

1. Mechanisms of RBC Aggregation

The mechanism of rouleaux formation is still poorly understood. Several theories have been proposed to explain the mechanism of RBC aggregation induced by macromolecules.[29,31] Among them, the model of red cell aggregation by macromolecular bridging of cell surface, proposed by Chien,[31] is generally more acceptable.

In Chien's model,[31] the terminal segments of the macromolecule are absorbed onto the surfaces of two adjacent red cells and the central segment occupies the intercellular space. Red cells suspended in a macromolecular solution are brought into close range by external forces, e.g., gravitational sedimentation or thermal agitation. The absorption of macromolecules on the cell surface provides the attractive force. On the other hand, because red cell surface is negatively charged mainly as a result of the presence of sialic acid, the interaction of surface potentials causes mutual repulsion among the cells. RBC aggregation occurs when the macromolecular bridging force overcomes the electrostatic repulsive force and mechanical shearing force. The net aggregating energy is stored as a change in membrane strain energy. Meanwhile, the discoid shape of red cell and the deformability of the normal cell membrane tend to facilitate the red cells to aggregate into rouleaux.

Blood has been treated both as a continuum and as a dense distribution of small scatterers in biomedical ultrasound literature.[2,3,5] This topic will be discussed in detail in Chapter 5.

III. SOLID TISSUE ORGANS

Ultrasound imaging studies are frequently employed to diagnose structural abnormalities of solid tissue organs in acoustically accessible parts of the body. Architectural alterations resulting from disease processes modify the scattering properties of the tissues. In turn, these changes may be recognized as abnormal features in the ultrasound images. During this pattern recognition process, the image properties of the tissues are either directly compared to adjacent normal tissues or indirectly compared to a memorized normal pattern. Although a sonographic–anatomic correlation may be observed, defining the histologic tissue features responsible for the altered ultrasonic appearance of a tissue is sometimes elusive. Many investigators have correlated histologic features with acoustic properties of normal and pathologic tissue states.[11-13,43-82]

Perhaps the simplest way to consider this issue is to divide the image features into two groups: (1) macroscopic structures represented as large inhomogeneities with physical dimensions similar to or greater than the sample volume dimensions and (2) microscopic structures represented as small inhomogeneities with physical dimensions less than the sample volume dimensions. Here, sample volume is defined by the axial resolution, lateral resolution, and slice thickness parameters. These physical dimensions are generally in the 1/2- to 3-mm size range at the focus for commercial clinical ultrasound systems.

Large inhomogeneities typically include organ and vessel boundaries and connective tissue planes. These divisions are generally visible to the unaided eye during inspection of cut specimens and occur where significant changes in acoustic impedance are present. The common link is a structural size which is similar to or greater than the dimensions of the sample volume and a structural character which strongly reflects sound energy. The best examples of image features in group one can be seen in bistable images from early ultrasound systems and in high threshold images from modern gray-scale systems. These features can be thought of as the skeleton and represent strongly reflective structures at the macroscopic level. The soft tissue features eliminated by threshholding are weakly reflective structures and are considered in the second group.

By contrast, the image features in group two are the product of weak sound interactions at the diffuse scattering level and are represented by varying gray levels. By definition, scattering occurs from structures significantly smaller than the sample volume dimensions. Small inhomogeneities produce small changes in acoustic impedance at the scatterer sites, whose depth differences are smaller than the axial resolution (pulse length) of the transducer. Wave fronts returning to the transducer from these sites are out of phase and build an interference pattern. The transducer transforms the sum of these wave front pressures into an electrical signal with varying amplitude peaks and valleys and many frequency components. When a sequence of these A-mode lines is organized into a B-mode two-dimensional image, a texture pattern of dots with varying degrees of brightness is generated. This interference phenomenon is similar to that encountered for images produced by laser light. The random dot pattern, observed in both the light and the sound images, is

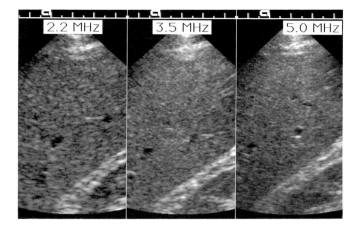

FIGURE 3. Images of the liver and the anterior aspect of the right kidney have been made using three different transducers with center frequencies of 2.2, 3.5, and 5 MHz. The dot size shifts from coarse to fine as the center frequency increases. This reflects the changing speckle pattern.

termed *speckle*; and its aggregate effect is often referred to as *texture*. However, since the electrical signal characteristics are not simply related to either the number or the location of the scatterer sites, the tissue texture pattern is in general not a true image of the histologic structure but rather an interference pattern that is determined to a great extent by the beam characteristics.[85-91]

In a two-dimensional image, speckle size has both axial and lateral dimensions. Axial speckle size is proportional to the bandwidth of the transducer (e.g., spatial pulse length) and is depth-independent since tissue attenuation which is frequency-dependent could affect the pulse spectrum and therefore bandwidth. Lateral speckle size is determined by the ultrasonic beam width which is a function of depth of penetration and focusing properties of the transducer. In Figure 3, images of liver made with 2.2, 3.5, and 5 MHz transducers show how speckle size decreases with increasing frequency, which improves both axial and lateral resolution.

From the preceding discussion, one can easily see that the texture observed in ultrasound images is a speckle pattern which is determined by the characteristics of the transducer and is therefore machine-dependent. In addition, attenuation of sound energy by intervening tissues negatively impacts speckle size. Since the texture pattern produced from diffuse scattering is corrupted by imaging system characteristics, tissue characterization by simple visual inspection and by simple computer analysis may not be as straightforward as one might think. However, sophisticated computer analysis using corrections for beam properties and attenuation applied to the radio frequency signal (prior to image processing stages) may yield clinically significant tissue characterization results. But, these methods are still largely experimental and not commercially available. In-depth discussions of speckle, texture, and sonographic transducer and equipment performance can be found in References 83 through 92 as well as Chapters 1 and 5 of this book.

Although the texture pattern exhibited by a tissue is corrupted by imaging system characteristics, it still carries an important tissue signature resulting from diffuse scattering. The diffuse scattering level from a tissue is dependent upon the size of scatterers, the density of scatterers, and the acoustic impedance difference between the scatterers and the surrounding medium. Details on this subject can be found in Chapters 4 and 5. These three parameters vary significantly among various tissue types, usually in complex ways. The frequency-dependent backscatter coefficient describes the aggregate effect of these parameters. A list of histologic features of tissues that may influence backscattering includes the connective tissue background, the size and density of cells, the physical arrangement of cells, the homogeneity or heterogeneity of cell populations, and the water content and distribution. Given this milieu of parameters and properties, designing even simple tissue models that can reflect reality is challenging. *A priori* prediction of backscatter response from biological tissues is analogous to a "black art". Fortunately, this multifaceted aspect of diffuse scattering provides a rich range of responses to interrogating sound energy and is the basis for our ability to depict differences among tissues using gray-scale ultrasound imaging, even though the mechanisms are not fully understood.

Visual analysis of patterns in images is not based upon observation of individual speckle. Rather, it is based upon the aggregate effect of speckle over a region of interest whose size may have physical dimensions of one millimeter to several centimeters. In an ultrasound image each small region of interest has a mean gray level which represents the backscattered sound intensity. From this pattern of gray levels emerge representations of physical structures whose sizes are greater than the spatial resolution, as determined by the sample volume dimensions. For example, a medullary pyramid of the kidney is a region of interest represented by aggregates of speckle with one mean gray level and histogram distribution. The adjacent cortex of the kidney is a second region of interest represented by aggregates of speckle with another mean gray level and histogram distribution (Figure 4). Thus, there is a direct anatomic–sonographic correlation at the macroscopic level. In contrast, there is only an indirect relationship between speckle and histology (at the microscopic level). However, the histology determines the diffuse scattering properties of each structure and, therefore, the mean gray levels observed in ultrasound images.

A disease process may alter the histology or gross architecture. Loss of cell populations, infiltration by a foreign cell population, increase in connective tissue content due to scarring, and alteration in lipid or protein or carbohydrate content of cells are a few examples of mechanisms. These changes may alter the diffuse scattering properties by modifying the size,

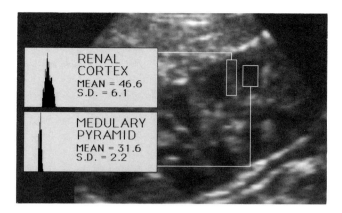

FIGURE 4. The darker medullary pyramids and the lighter cortex are macroscopic structures which are much larger than the sample volume dimensions. Histograms are used to measure the first order characteristics of mean gray level and standard deviation. These values describe the characteristics of the speckle pattern and are not specifically characteristic of the underlying histology.

density, and distribution of scatterers, and the acoustic impedance differences encountered by the beam of sound energy. For example, a focal mass may appear hyperechoic, hypoechoic, or isoechoic with respect to surrounding normal tissues; or, a homogeneous tissue may become heterogeneous. The gray levels may change in uniform or non-uniform ways or perhaps not at all. Not all disease processes (e.g., biochemical enzyme inhibition) will alter the histology. Not all changes in histology will result in diffuse scattering changes. Therefore, it is possible to observe no change in the ultrasound appearance of an organ, even though physiologic function may change dramatically.

A. MYOCARDIUM

The structure of the myocardium is relatively simple compared to other soft tissues. It consists of predominantly cardiac muscle fibers or cells, interspersed among which is a network of blood vessels and bile ducts.[39] Figure 5 is a light micrograph of the bovine myocardium stained with hematoxylin and eosin at a magnification of 312. The myocardial fibers are more or less cylindrical in shape with a diameter between 10 to 30 μm depending upon the animal species. The cardiac muscle fibers occupy approximately 90% of the volume of the myocardium. Each cardiac muscle cell has one or two nuclei usually located near the mid-portion of the cell. The cardiac muscle cells are interconnected via junctions known as intercalated discs that have areas of low electrical resistance allowing rapid transmission of contractile excitation from one fiber to the other.

Previous investigations supported by experimental data[9,12,13] suggested that myocardium may be treated as densely packed small cylindrical scatterers

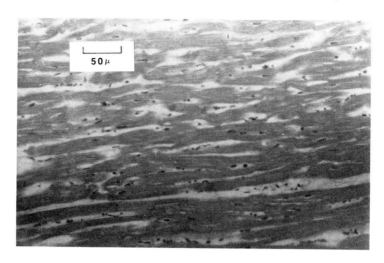

FIGURE 5.   Light micrograph of H and E stained normal bovine heart at a magnification of 312. Individual cardiac muscle cells are clearly seen.

which have a 3rd power frequency dependence.[40] As such, myocardium is considered an anisotropic medium, i.e., its scattering properties are dependent upon tissue orientation relative to the incident direction of the wave.[41]

B. LIVER

The liver is the largest gland and the most biochemically active organ in the body. It is highly perfused with blood from the portal vein (80%) and the hepatic artery (20%).[39,42] Its functions include carbohydrate storage in the form of glycogen, ketone body formation, bile formation, detoxification of drugs and toxins, production of plasma globular proteins, urea formation, inactivation of polypeptide hormones, metabolism of fat, and maintenance of blood glucose levels between meals. It is covered by a thin connective tissue capsule.

Structurally and functionally the liver is divided into lobules. The liver lobules are a polygonal mass of tissue composed of a central vein surrounded by plates of hepatocytes. These plates extend radially from the central vein to the periphery. At the corners of the lobule the portal triads are found. The portal triads contain an arteriole, a venule, a bile duct, and lymphatic vessels. All of these structures are surrounded by a sheath of connective tissue. In general, liver has three to six portal triads per lobule. Blood flows from the portal triads via the liver sinusoids among the plates toward the central vein.

In humans and dogs the lobules are in close proximity to one another. However, in the pig the lobules are separated by a layer of connective tissue. This layer is so thick that it is plainly visible with the unaided eye.

In Figure 6 histologic sections at increasing magnification demonstrate the physical sizes and relationship of component structures in monkey liver.

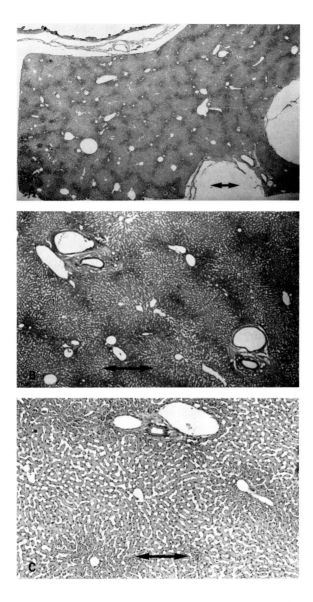

FIGURE 6. Morphology of monkey liver. (A) Double arrow is 1 mm. Low power light micrograph shows lobular architecture of the liver. Each central lobule is partially outlined by a darker ring and contains a central vein. Essentially all vessels are less than 0.5 mm diameter and might act as diffuse scattering sites. (B) Double arrow is 0.5 mm. Photomicrograph at 4× power shows two portal triads which are about 0.5 mm diameter. The bile duct, hepatic artery, and portal vein are surrounded by a lighter connective tissue which might act as a diffuse scatterer. (C) Double arrow is 0.2 mm. Photomicrograph at 10× power shows the hepatocytes and sinusoids (white channels). They are too small to individually act as scattering sites. A portal triad is seen at top center.

At low power (Panel A) this 10 by 6 mm section shows two large vessels (about 2 mm diameter), which might be resolved by an ultrasound imaging system. The many small liver lobules, each with its central hepatic vein, are slightly less than 1 mm in diameter. Dark boundaries outline each lobule but are not necessarily acoustically significant. The many small vessels are only a few tenths of a millimeter in diameter and are potential sites of diffuse scattering events and sources of ultrasonic speckle. At $4 \times$ power, the portal triads are also only a few tenths of a millimeter diameter each (Panel B), and the cellular network of the venous sinusoids are becoming visible. At $10 \times$ power, the individual cells are now evident (Panel C). Aggregates of hepatocytes outlining sinuoids possibly could act as scattering sites at the Rayleigh scattering level. That there are few sonographically resolvable structures is readily evident; however, there are many structures smaller than the dimensions of the sample volume that could contribute to diffuse scatttering.

Which anatomical components in liver are responsible for ultrasonic scattering is an unresolved issue. Some investigators[43] suggest that the connective tissue network is the most likely source of ultrasonic scattering whereas others[13] propose hepatic cells. Analytically liver has been treated mostly as an isotropic continuum presumably due to its anatomic complexity.[4,44]

C. SPLEEN

Spleen is a large lymphoid tissue.[39] Its major function is to filter blood, thus acting as a phagocytic site for blood cells and a defense against microorganisms that enter the blood. It also acts as a site of antibody formation since it is in contact with many antigens carried in the blood.

Like the liver, the spleen is covered by a capsule, made of dense connective tissue. The capsule of the spleen gives off supporting connective tissue trabeculae.

In dogs the splenic trabeculae contain much smooth muscle. Thus, in stressful situations as well as under hormonal influence, the spleen can act as a blood reservoir, i.e., upon contraction the blood within the spleen can be put into circulation. This is not very relevant in humans since the spleen does not contain as much smooth muscle as the dog spleen.

The splenic parenchyma is composed mostly of white and red pulp supported by a connective tissue network rich in reticular fibers. Specifically, the white pulp is primarily composed of T and B lymphocytes forming a "sleeve" surrounding the central arteries. In turn the white pulp is surrounded by the red pulp which is made up of reticular tissue (splenic cords) surrounding the endothelial lined sinusoids. The splenic cords are continuous throughout and of irregular structure. They are made up of reticular cells, fixed and wandering macrophages, monocytes, lymphocytes, and plasma cells.[39,42] Blood in the spleen flows from the splenic artery through several branches to the central arteries and then the pulp arteries, passes through the sinusoids, and exits via the trabecular veins.

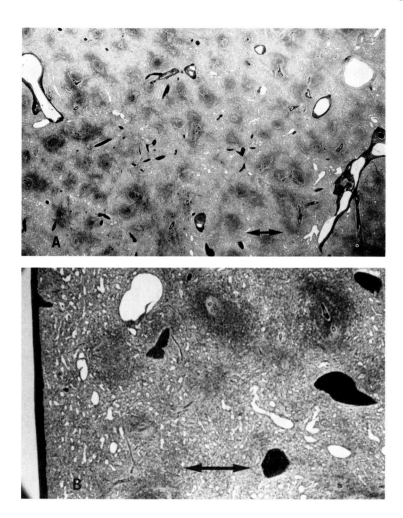

FIGURE 7. Morphology of human spleen. (A) Double arrow is 1 mm. Low power photomicrograph shows fewer blood vessels (voids) per unit area when compared to liver. Dark gray patches are white pulp which surround central arterioles and have diameters of about 0.1 mm. The light gray background is red pulp. (B) Double arrow is 0.5 mm. Photomicrograph at 4× power shows two dark gray patches of white pulp in the upper right half of the image. Four clumps of black stained connective tissue are several tenths of a millimeter diameter and might act as diffuse scattering sites.

In Figure 7 histologic sections of the spleen have been prepared with reticulum stain which preferentially identifies connective tissue. This is a 10 by 6 mm sample taken from the periphery of the spleen. There are no ultrasonically resolvable structures. The dark patches are white pulp which surround central arterioles and have diameters of about one tenth of a millimeter.

The white pulp is surrounded by red pulp (light gray background). Some larger vessels of a few tenths of a millimeter diameter are noted. Very little experimental data exist on ultrasonic scattering in spleen.[13] The likely scattering components are the connective network and blood vessels. Another possible candidate would be the white pulp (including the central artery) if acoustic properties of the white pulp and the red pulp are sufficiently different such that a boundary can be defined.

D. KIDNEY

The major function of the kidney is the maintenance of homeostasis, in other words, to maintain fluid and electrolyte balance and to remove metabolic wastes.[39,42] Its major functional unit is the nephron which is a long tortuous cylinder of cells whose primary function is to filter blood plasma. The kidney is surrounded by several layers of fat and is enclosed by a thin collagenous capsule. Its parenchyma is divided into two major sections; the outermost known as the cortex, and the innermost known as the medulla (renal pyramids).

Anatomically the cortex consists of renal corpuscles and tubules, and blood vessels. The renal corpuscles consist of the glomeruli which are dense, rounded structures enclosed by Bowman's capsule. The two structures are separated by Bowman's space. The mean diameter of the renal corpuscles is between 100 and 200 μm. The medulla is primarily composed of the loops of Henle, collecting ducts, and the vasa recta.

In Figure 8 histologic sections at increasing magnification demonstrate the physical size relationships of component structures of the newborn human kidney. At low power (Panel A) this 10 by 6 mm section shows the clearly different histologic patterns for the cortex and medulla. Intuitively, the more homogeneous character of the medulla should have fewer scatterer sites; hence, the hypoechoic nature of the medullary pyramid is observed in sonograms. The obvious heterogeneous character of the cortex should provide an abundance of scattering sites; hence, diffuse scattering should be greater than the medulla, and the cortex should be hyperechoic. The arcuate artery and the calyx are the only structures at the 0.5 mm size. Both are potentially large enough to be resolved on sonograms and often appear as bright specular echoes. At 4× power, the heterogeneous appearance of the cortex (Panel B) and the more homogeneous appearance of the medulla (Panel C) are significant for the renal corpuscles containing glomeruli seen in the cortex. The tubule network background is similar in both. The size range of the renal corpuscles at 0.1 to 0.2 mm is readily apparent (Panel D).

Structurally and functionally the kidneys are very similar in humans and animals. In dogs, particularly, the cortex is approximately 1 cm thick and the medulla is primarily represented by five to six medullary pyramids.

Kidney cortex has been shown to be ultrasonically anisotropic.[45,46] Recent results which will be described in Chapter 4 suggested that ultrasonic scattering

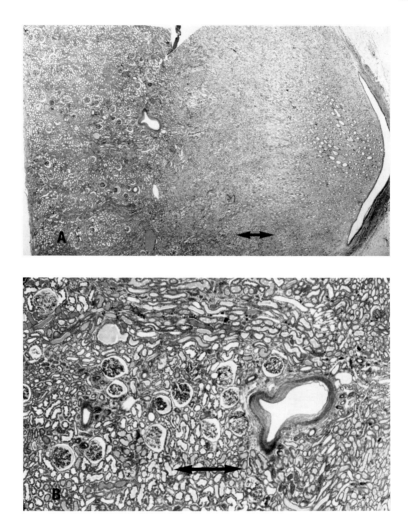

FIGURE 8. Morphology of newborn human kidney. (A) Double arrow is 1 mm. Low power photomicrograph shows the rather homogeneous medullary pyramid in the right half and heterogeneous cortex in the left half. The lucent crescent at the extreme right is the calyx which normally contains urine. (B) Double arrow is 0.5 mm. Photomicrograph at 4× power shows the heterogeneous pattern of tubules and renal corpuscles containing glomeruli which might act as diffuse scatterers. The arcuate artery is 0.5 mm diameter and may produce a specular echo.

in kidney cortex can be modeled as a two-component scattering medium. The two types of scatterers involved are the renal corpuscles, and the renal tubules and the blood vessels.

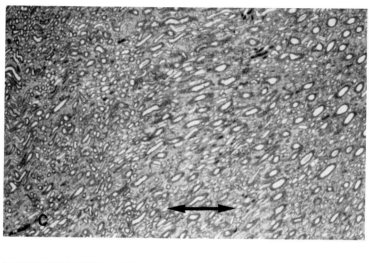

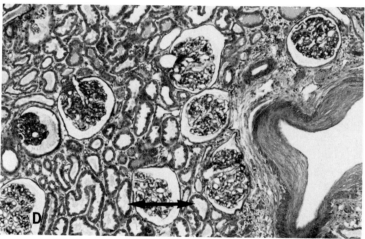

FIGURE 8 (continued).   (C) Double arrow is 0.5 mm. Photomicrograph at 4× power shows the more uniform pattern of tubules in the medullary pyramid. Most structures are under 0.1 mm diameter. (D) Double arrow is 0.2 mm. Photomicrograph at 10× power shows the detail of renal corpuscles which have diameters of 0.1 to 0.2 mm. Bowman's capsule surrounds a space which contains the glomerular tuft. Tubules are seen in between.

REFERENCES

1. **Chivers, R. C.**, The scattering of ultrasound by human tissues — some theoretical models, *Ultras. Med. Biol.*, 3, 1, 1977.
2. **Atkinson, P. and Berry, M. V.**, Random noise in ultrasonic echoes diffracted by blood, *J. Phys. A: Math. Nucl. Gen.*, 7, 1293, 1974.
3. **Angelsen, B. J.**, A theoretical study of the scattering of ultrasound from blood, *IEEE Trans. Biomed Eng.*, BME-27, 61, 1980.
4. **Lizzi, F. L., Greenbaum, M., Feleppa, E. J., and Elbaum, M.**, Theoretical framework for spectrum analysis in ultrasonic tissue characterization, *J. Acoust. Soc. Am.*, 73, 1366, 1983.
5. **Lucas, R. J. and Twersky, V.**, Inversion of ultrasonic scattering data for red cell suspensions under different flow conditions, *J. Acoust. Soc. Am.*, 82, 794, 1987.
6. **Fields, S. and Dunn, F.**, Correlation of echographic visualization of tissue with biological composition and physiologic state, *J. Acoust. Soc. Am.*, 54, 809, 1973.
7. **Goss, S. A. and O'Brien, W. D.**, Dependence of the ultrasonic properties of biological tissues on constituent proteins, *J. Acoust. Soc. Am.*, 67, 1041, 1980.
8. **O'Donnell, M., Mimbs, J. W., and Miller, J. G.**, The relationship between collagen and ultrasonic attenuation in myocardial tissue, *J. Acoust. Soc. Am.*, 65, 512, 1979.
9. **O'Donnell, M., Mimbs, J. W., and Miller, J. G.**, Relationship between collagen and ultrasonic backscatter in myocardial tissue, *J. Acoust. Soc. Am.*, 69, 580, 1981.
10. **Greenleaf, J. F.**, *Tissue Characterization with Ultrasound*, CRC Press, Boca Raton, FL, 1986.
11. **Bamber, J. C., King, J. A., and Hill, C. R.**, Acoustic properties of normal and cancerous liver. II. Dependence on tissue structure, *Ultras. Med. Biol.*, 7, 135, 1979.
12. **Shung, K. K.**, Ultrasonic characterization of biological tissues, *ASME J. Biomech. Eng.*, 107, 309, 1985.
13. **Fei, D. Y. and Shung, K. K.**, Ultrasonic backscatter from mammalian tissues, *J. Acoust. Soc. Am.*, 78, 871, 1985.
14. **Borders, S. H., Fronek, A., Kemper, W. S., and Franklin, D.**, Ultrasonic energy backscattered from blood: an experimental determination of the variation of sound energy with hematocrit, *Ann. Biomed. Eng.*, 8, 83, 1978.
15. **Yuan, Y. W. and Shung, K. K.**, Ultrasonic backscatter from flowing whole blood. I. Dependence on shear rate and hematocrit, *J. Acoust. Soc. Am.*, 84, 52, 1988.
16. **Yuan, Y. W. and Shung, K. K.**, Ultrasonic backscatter from flowing whole blood. II. Dependence on frequency and fibrinogen, *J. Acoust. Soc. Am.*, 84, 1159, 1988.
17. **Miale, J. B.**, *Laboratory Medicine — Hematology*, C. V. Mosby, St. Louis, 1977.
18. **Platt, W. R.**, *Color Atlas and Textbook of Hematology*, J. B. Lippincott, Philadelphia, 1979.
19. **Oscar, W. and Schalam**, *Veterinary Hematology*, Lea & Febiger, Philadelphia, 1961.
20. **Agar, N. S. and Board, P. G.**, *Red Blood Cells of Domestic Mammals*, Elsevier, Amsterdam, 1983.
21. **Albritton, E. C.**, *Standard Values in Blood*, W. B. Saunders, Philadelphia, 1953.
22. **Harris, J. W.**, *The Red Cell*, Harvard University Press, Cambridge, MA, 1965.
23. **Wintrobe, M. M.**, *Clinical Hematology*, 7th ed., Lea & Febiger, Philadelphia, 1974.
24. **Ponder, E.**, *Hemolysis and Related Phenomena*, Churchill Livingstone, London, 1948.
25. **Whitmore, E. L.**, *Rheology of the Circulation*, Pergamon Press, Hedington Hill Hall, Oxford, U.K., 1968.
26. **Eylar, E. H., Madoff, M. A., Brody, O. V., and Oncley, J. L.**, The contribution of sialic acid to the surface charge of the erythrocytes, *J. Biol. Chem.*, 237, 1992, 1962.
27. **Cook, G. M. W., Heard, D. H., and Seaman, G. V. F.**, Sialic acid and the electrokinetic change of the human erythrocyte, *Nature*, 191, 44, 1961.
28. **Tanford, C.**, *Physical Chemistry of Macromolecules*, John Wiley & Sons, New York, 1965.

29. **Fahraeus, R.,** The suspension stability of blood, *Physiol. Rev.,* 9, 241, 1921.
30. **Schmid-Schonbein, H., Gaeghtgens, P., and Hirsch, H.,** On the shear rate dependence of red cell aggregation in vitro, *J. Clin. Invest.,* 47, 1447, 1968.
31. **Chien, S.,** Biophysical behavior of red cells in suspension, in *The Red Blood Cell,* Vol. II, 2nd ed., Surgenor, D. M., Ed., Academic Press, New York, 1975.
32. **Usami, S. and Chien, S.,** Optical reflectometry of red cell aggregation under shear flow, 7th Eur. Conf. Microcirc. Proc., 91, 1972.
33. **Chien, S., Usami, S., Dellenback, R. J., Gregersen, M. I., Nanninga, L. B., and Guest, M. M.,** Blood viscosity: influence of erythrocyte aggregation, *Science,* 157, 829, 1967.
34. **Charm, S. E. and Kurland, G. S.,** *Blood Flow and Microcirculation,* John Wiley & Sons, New York, 1974.
35. **Chien, S., Usami, S., Dellenback, R. J., and Bryant, C. A.,** Comparative hemorheology-hematological implications of species difference in blood viscosity, *Biorheology,* 8, 33, 1971.
36. **Brooks, D. E., Goodwin, J. W., and Seaman, G. V. F.,** Interaction among erythrocytes under shear, *J. Appl. Physiol.,* 28, 172, 1970.
37. **Cokelet, G. R., Meiselman, H. J., and Brooks, D. E.,** *Erythrocyte Mechanics and Blood Flow,* Alan R. Liss, New York, 1980.
38. **Chien, S. and Jan, K. M.,** Ultrastructural basis of the mechanism of Rouleaux formation, *Microvasc. Res.,* 5, 155, 1973.
39. **Wheater, P. R., Burkitt, H. G., and Daniels, V. G.,** *Functional Histology,* Churchill Livingstone, Edinburgh, 1979.
40. **Morse, P. M. and Ingard, K. U.,** *Theoretical Acoustics,* McGraw-Hill, New York, 1968.
41. **Mottley, J. G. and Miller, J. G.,** Anisotropy of the ultrasonic backscatter of myocardial tissue: theory and measurements in vitro, *J. Acoust. Soc. Am.,* 83, 755, 1988.
42. **Junqueira, L. C., Carneiro, J., and Long, J.,** *Basic Histology,* Lange Medical, Los Altos, CA, 1986.
43. **Nicholas, D.,** Evaluation of backscattering coefficients for excised human tissues: results, interpretations, and associated measurements, *Ultras. Med. Biol.,* 8, 17, 1982.
44. **Campbell, J. A. and Waag, R. C.,** Measurement of calf liver ultrasonic differential and total scattering cross sections, *J. Acoust. Soc. Am.,* 75, 603, 1984.
45. **Rubin, J. M., Carson, P. L., and Meyer, C. R.,** Anisotropic ultrasonic backscatter from renal cortex, *Ultras. Med. Biol.,* 14, 507, 1988.
46. **Insana, M. F., Wagner, R. F., Brown, D. G., and Hall, T. J.,** Describing small-scale structure in random media using pulse echo ultrasound, *J. Acoust. Soc. Am.,* 87, 179, 1990.
47. **Hricak, H., Cruz, C., Romanski, R., et al.,** Renal parenchymal disease: sonographic-histologic correlation, *Radiology,* 144, 141, 1982.
48. **Choyke, P. L., Grant, E. G., Hoffer, F. A., et al.,** Cortical echogenicity in the hemolytic uremic syndrome: clinical correlation, *J. Ultras. Med.,* 7, 439, 1988.
49. **Graif, M., Shohet, I., Stauss, S., Yahav, J., and Itzchak, Y.,** Hemolytic uremic syndrome: sonographic-clinical correlation, *J. Ultras. Med.,* 3, 563, 1984.
50. **Hayden, C. K., Santa-Cruz, F. R., and Amparo, E. G.,** Ultrasonographic evaluation of the renal parenchyma in infancy and childhood, *Radiology,* 152, 413, 1984.
51. **Hricak, H., Slovis, T. L., Callen, C. W., et al.,** Neonatal kidneys: sonographic anatomic correlation, *Radiology,* 147, 699, 1983.
52. **Haller, J. O., Berdon, W. E., and Friedman, A. P.,** Increased renal cortical echogenicity: a normal finding in neonates and infants, *Radiology,* 142, 173, 1982.
53. **Scheible, W. and Leopold, G. R.,** High-resolution real-time ultrasonography of neonatal kidneys, *J. Ultras. Med.,* 1, 133, 1982.

54. **Charboneau, J. W., Hattery, R. R., Ernest, E. C., et al.**, Spectrum of sonographic findings in 125 renal masses other than benign simple cyst, *Am. J. Radiol.*, 140, 87, 1983.

55. **Hartman, D. S., Goldman, S. M., Friedman, A. C., et al.**, Angiomyolipoma: ultrasonic-pathologic correlation, *Radiology*, 139, 451, 1981.

56. **Coleman, B. G., Arger, P. H., and Mulhern, C. B.**, Gray-scale sonographic spectrum of hypernephromas, *Radiology*, 137, 757, 1980.

57. **Subramanyam, B. R., Raghavendra, N., and Madamba, M. R.**, Renal transitional cell carcinoma: sonographic and pathologic correlation, *J. Clin. Ultras.*, 10, 203, 1982.

58. **Blei, C. L., Hartman, D. S., Friedman, A. C., et al.**, Papillary renal cell carcinoma: ultrasonic/pathologic correlation, *J. Clin. Ultras.*, 10, 429, 1982.

59. **Hricak, H., Cruz, C., Eyler, W. R., et al.**, Acute post-transplantation renal failure: differential diagnosis by ultrasound, *Radiology*, 139, 441, 1981.

60. **Hricak, H., Romanski, R. N., and Eyler, W. R.**, The renal sinus during allograft rejection: sonographic and histopathologic findings, *Radiology*, 142, 693, 1982.

61. **Fried, A. M., Woodring, J. H., Loh, F. K., et al.**, The medullary pyramid index: an objective assessment of prominence in renal transplant rejection, *Radiology*, 149, 7871, 1983.

62. **Slovis, T. L., Babcock, D. S., and Hricak, H.**, Renal transplant rejection: sonographic findings in children, *Radiology*, 153, 659, 1984.

63. **Tanaka, S., Kitamura, T., and Imaoka, S.**, Hepatocellular carcinoma: sonographic and histologic correlation, *Am. J. Radiol.*, 140, 701, 1983.

64. **Sheu, J. C., Sung, J. L., Chen, D. S., et al.**, Ultrasonography of small hepatic tumors using high-resolution linear-array real-time instruments, *Radiology*, 150, 797, 1984.

65. **Marchal, G. J., Pylyser, K., Tshibwabwa-Tumba, E. A., et al.**, Anechoic halo in solid liver tumors: sonographic microangiographic and histologic correlation, *Radiology*, 156, 479, 1985.

66. **King, D. L., Lizzi, F. L., Feleppa, E. J., et al.**, Focal and diffuse liver disease studied by quantitative microstructural sonography, *Radiology*, 155, 457, 1985.

67. **Scatarige, J. C., Scott, W. W., Donovan, P. J., Siegelman, S. S., and Sanders, R. C.**, Fatty infiltration of the liver: ultrasonographic and computed tomographic correlation, *J. Ultras. Med.*, 3, 9, 1984.

68. **Taylor, K. J. W., Gorelick, F. S., Rosenfield, A. T., et al.**, Ultrasonography of alcoholic liver disease with histological correlation, *Radiology*, 141, 157, 1981.

69. **DiLelio, A., Cestari, C., Lomazzi, A., et al.**, Cirrhosis: diagnosis with sonographic study of the liver surface, *Radiology*, 172, 389, 1989.

70. **Grossman, H., Ram, P. C., Coleman, R. A., et al.**, Hepatic ultrasonography in type I glycogen storage disease (von Gierke Disease), *Radiology*, 141, 753, 1981.

71. **Bowerman, R. A., Samuels, B. I., and Silver, T. M.**, Ultrasonographic features of hepatic adenomas in type I glycogen storage disease, *J. Ultras. Med.*, 2, 51, 1983.

72. **Sommer, F. G., Gregory, P. B., Fellingham, L. L., et al.**, Measurement of attenuation and scatterer spacing in human liver tissue: preliminary results, *J. Ultras. Med.*, 3, 557, 1984.

73. **Sommer, F. G., Joynt, L. F., Carroll, B. A., and Macovski, A.**, Ultrasonic characterization of abdominal tissues via digital analysis of backscattered waveforms, *Radiology*, 141, 811, 1981.

74. **Sommer, F. G., Joynt, L. F., Hayes, D. L., and Macovski, A.**, Stochastic frequency-domain tissue characterization: application to human spleens "in vivo", *Ultrasonics*, 20, 82, 1982.

75. **Sommer, F. G., Hoppe, R. T., Fellingham, L., et al.**, Spleen structure in Hodgkin disease: ultrasonic characterization, *Radiology*, 153, 219, 1984.

76. **Marks, W. M., Filly, R. A., and Callen, P. W.**, Ultrasonic evaluation of normal pancreatic echogenicity and its relationship to fat deposition, *Radiology*, 137, 475, 1980.

77. **Shawker, T. H., Linzer, M., and Hubbard, V. S.,** Chronic pancreatitis: the diagnostic significance of pancreatic size and echo amplitude, *J. Ultras. Med.,* 3, 267, 1984.
78. **Carroll, B. A. and Gross, D. M.,** High-frequency scrotal sonography, *Am. J. Radiol.,* 140, 511, 1983.
79. **Dahnert, W. F., Hamper, U. M., Eggleston, J. C., et al.,** Prostatic evaluation by transrectal sonography with histopathologic correlation: the echogenic appearance of early carcinoma, *Radiology,* 158, 97, 1986.
80. **Linzer, M. and Norton, S. J.,** Ultrasonic tissue characterization, *Annu. Rev. Biophys. Bioeng.,* 11, 303, 1982.
81. **Davis, P. L., Filly, R. A., and Goerke, J.,** In vitro demonstration of an echogenic emulsion: relationship of lipid particle size to echo detection, *J. Clin. Ultras.,* 9, 263, 1981.
82. **Filly, R. A., Sommer, F. G., and Minton, M. J.,** Characterization of biological fluids by ultrasound and computed tomography, *Radiology,* 134, 167, 1980.
83. **Jaffe, C. C. and Harris, D. J.,** Sonographic tissue texture: influence of transducer focusing pattern, *Am. J. Radiol.,* 135, 343, 1980.
84. **Wells, P. N. T. and Halliwell, M.,** Speckle in ultrasonic imaging, *Ultrasonics,* 19, 225, 1981.
85. **Flax, S. W., Glover, G. H., and Pelc, N. J.,** Textural variations in B-Mode ultrasonography: a stochastic model, *Ultras. Imag.,* 3, 235, 1981.
86. **Jaffe, C. C., Harris, D. J., Taylor, K. J. W., et al.,** Sonographic transducer performance cannot be evaluated with clinical images, *Am. J. Radiol.,* 137, 1239, 1981.
87. **Jones, J. P. and Kimme-Smith, C.,** An analysis of the parameters affecting texture in a B-Mode ultrasonogram, *Acoust. Imag.,* 10, 295, 1982.
88. **Smith, S. W.,** A contrast-detail analysis of diagnostic ultrasound imaging, *Med. Phys.,* 9, 4, 1982.
89. **Sommer, F. G. and Sue, J. Y.,** Imaging processing to reduce ultrasonic speckle, *J. Ultras. Med.,* 2, 413, 1983.
90. **Kimme-Smith, C. and Jones, J. P.,** The relative effects of system parameters on texture in gray-scale ultrasonograms, *Ultras. Med. Biol.,* 10, 299, 1984.
91. **Thijssen, J. M. and Oostervel, B. J.,** Texture in tissue echograms: speckle or information?, *J. Ultras. Med.,* 9, 215, 1990.
92. **Alasaarela, E. and Koivukangas, J.,** Evaluation of image quality of ultrasound scanners in medical diagnostics, *J. Ultras. Med.,* 9, 23, 1990.

Chapter 4

ACOUSTIC SCATTERING THEORY APPLIED TO SOFT BIOLOGICAL TISSUES

Michael F. Insana and David G. Brown

TABLE OF CONTENTS

0-8493-6568-6/93/$0.00 + $.50

I. INTRODUCTION

Interest in quantitative ultrasound scattering measurements in biological tissues stems from the belief that only a fraction of the total information available in the echo signal is visible in the gray-scale ultrasound image. Much information of potential diagnostic significance concerning tissue characteristics can be made available only through sophisticated signal processing. A more fundamental understanding of the basic acoustic interactions in tissue is prelude to this processing, which holds the promise for improvement in imaging technology and enhancement in the diagnostic ultility of the modality.

The complexity and diversity of biological tissues make it unlikely that a rigorous and generally applicable theoretical description of medical ultrasound imaging will evolve. Images are generated by complex interactions among the incident pressure field, the geometry of the detector, and fluctuations in mechanical properties within tissues. Boundaries between tissues that are large compared to the wavelength of sound produce strong specular reflections that usually dominate the image. Small diffuse structures produce scattered waves that coherently interfere at the detector, generating a speckle pattern in the image that is characteristic both of the instrumentation and of the tissue. As a result of this complexity, scattering from tissues must remain an empirical science, but one requiring the guidance of a realistic theoretical treatment. Understanding scattering mechanisms in tissues at a fundamental level requires precise specification of the tissue characteristics and acoustic pressure fields. Differences among the various approaches to scattering measurements found in the literature are related to approximations and assumptions regarding these specifications.

In this chapter we review the basic equations of acoustic scattering theory. Our emphasis will be on providing an intuitive "feel" for the underlying phenomena being described, so some of our arguments will be simplified forms of more rigorous arguments referenced. The basic characteristics (if not the precise details) of scattering follow naturally from a few easily understood physical principles, and the mathematical complexity of the derivations tends to obscure the simplicity of these relationships. On the other hand, we will work out in detail some of the calculations which are glossed over briefly in the standard treatments of the subject in order to engender a higher degree of comfort with the final results. Special attention is given to the approximations and assumptions made when applying the results to soft tissues.

II. DEFINITIONS AND ASSUMPTIONS

The interactions between ultrasonic waves and soft biological tissues may be divided into absorption and scattering processes. The total energy lost to those two processes is the attenuation. Absorption is the transformation of acoustic energy into thermal energy, while scattering is the re-radiation of acoustic waves with properties different from those incident. The term "scattering" as used in this chapter refers to elastic scattering, as occurs in those interactions for which the energy of the wave is conserved, e.g., reflection, refraction, and diffraction. Mathematically, acoustic scattering may be expressed as a boundary value problem, where the scatter field is obtained using wave equations and matching boundary conditions at the surface of the "scatterer", defined as a spatial fluctuation in the density ρ and/or compressibility κ of the medium. Therefore, to achieve a basic understanding of acoustic scattering in tissues requires considerable knowledge of the spatial variation in ρ and κ in tissues. Where that knowledge is incomplete, it is necessary to make assumptions and approximations. The following discussion, although far from complete, outlines some of the basic assumptions that must be considered when discussing the propagation and scattering of acoustic waves in tissues.

Soft tissues are modeled as *fluids* containing either *discrete* or *continuously varying* inhomogeneities in density and compressibility. In either case, the inhomogeneities may be randomly positioned, regularly positioned, or somewhere in between. Discrete scattering media are characterized by the average number, average dimension, and average separation of inhomogeneities. Continuously varying scattering media, on the other hand, are characterized by their correlation function, which expresses a characteristic dimension (correlation length) of the scatterers and the period of any regular structure present. Considering the lack of knowledge regarding the three-dimensional architecture of tissue, the *inhomogeneous continuum model* for scattering processes is probably the most realistic choice at this point in time. In this model a fluid is assumed to be a continuous medium characterized by steady state values of density and compressibility, but containing in addition inhomogeneities of differing equilibrium density and/or compressibility from which scattering of acoustic waves takes place. At common diagnostic medical ultrasound frequencies (5 MHz: 0.3 mm wavelength) even a cube of water one thousandth of a wavelength on a side contains over 10^9 molecules, so that quantum effects, e.g., molecular density fluctuations, may be safely ignored.

The structure or morphology of tissue is assumed to consist of ''small'' random inhomogeneities. Specifically, we assume that density and compressibility are *random* functions of position, and that the fluctuations are *small in magnitude* so as to produce low-amplitude scatter waves and *small in size* as compared to the entire scattering volume. Histology reveals that the supporting structures or stroma in many tissues are microscopic variations in density and compressibility which may be considered randomly positioned. Also, there is significant evidence to suggest that soft tissues are weakly scattering media. Measurements by two investigators[8,28] show that scattering accounts for only 2 to 10% of the total attenuation in liver. In one of these cases,[8] the total scattering cross-section per unit volume was $1.7 \times 10^{-3} f^{1.3}$ cm^{-1} whereas the attenuation coefficient was $5.3 \times 10^{-2} f^{1.23}$ cm^{-1}, a ratio of 30:1 at 2 MHz.

We require two important statistical properties of the random medium: we assume that the random process is *stationary* in time and *homogeneous* in space. (The term ''stationary'' may be applied to both temporally and spatially dependent processes, but, like Ishimaru,[21] we will use the term ''homogeneous'' to specify processes that are stationary over space.) If the mean value of a process $\psi(t)$ is independent of time and if its correlation properties, e.g., the joint second moment $\langle \psi(t_1)\psi^*(t_2) \rangle$, do not depend on the time that any one event is recorded but only on the difference in time between the recordings of any two events, the process is stationary to second order, or *weakly stationary*. Since the following analysis is limited to first- and second-order statistical properties, the condition that the medium is weakly stationary is sufficient for our purposes. In actual practice, time varying factors, such as scattering from flowing blood, are negligible as compared to scattering from tissue parenchyma, and therefore the assumption of a weakly stationary process is a good one. Similarly, if the mean value of a process is independent of position and if the correlation properties are a function of relative rather than absolute position, the process is statistically homogeneous. All tissues are statistically inhomogeneous at some scale, but for pulsed acoustic beams it is sufficient to assume that tissues are *locally homogeneous,* i.e., homogeneous over a region several times the volume of tissue occupied by the interrogating pulse of sound. (Note that a statistically homogeneous, random *scattering* medium must have inhomogeneities in density and compressibility.)

The treatment that we develop is strictly valid only for waves traveling through tissue *without loss*. Since substantial absorption losses do occur in tissues — typical values for attenuation at 5 MHz result in a factor of two loss of the incident intensity for every centimeter of tissue traversed — we modify the approach to include attenuation by introducing a complex wave number,[22] as discussed in Section III.D.1. In practice, losses are accounted for by applying the average attenuation estimate throughout the scattering volume, thereby making the further approximation that *attenuation is uniform*. In addition, it is known that tissues conduct sound as a viscous fluid that supports shear waves.[16] However, any shear waves generated are damped locally and therefore may be included in the attenuation coefficient. Ideally, absorption should be included in the equations governing acoustic wave propagation, and recently a detailed treatment of this subject has been published.[35]

Unless otherwise noted, we assume that tissues are *isotropic*. For isotropic tissues, scattering is independent of the relative orientation of the transducer and tissue volume. Several soft tissues with regular structure are known to be highly anisotropic with respect to attenuation and backscatter, e.g., myocardium,[32,33] skeletal muscle,[36,39] and kidney,[20,46] whereas tissues with less regular structure are considered to be isotropic, e.g., liver.[8,33] It often is possible, however, to assume that such anisotropic tissues are isotropic within a plane. For example, Levinson[25] showed that the angular variations in the speed of sound of muscle could be explained by a transverse isotropy model that describes tissue structures as having symmetry about the longitudinal axis and for which the elastic properties are isotropic within planes perpendicular to that axis. To avoid confusion in the literature due to an incorrect assumption of isotropy, it is essential that investigators specify the direction of the sound beam relative to standard anatomical landmarks when reporting measurements for tissues.

We assume that the *speed of sound is constant* over the range of frequencies analyzed. Although dispersion exists in all attenuating media, it is often small in biological materials. O'Donnell et al.[43] described a relationship between attenuation and phase velocity that predicted that the speed of sound varied by 3 parts in 1500 between 1 and 10 MHz, when the attenuation increases linearly with frequency. The relations were verified experimentally when they correctly predicted the frequency dependence of the speed of sound in hemoglobin solutions. Under most measurement conditions, it is reasonable to assume a constant speed of sound.

We assume that the *incident pressure field is known* throughout the scattering volume. The simplest solutions to the scattering equations are obtained for incident plane waves. Under practical conditions, however, it is necessary to include properties of the transducer beam in the scattering equations. Two general approaches to the problem of accounting for transducer sensitivity and beam directivity are found in the literature. The *first* and most popular approach assumes the transducer beam may be modeled by an appropriate mathematical equation, which allows the integral equations for scattering to be solved in closed form.[7,19,26,40,50] The principal disadvantage of this approach is to limit measurements to the focal region for focused transducers or far field for unfocused transducers. A variation of this method[37] involves reducing the data using the directivity function or beam diffraction pattern *measured* under experimental conditions. The *second* approach[29] requires fewer assumptions regarding the experimental geometry. Accurate models of the incident pressure field are introduced into the integral equations for scattering, and a solution is found numerically. Although this second approach yields range-independent results, a large amount of time is required to reduce the data. When scanning tissues in the body, all of these methods are susceptible to random distortions of the incident pressure field due to inhomogeneities in overlying tissues. Regardless of how accurately the beam is measured or modeled, it is impossible to anticipate every beam-distorting situation encountered, and aberrations from overlying tissues can introduce a large source of error in scattering measurements.[54] Consequently, the first approach, which is the most restrictive in terms of the measurement geometry but the simplest in terms of the ease and speed of data reduction, is the most popular, particularly when making measurements *in vivo* where often there are significant random distortions of the incident pressure field.

In the following sections, several of these topics are developed further, and additional restrictions imposed by the use of a Green's function approach to solve the inhomogeneous wave equation are discussed. Further assumptions and approximations involving the measurement of scattering cross-sections are discussed in later chapters.

III. FUNDAMENTALS OF WAVE PROPAGATION IN HOMOGENEOUS CONTINUA

Fluids have elasticity (compressibility κ) and inertia (mass density ρ), the two characteristics required for wave phenomena in a spatially distributed physical system whose elements are coupled. Elasticity implies that any deviation from the equilibrium state of the fluid will tend to be corrected; inertia implies that the correction will tend to overshoot, producing the need for a correction in the opposite direction and hence allowing for the possibility of propagating phenomena — acoustic (pressure) waves.

The description of these phenomena requires just three equations: (1) the equation of continuity or (differential) equation for the conservation of mass, (2) Euler's equation or (differential) equation for the conservation of momentum, and (3) the (differential) equation of state relating changes in pressure and density. A fourth, the differential equation of conservation of energy, may also be needed if it is necessary to account for heat conduction; however, we will make the adiabatic assumption of zero heat flow (and therefore that this is a constant entropy or "isentropic" process). The validity of this assumption is discussed below.

A. EQUATION OF CONTINUITY

For conservation of mass, the change in the amount of mass within a volume must equal the difference between the mass entering and leaving the volume. Consider a volume element ΔV extending in the x direction from x to $x + \Delta x$. The mass entering at x during the time Δt is the product of the x component of velocity u_x, times the density at x, times the area $\Delta A = \Delta y \Delta z$, time Δt. The same relationship holds for $x + \Delta x$, except that here a positive velocity corresponds to a loss of mass. Finally, the total gain in mass of the volume element is simply $\Delta \rho$ times the volume. Therefore we have

$$\Delta \rho \Delta x \Delta y \Delta z = (u_x(x)\rho(x) - u_x(x + \Delta x)\rho(x + \Delta x))\Delta y \Delta z \Delta t$$

or in the limit as the differences (Δ) go to zero

$$\left. \frac{\partial \rho}{\partial t} \right|_x = - \frac{\partial(u_x \rho)}{\partial x} \tag{1}$$

If we now generalize to include the other two dimensions, recalling the definition of the divergence of a vector $\mathbf{a}$, $\nabla \cdot \mathbf{a} = \partial a_x/\partial x + \partial a_y/\partial y + \partial a_z/\partial z$, we obtain

$$\frac{\partial \rho}{\partial t} + \nabla \cdot (\rho \mathbf{u}) = 0 \tag{2}$$

Finally, note that $\nabla \cdot (\alpha \mathbf{a}) = \alpha \nabla \cdot \mathbf{a} + (\mathbf{a} \cdot \nabla)\alpha$, where $\mathbf{a} \cdot \nabla = a_x \partial/\partial x + a_y \partial/\partial y + a_z \partial/\partial z$, α is an arbitrary scalar, and $\mathbf{a}$ is an arbitrary vector function of the coordinates:

$$\frac{\partial \rho}{\partial t} + (\mathbf{u} \cdot \nabla)\rho + \rho \nabla \cdot \mathbf{u} = 0$$

$$\frac{D\rho}{Dt} + \rho \nabla \cdot \mathbf{u} = 0 \tag{3}$$

where D/Dt is the total or "Lagrangian" derivative

$$\frac{D\alpha}{Dt} = \frac{\partial \alpha}{\partial t} + \frac{\partial \alpha}{\partial x}\frac{dx}{dt} + \frac{\partial \alpha}{\partial y}\frac{dy}{dt} + \frac{\partial \alpha}{\partial z}\frac{dz}{dt}$$

$$= \left(\frac{\partial}{\partial t} + \mathbf{u} \cdot \nabla \right)\alpha \tag{4}$$

The total derivative is the change in some quantity for an element which is moving with the *local* velocity **u**.

B. EULER'S EQUATION

The derivation of Euler's equation is somewhat more complicated, but follows from the same general considerations. After all, the derivation of Equation 1 is valid not just for mass, but for the conservation of any quantity carried along with the fluid. Thus, for the conservation of momentum (ignoring any forces for the time being) we have

$$\Delta(\rho\mathbf{u})\Delta x\Delta y\Delta z = (u_x(x)\rho(x)\mathbf{u}(x) - u_x(x + \Delta x)$$

$$\times \rho(x + \Delta x)\mathbf{u}(x + \Delta x))\Delta y\Delta z\Delta t$$

$$\left.\frac{\partial(\rho\mathbf{u})}{\partial t}\right|_x = -\frac{\partial(u_x\rho\mathbf{u})}{\partial x} \tag{5}$$

$$\frac{\partial(\rho\mathbf{u})}{\partial t} = -\nabla \cdot (\mathbf{u}(\rho\mathbf{u}))$$

The last equation is the three-dimensional generalization of the former, with the dot product acting on the first vector of the dyad $\mathbf{u}(\rho\mathbf{u})$ but the differentiation applying to the entire dyad (as seen clearly in the one-dimensional version of the equation). The respective derivatives may be expanded to give the following result:

$$\rho\frac{\partial\mathbf{u}}{\partial t} + \mathbf{u}\frac{\partial\rho}{\partial t} = -\rho(\mathbf{u} \cdot \nabla)\mathbf{u} - \mathbf{u}\nabla \cdot (\rho\mathbf{u}) \tag{6}$$

Since by Equation 2 the second terms on the left and right cancel, this becomes

$$\rho\frac{D\mathbf{u}}{Dt} = 0 \tag{7}$$

Of course, momentum is not conserved: the time rate of change of momentum is not zero but equal to the sum of forces acting on the element of the fluid. In particular, a pressure p acting at x represents a force $p(x)\Delta y\Delta z$, whereas one at $x + \Delta x$ is a countervailing force $p(x + \Delta x)\Delta y\Delta z$, yielding a net force per unit volume $-\partial p/\partial x$, or in three dimensions $-\nabla p$ and

$$\rho\frac{D\mathbf{u}}{Dt} = -\nabla p \tag{8}$$

Other effects such as viscosity of the medium and external forces such as gravity are implemented as terms on the right-hand side of Equation 8.

We are assuming that all such terms are sufficiently small that they do not significantly affect our result and may be safely ignored.

C. EQUATION OF STATE

The equation of state of a fluid is required in order to relate the pressure and density of the fluid and thereby eliminate one of these variables from the above equations. In general, any "state" variable of the fluid, e.g., volume, pressure, or entropy, may be expressed as a function of any other two. Therefore, if we want to eliminate the density from our equations, and if we asssume that the acoustic process is adiabatic, (heat flow $\delta q = 0$, implying that entropy s is constant since $\delta q = T\delta s$), we will choose $\rho = \rho(p,s) = \rho(p)$. In actual practice we consider fluctuations in the medium $\delta\rho$ about the mean value ρ_o as a Taylor series expansion in the fluctuations about the mean pressure for the adiabatic case:

$$\rho = \rho_o + \delta\rho \tag{9}$$

where

$$\delta\rho = \left.\frac{\partial\rho}{\partial p}\right|_{\delta s=0} \delta p + \left.\frac{\partial^2\rho}{\partial p^2}\right|_{\delta s=0} \frac{(\delta p)^2}{2} + \cdots \tag{10}$$

How valid is the adiabatic assumption? Pierce[44] states that there is a transition frequency f_{TC}, which for water is on the order of 2×10^{12} Hz, and below which the adiabatic approximation can be expected to hold. Thus frequencies used for diagnostic ultrasound, typically less than 10^7 Hz, are well below f_{TC}. Note that the process is adiabatic at *low frequencies* because long-wavelength pressure fields don't provide steep enough temperature gradients for significant heat flow to occur. Conversely, the process is nonadiabatic at *high frequencies* with steep temperature gradients, even though it might seem that there would not be "sufficient time" for heat to flow.

D. WAVE PROPAGATION IN A HOMOGENEOUS MEDIUM

Next we combine the three fundamental equations to obtain first order (linear) equations describing the system. The validity of neglecting second order effects is discussed in Appendix A.

Consider a fluid whose equilibrium parameters (ρ_o, p_o, $\mathbf{u_o}$) describe a quiescent medium, so that $\mathbf{u_o} = 0$ and the respective time derivatives are zero ($\partial(\cdot)/\partial t = 0$), and which is also a homogeneous medium, so that the spatial derivatives are zero ($\partial(\cdot)/\partial x_i = 0$). We subject the fluid to a disturbance which we characterize as a perturbation on the equilibrium state:

$$\rho = \rho_o + \delta\rho$$

$$p = p_o + \delta p$$

$$\mathbf{u} = \delta\mathbf{u} \tag{11}$$

Inserting these expansions into the three fundamental equations, it is trivial to extract the linear equations from the exact expressions governing the system. Here the linear terms are those containing only one of the differential factors $\delta\rho$, δp, or $\delta\mathbf{u}$. For example, in the continuity equation, Equation 2, we have

$$\frac{\partial(\rho_o + \delta\rho)}{\partial t} + \nabla \cdot ((\rho_o + \delta\rho)\delta\mathbf{u}) = 0 \tag{12}$$

Neglecting the second order term $\nabla \cdot (\delta\rho\ \delta\mathbf{u})$, we obtain the following first order equation:

$$\frac{\partial(\delta\rho)}{\partial t} = -\rho_o\nabla \cdot \delta\mathbf{u} \tag{13}$$

Similarly, for Euler's equation, Equation 8, we obtain

$$(\rho_o + \delta\rho)\left(\frac{\partial(\delta\mathbf{u})}{\partial t} + \delta\mathbf{u} \cdot \nabla(\delta\mathbf{u})\right) = -\nabla(\delta p) \tag{14}$$

$$\rho_o\frac{\partial(\delta\mathbf{u})}{\partial t} = -\nabla(\delta p) \tag{15}$$

where we have neglected the higher order terms $\rho_o(\delta\mathbf{u}\cdot\nabla(\delta\mathbf{u})$, $\delta\rho\ \partial(\delta\mathbf{u})/\partial t$, and $\delta\rho(\delta\mathbf{u}\cdot\nabla(\delta\mathbf{u}))$. Finally, the equation of state, Equation 10, gives the relationship between increments in density and pressure:

$$\delta\rho = \left.\frac{\partial\rho}{\partial p}\right|_{\delta s = 0} \delta p$$

$$= \kappa_o\rho_o\delta p \tag{16}$$

where we have used the first order equation to define the average compressibility κ_o. We are now in a position to combine the linear acoustic equations to derive the first order acoustic wave equation. From this point forth we will focus our attention on the linear equations, and will redefine p to mean δp and $\mathbf{u}$ to mean $\delta\mathbf{u}$.

1. How Good is the Linear Approximation?

At first glance there would seem to be a serious problem with labeling the pressure changes introduced by medical diagnostic transducers as small perturbations, since typical peak pressure values range from 10 to 100 atm,[11] and any expansion in $p/p_o \sim 10$ to 100 would seem unlikely to converge (Appendix A). Fortunately, however, p_o (unlike ρ_o) does *not* appear in the equations, and is thus irrelevant to the normalization — so that the ratio of fundamental significance is $\delta\rho/\rho_o$. (Note that this is in contradiction to the argument of Morse and Ingard[31], page 233.) In Appendix A we show that $\delta\rho/\rho_o$ is on the order of one part in a hundred, so that second order effects are generally small in comparison with those of first order.

This does not mean that second order effects are unimportant. Higher order terms lead to the conversion of energy into harmonic frequencies, particularly over long transmission paths.[34,51] Note how the second order term in the equation of state contains $(\delta p)^2$,

$$\delta\rho = \left.\frac{\partial\rho}{\partial p}\right|_{\delta s=0} \delta p + \left.\frac{\partial^2\rho}{\partial p^2}\right|_{\delta s=0} \frac{(\delta p)^2}{2} \tag{17}$$

and leads naturally to the generation of these higher harmonics. Moreover, nonlinear characteristics have been suggested as tissue signatures by several investigators.[12,48] Since the transducers and associated electronics are insensitive in the steady state to the higher harmonics,[45] we might tend to ignore these effects when present, but would find that the energy at high frequencies is rapidly attenuated and the resulting losses unaccounted for. For our purposes of characterizing scatter, however, we will not consider higher order terms. Nonlinear effects are minimized in laboratory measurements by keeping the incident pressure intensity just high enough to provide sufficient signal strength.

The first order equations may be easily combined to eliminate **u** by taking the time derivative of Equation 13 and the divergence of Equation 15. Then, using Equation 16 to eliminate $\delta\rho$, we obtain the homogeneous wave equation

$$\nabla^2 p - \rho_o\kappa_o \frac{\partial^2 p}{\partial t^2} = 0 \tag{18}$$

with solutions of the form $p = f(\mathbf{k}\cdot\mathbf{r} - \omega t)$, where $\nabla^2 p = k^2 f''$ and $\partial^2 p/\partial t^2 = \omega^2 f''$ (f'' is the second derivative of f with respect to its argument). The parameters k and ω must satisfy the relationship $(\omega/k)^2 = c^2 = 1/(\rho_o\kappa_o)$, where c is the speed of propagation in the (lossless) medium, ω is the frequency, and κ is the wave number. As noted earlier, we can account for losses in acoustic energy as the wave travels through an attenuating medium by adopting a complex wave number, $k = \omega/c + i\alpha$, where α is the attenuation coefficient characteristic of the medium. Using this expression for k, it is easy to show that the loss of acoustic energy experienced by a plane or spherical wave traveling a distance r through an attenuating medium is given the factor $\exp(-\alpha r)$.

2. Plane and Spherical Waves

Two solutions of particular importance are those for a plane wave, $p = A \exp i(\mathbf{k \cdot r} - \omega t)$, and for a spherical wave, $p = A/r \exp i(kr - \omega t)$. Note how, once given the pressure, we may use the three first order equations, Equations 13, 15, and 16, to solve immediately for $\rho, \mathbf{u}$, and such derived quantities as the impedance $z = p/u$, and intensity, i.e., the power per unit area carried by the wave, $I = \frac{1}{2} Re(p^* u)$, where $Re(\cdot)$ is the real part of the quantity and p^* is the complex conjugate of p.[†]

For example, for the plane wave, $-\nabla p = -ikp$, and assuming $\mathbf{u}(\mathbf{r},t) = \mathbf{u}(\mathbf{r}) \exp(-i\omega t)$ then $\rho_o \partial \mathbf{u}/\partial t = -i\rho_o \omega \mathbf{u}$ so that by Equation 15,

$$-i\rho_o \omega \mathbf{u} = -i\omega p \mathbf{a_k}/c$$

$$z = \frac{p}{u} = \rho_o c \tag{19}$$

where the unit vector $\mathbf{a_k} = \mathbf{k}/k$, and $I = \frac{1}{2} Re(p^* p/z) = \frac{1}{2}|p|^2 Re(1/z) = |p|^2/2\rho_o c$.

Similarly, for the spherical wave, $-\nabla p = -\mathbf{a_r} \partial p/\partial r = \mathbf{a_r}(1/r - ik)p$, where $\mathbf{a_r} = \mathbf{r}/r$, and

$$-i\rho_o \omega \mathbf{u} = (1/r - ik)p\mathbf{a_r}$$

$$z = \frac{p}{u} = \frac{\rho_o c}{1 + i/kr}$$

$$= \rho_o c \cos \alpha e^{-i\alpha}$$

$$Re\left(\frac{1}{z}\right) = \frac{\cos \alpha}{\rho_o c \cos \alpha} = \frac{1}{\rho_o c} \tag{20}$$

where $kr = \cot\alpha$. In spite of the complicated form for z, $I = |p|^2/2\rho_o c$ just as for the plane wave. Note that if we assume solutions of the form $p(\mathbf{r}) \exp(-i\omega t)$, then the wave equation, Equation 18, is transformed into the Helmholtz equation

$$\nabla^2 p + k^2 p = 0 \tag{21}$$

so that only the spatial dependence need be considered.

[†]   There are (at least) two conventions for defining acoustic intensity. First, $I = \frac{1}{2} Re(p^* u)$ where p is the peak pressure. Second, $I = Re(p_a^* u)$ where p_a is the time-average pressure. In this chapter we use the first convention.

2. The Green's Function Approach

The Green's function approach is a standard technique for solving the inhomogeneous Helmholtz equation for p_ω which involves rewriting the differential equation as an integral equation for pressure.[30] This approach is attractive because the solutions are physically descriptive. As shown below, the net scattered pressure is obtained by summing the individual scattered waves resulting from interactions between the pressure field and elements of an extended medium. The Green's function $G(\mathbf{r}|\mathbf{r_o})$ describes the observed pressure field at the detector's location $\mathbf{r}$ resulting from a point source of scattering located at $\mathbf{r_o}$. A source may be considered a "point source" if its size is much smaller than the wavelength of sound. Our medium consists of a *deterministic*, continuous distribution of inhomogeneities in an otherwise acoustically uniform material, and may be treated as a collection of point scattering sources. If the incident pressure field and the sources are known, then the total field at the detector can be determined.

The Green's function is a solution to the inhomogeneous differential equation

$$\nabla^2 G_\omega(\mathbf{r}|\mathbf{r_o}) + k^2 G_\omega(\mathbf{r}|\mathbf{r_o}) = -\delta(\mathbf{r} - \mathbf{r_o}) \tag{34}$$

for a point source at $\mathbf{r_o}$ radiating out into an unbounded medium, and $\delta(\mathbf{r} - \mathbf{r_o})$ is the Dirac delta function for three dimensions. (Derivatives are with respect to the observer's coordinate system, $\mathbf{r}$.) The simple harmonic scattering sources described in Equation 34 "generate" spherical pressure waves of the form

$$G_\omega(\mathbf{r}|\mathbf{r_o}) = \frac{1}{4\pi|\mathbf{r} - \mathbf{r_o}|} e^{ik|\mathbf{r} - \mathbf{r_o}|} \tag{35}$$

This well-known *Green's function in free space* is a solution to Equation 34 that satisfies the condition that $G = 0$ on the surface of the scattering volume located at infinity (unbounded medium). An important property of Green's functions is the principle of reciprocity, which states that the value of the function remains the same when the source and detector are interchanged, i.e., $G_\omega(\mathbf{r}|\mathbf{r_o}) = G_\omega(\mathbf{r_o}|\mathbf{r})$.

The first step towards obtaining an integral expression for p_ω is to multiply Equation 32 by G_ω and Equation 34 by p_ω and subtract, to give

$$G_\omega(\mathbf{r}|\mathbf{r_o})\nabla_o^2 p_\omega(\mathbf{r_o}) - p_\omega(\mathbf{r_o})\nabla_o^2 G_\omega(\mathbf{r}|\mathbf{r_o}) = p_\omega(\mathbf{r_o})\delta(\mathbf{r} - \mathbf{r_o}) - f_\omega(\mathbf{r_o})G_\omega(\mathbf{r}|\mathbf{r_o}) \tag{36}$$

The symmetric properties of G_ω and δ enabled us to interchange $\mathbf{r}$ and $\mathbf{r_o}$. (∇_o refers to differentiation with respect to the coordinate system of the source.) Next, we integrate over volume elements dv_o located at $\mathbf{r_o}$ of a volume

V containing the scattering sources. Recognizing that $p_\omega(\mathbf{r}) = \int p_\omega(\mathbf{r_o})\delta(\mathbf{r} - \mathbf{r_o})dv_o$, we find that

$$p_\omega(\mathbf{r}) = \int_V [G_\omega(\mathbf{r}|\mathbf{r_o})\nabla_o^2 p_\omega(\mathbf{r_o}) - p_\omega(\mathbf{r_o})\nabla_o^2 G_\omega(\mathbf{r}|\mathbf{r_o})] \, dv_o \qquad (37)$$
$$+ \int_V f_\omega(\mathbf{r_o})G_\omega(\mathbf{r}|\mathbf{r_o}) \, dv_o$$

The first volume integral may be rewritten as an integral over the surface S enclosing the scattering volume by applying Green's theorem (a special case of Gauss' theorem which states that $\int_V \nabla \cdot \mathbf{F}dv = \int_s \mathbf{F} \cdot \hat{\mathbf{n}}ds$).[32] That is, since

$$G\nabla^2 p - p\nabla^2 G = \nabla \cdot (G\nabla p - p\nabla G) \qquad (38)$$

then

$$p_\omega(\mathbf{r}) = \int_S [G_\omega(\mathbf{r}|\mathbf{r_o})\nabla_o p_\omega(\mathbf{r_o}) - p_\omega(\mathbf{r_o})\nabla_o G_\omega(\mathbf{r}|\mathbf{r_o})] \cdot \hat{\mathbf{n}} \, ds_o \qquad (39)$$
$$+ \int_V f_\omega(\mathbf{r_o})G_\omega(\mathbf{r}|\mathbf{r_o}) \, dv_o$$

where $\hat{\mathbf{n}}$ is the unit vector normal to the surface, pointing away from the scattering volume. Morse and Ingard[31] (page 321) provide some intuition regarding the physics of Equation 39. They explain that since $G_\omega(\mathbf{r}|\mathbf{r_o})$ is the field at $\mathbf{r}$ due to a point source at $\mathbf{r_o}$, Equation 39 describes the total field at $\mathbf{r}$ as the sum of fields from each elementary source $f_\omega dv_o$ plus the waves reflected by the boundary surface.

The surface integral is simplified by applying the assumption of an unbounded medium, e.g., let V be spherical with outer surface radius $\mathfrak{R}$ such that $\mathfrak{R} \to \infty$. The scattering contribution due to the surface integral becomes negligible as $\mathfrak{R} \to \infty$ but the incident pressure field, p_i, remains constant, and by Huygen's principle integrates to p_i for any observation point $\mathbf{r}$ within the volume bounded by the surface. Thus, Equation 39 may be written as

$$p_\omega(\mathbf{r}) = p_i(\mathbf{r}) + \int_V f_\omega(\mathbf{r_o})G_\omega(\mathbf{r}|\mathbf{r_o}) \, dv_o$$
$$= p_i(\mathbf{r}) + \underbrace{\int_V [k^2\gamma_\kappa(\mathbf{r_o})p_\omega(\mathbf{r_o}) - \nabla_o \cdot (\gamma_\rho(\mathbf{r_o})\nabla_o p_\omega(\mathbf{r_o}))]G_\omega(\mathbf{r}|\mathbf{r_o}) \, dv_o}_{p_s(\mathbf{r})} \qquad (40)$$

Is the assumption of an unbounded medium realistic? Usually it is, since under measurement conditions electronic gates and transducer beams limit the incident pressure field, such that the contribution from real tissue boundaries outside these limits is negligible. Hence, the assumption that the medium is unbounded does not necessarily mean that the volume need be infinite in extent.

Equation 40 states that the total field observed is the sum of the incident and scattered fields. The scattered field at point $\mathbf{r}$ is the coherent sum of waves scattered from point sources positioned at $\mathbf{r}_o$ with magnitude $f_\omega dv_o$. The observed field from each point source is given by $G_\omega(\mathbf{r}|\mathbf{r}_o)$.

3. The Scattered Pressure Field

We now focus our attention on the scattered pressure field, p_s. First, we expand the second term in the integral expression for p_s using the product rule for differentiation and find that

$$\int_V G_\omega \nabla_o \cdot [\gamma_\rho \nabla_o p_\omega] \, dv_o = \int_V \nabla_o \cdot [G_\omega \gamma_\rho \nabla_o p_\omega] \, dv_o - \int_V \gamma_\rho \nabla_o p_\omega \cdot \nabla_o G_\omega \, dv_o \quad (41)$$

Gauss' theorem allows us to replace the first integral on the right of Equation 41 with a surface integral over the component of the integrand normal to the surface. Since the surface is outside the scattering inhomogeneities, $\gamma_\rho = 0$ and this integral is also zero. Thus, substituting Equation 41 into Equation 40 results in the following expression for the scattered pressure:

$$p_s(\mathbf{r}) = \int_V [k^2 \gamma_\kappa(\mathbf{r}_o) p_\omega(\mathbf{r}_o) G_\omega(\mathbf{r}|\mathbf{r}_o) + \gamma_\rho(\mathbf{r}_o) \nabla_o p_\omega(\mathbf{r}_o) \cdot \nabla_o G_\omega(\mathbf{r}|\mathbf{r}_o)] \, dv_o \quad (42)$$

''Small'' scattering sources give rise to a monopole contribution from the compressibility source term and a dipole contribution from the density source term. That is, scattered waves from fluctuations in compressibility radiate spherically from the source, whereas fluctuations in density produce a dipole (''$\cos\Theta$'') radiation pattern, as shown below. Since the Green's function describes spherically divergent waves, $\nabla_o G_\omega$ is a vector normal to the surface of a sphere surrounding each source. The force on the scatterer is $\nabla_o p_\omega dv_o$. The density term, therefore, is the component of $\nabla_o G_\omega$ in the direction of the force, which accounts for the $\cos\Theta$ radiation pattern. The dipole contribution is a consequence of the scattering particle with nonzero γ_ρ moving back and forth with respect to the surrounding medium, whereas the monopole contribution results from radial excitation with no relative motion, as described in Section IV.A. For *extended sources*, Equation 42 is an integration over dipole contributions that vary in magnitude and direction, so that the result may not be a simple dipole field. Encouraged by the lack of a clear dipole radiation pattern for tissues and early results indicating that γ_κ was an order of magnitude greater than γ_ρ,[15] some investigators have simplified Equation 42 by arguing that γ_ρ is negligible for tissues. In general, however, this assumption is not valid: more recent studies[38,55] have shown that density variations contribute significantly to scattering from liver and other tissues. Therefore scattering from fluctuations in both density and compressibility must be considered for tissue applications.

4. Observation at Large Distances

If the observation distance is large compared to the size of the volume occupied by scatterers (e.g., Figure 1, where $r \gg r_o$), the source-to-observer distance $|\mathbf{r} - \mathbf{r_o}|$ may be expanded as follows:

$$|\mathbf{r} - \mathbf{r_o}| = r\left(1 + \frac{2}{r}\mathbf{r_o} \cdot \hat{\mathbf{o}} + \left(\frac{r_o}{r}\right)^2\right)^{1/2}$$

$$\approx r - \mathbf{r_o} \cdot \hat{\mathbf{o}} \tag{43}$$

where $\hat{\mathbf{o}} = \mathbf{r}/r$ is the unit vector in the direction of the observer. The Green's function in Equation 42 and its gradient may therefore be approximated by the expressions

$$G_\omega(\mathbf{r}|\mathbf{r_o}) \approx \frac{1}{4\pi r} e^{ikr} e^{-i\mathbf{k_s} \cdot \mathbf{r_o}}$$

$$\nabla_o G_\omega(\mathbf{r}|\mathbf{r_o}) \approx -ik^2 \frac{\hat{\mathbf{o}}}{k} G_\omega(\mathbf{r}|\mathbf{r_o}) \tag{44}$$

where $\mathbf{k_s} = k\hat{\mathbf{o}}$ is the wave number vector in the direction of the observer. Thus, far from the inhomogeneities, we observe a scattered pressure given by

$$p_s(\mathbf{r}) \approx \frac{k^2 e^{ikr}}{4\pi r} \int_V \left[\gamma_\kappa(\mathbf{r_o})p_\omega(\mathbf{r_o}) - i\gamma_\rho(\mathbf{r_o})\left(\nabla_o p_\omega(\mathbf{r_o}) \cdot \frac{\hat{\mathbf{o}}}{k}\right)\right] e^{-i\mathbf{k_s} \cdot \mathbf{r_o}} \, dv_o \tag{45}$$

Notice that the scattered wave that travels to the observer is spherical ($\exp(ikr)/r$) and has an amplitude proportional to the three-dimensional spatial Fourier transform of the scattering sources.

5. Incident Plane Waves and the Born Approximation

As seen by Equation 45, we must determine the *total* pressure field within the scattering medium before we can determine the scattered field. Towards this goal, we make two assumptions. First, we assume that the incident pressure is a plane wave. Although the plane-wave assumption is inappropriate for most experimental situations, as discussed in great detail by many investigators, e.g., References 7, 19, 26, 29, 37, 40, 42, 50, 52 and in other chapters of this book, it is used in the following analysis to simplify the discussion. The incident pressure field, therefore, is given by $p_i(\mathbf{r}) = P$

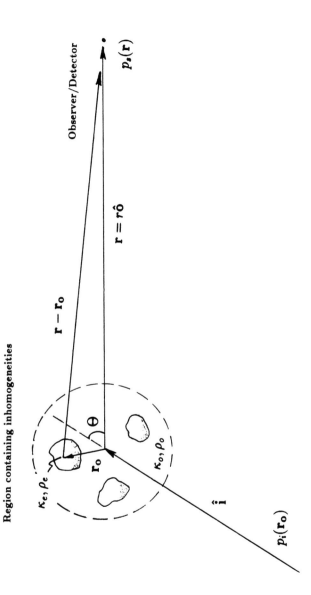

FIGURE 1. The scattering geometry.

$\exp(i\mathbf{k}_i \cdot \mathbf{r})$ and $\nabla p_i = i\mathbf{k}_i p_i$, where $\mathbf{k}_i = k\hat{\mathbf{i}}$ is the wave number vector in the direction of propagation of the incident field (Figure 1). Although we have not explicitly shown the time dependence, it is understood that $p_i(\mathbf{r},t) = p_i(\mathbf{r})\exp(-i\omega t)$. Second, we assume that the scattered wave amplitude is much smaller than that of the incident pressure (Born approximation), such that the total field, p_ω, can be replaced by the field incident. This requires that γ_κ and γ_ρ be small, so that the scattered wave is small in comparison to the incident wave. In addition, however, the Born approximation is a "single scatter" approximation, and thus fails, in a strict sense, when the wavelength is on the order of or less than the dimensions of the scatterer, since then resonances and other multiple scatter effects become important.

Combining the Born approximation and plane-wave assumption with Equation 45 we write

$$
\begin{aligned}
p_s(\mathbf{r}) &\simeq \frac{k^2 e^{ikr}}{4\pi r} \int_V \left[\gamma_\kappa(\mathbf{r}_o)p_i(\mathbf{r}_o) - i\gamma_\rho(\mathbf{r}_o)\left(\nabla_o p_i(\mathbf{r}_o) \cdot \frac{\hat{\mathbf{o}}}{k} \right) \right] e^{-i\mathbf{k}_s \cdot \mathbf{r}_o}\, dv_o \\
&= \frac{Pe^{ikr}}{r} \frac{k^2}{4\pi} \int_V [\gamma_\kappa(\mathbf{r}_o) + \gamma_\rho(\mathbf{r}_o)(\hat{\mathbf{i}} \cdot \hat{\mathbf{o}})] e^{-i\mathbf{K} \cdot \mathbf{r}_o}\, dv_o
\end{aligned}
\tag{46}
$$

where $\mathbf{K} = \mathbf{k}_s - \mathbf{k}_i = k(\hat{\mathbf{o}} - \hat{\mathbf{i}})$ is the *scattering vector* of magnitude

$$
\begin{aligned}
K &= k(|\hat{\mathbf{o}}|^2 + |\hat{\mathbf{i}}|^2 - 2(\hat{\mathbf{i}} \cdot \hat{\mathbf{o}}))^{1/2} \\
&= 2k\left(\frac{1 - \cos \Theta}{2} \right)^{1/2} \\
&= 2k \sin \Theta/2
\end{aligned}
\tag{47}
$$

and Θ is the scattering angle determined by the directions of the incident field and the observer in the coordinate system illustrated by Figure 1: $\cos\Theta = (\hat{\mathbf{i}} \cdot \hat{\mathbf{o}})$. From Equation 46, the scattering pressure is again seen to be proportional to a Fourier transform of the scattering sources, but now with respect to $\mathbf{K}$.

It is important to note that the frequency dependence of scattering from inhomogeneities in compressibility is the same as scattering from inhomogeneities in density. This will always be true for weak scattering of plane waves, where the density source term in Equation 42,

$$
\gamma_\rho \nabla_o p_\omega \cdot \nabla_o G_\omega = k^2 \gamma_\rho(\hat{\mathbf{i}} \cdot \hat{\mathbf{o}})p_i G_\omega
$$

and the compressibility term are both proportional to k^2. This plane-wave result may be extended to include most experimental conditions: it also holds true for scattering from focused and nonfocused circular transducers, as shown by Ueda and Ichikawa[52] and Lizzi et al.,[26] except in the extreme near field.

The similarity in frequency dependence for the two source terms becomes important in the discussion of correlation functions below.

We have argued that the Born approximation is strictly valid only when the medium is weakly scattering and the size of the scatterers is much smaller than the wavelength. However, in many experimental situations with scatterers of size on the order of the wavelength we find agreement between experiment and theory using the Born approximation, e.g., References 9 and 19. This implies that in these media the effects of multiple scattering are negligibly small, and the Born approximation holds for finite-size scatterers. There is currently no definitive evidence to evaluate the validity of the Born approximation in biological tissues. However, it has been used, at least implicitly, in the analysis of all scattering measurements made for tissues, in spite of the fact that there are structures in tissues on the order of the wavelength that scatter sound.

It is convenient for the remaining discussion to express the scattered pressure, Equation 46, in terms of the angle distribution factor for scattered waves, Φ[31] i.e.,

$$p_s(\mathbf{r}) = P \frac{e^{ikr}}{r} \Phi(\mathbf{K}) \tag{48}$$

$$\Phi(\mathbf{K}) = \frac{k^2}{4\pi} \int_V \gamma(\mathbf{r_o}, \Theta) e^{-i\mathbf{K} \cdot \mathbf{r_o}} \, dv_o \tag{49}$$

where $\gamma(\mathbf{r_o}, \Theta) = \gamma_\kappa(\mathbf{r_o}) + \gamma_\rho(\mathbf{r_o})\cos\Theta$. Equations 48 and 49 provide considerable detail about the physics of the scattered field. First, observing the field far from the inhomogeneity, the scattered field behaves as a spherical wave with an angular-dependent amplitude $\phi(\mathbf{K})$. Second, the compressibility and density contributions to p_s are identical except for their relative weights γ_κ and $\gamma_\rho \cos\Theta$, under the conditions of the Born approximation. Therefore, by fixing Θ at one scattering angle, e.g., $\Theta = 180°$ (backscatter), the medium may be characterized by a single function $\gamma(\mathbf{r})$. Note that the backscattered pressure field for small density and compressibility fluctuations is proportional to the relative fluctuation in the plane wave acoustic impedance z, i.e., $\gamma = \Delta\kappa/\kappa - \Delta\rho/\rho = -2\Delta z/z$, since $z = \rho c = \sqrt{\rho/\kappa}$. Only by measuring scattering as a function of Θ can the compressibility and density terms be separated. The backscatter example is an important one for *in vivo* tissue measurements, because access to tissues for scattering measurements at other angles is severely limited in the body.

To summarize, Equations 48 and 49 express the scattered pressure field that results from the interaction between plane waves and a small well-defined region of scattering inhomogeneities positioned inside an unbounded medium. We have assumed that the detector (observer) is far from the region as compared to the size of the detector, and that the Born approximation is valid, which means small relative fluctuations and, strictly speaking, wavelengths long in comparison to the dimensions of the inhomogeneities.

6. Scattering Cross-Sections

The incident power Π lost to interactions with the medium divided by the incident intensity I_i is defined as the *total cross-section* of the medium that interacts with the incident intensity.[31] The total acoustic cross-section, σ_t, is the sum of absorption and scattering cross-section, i.e., $\sigma_t = \sigma_a + \sigma_s$, and has units of area. Cross-sections provide an important interface between theory and measurement, and therefore are essential to understanding basic acoustic interactions. In general, cross-sections are used to relate measurements of acoustic power to the structure and composition of the medium as modeled by theory.

Scattering interactions in tissues may be characterized by the appropriate scattering cross-section. For example, when a finite-aperture detector is placed far from the scattering volume, as in the discussion above, only a fraction of the total scattered power is sampled. The relevant measure in this example is the differential scattering cross-section per unit solid angle, $\sigma_{ds} = d\sigma_s/d\Omega$, or simply the *differential scattering cross-section*. This quantity is found from the ratio of the differential scattering power and the incident intensity. An expression for power is obtained using Equation 48 and the fact that p_s is a spherical wave, so that $d\Pi = I_s r^2 d\Omega$, where I_s is the scattered intensity:

$$d\Pi(\mathbf{K}) = \frac{|p_s|^2}{2\rho_o c} r^2 \, d\Omega$$

$$= \frac{P^2 \, d\Omega}{2\rho_o c} |\Phi(\mathbf{K})|^2 \tag{50}$$

Recalling that $p_i(\mathbf{r}) = P \exp(i\mathbf{k_i} \cdot \mathbf{r})$, the incident plane wave intensity is therefore $I_i = |p_i|^2/2\rho_o c = P^2/2\rho_o c$. Consequently, the differential scattering cross-section is

$$\sigma_{ds} = \frac{d\Pi}{I_i \, d\Omega} = |\Phi(\mathbf{K})|^2 \tag{51}$$

where the quantity $|\Phi|^2$ describes the angular distribution of scattered power. In general σ_{ds} is a function of the vector $\mathbf{K}$, but, as we show later, for isotropic scattering σ_{ds} may be expressed as a function of its magnitude K.

The total scattering cross-section is obtained by integrating over the surface of a sphere of radius R:

$$\sigma_s = 2\pi \int_0^\pi |\Phi(\Theta)|^2 \sin \Theta d\Theta \tag{52}$$

The total scattering cross-section may be described in physical terms as the area normal to the incident beam which intercepts an amount of incident power equal to the scattered power.[30]

Thus far, we have considered scattering from deterministic media, i.e., media where we know the function $\gamma(\mathbf{r})$ precisely. However, to describe scattering in biological tissues we must extend the discussion to include random media with many scattering sites. In such media, the scattered power received at the detector increases with V and, for incoherent scattering, the increase is linear, as shown below. (We now redefine V to be the volume of the medium that participates in generating the differential power $d\Pi$.) The relevant cross-section relating measurements of received power at the detector and characteristics of a random medium is given by the *differential scattering cross-section per unit volume*, $\sigma_d = |\Phi(\mathbf{K})|^2/V$. Of particular interest is the *differential backscattering cross-section per unit volume*, also known as the *backscatter coefficient*, $\sigma_b = |\Phi(2k)|^2/V$. (Since $\Theta = \pi$ for backscatter, $K = 2k$.) The traditional units of the backscatter coefficient are $cm^{-1} sr^{-1}$. For isotropic random media, σ_b is a function of one experimental variable — frequency, which may be expressed as either spatial (k) or temporal (f) frequency, since $k = 2\pi f/c$.

C. SCATTERING FROM RANDOM CONTINUA

Consider plane waves traveling through a medium characterized by its fluctuations in density and compressibility, $\gamma(\mathbf{r},\Theta)$, where the vector $\mathbf{r}$ identifies the position in the medium and Θ identifies the scattering angle. Let γ be a continuous random function of position throughout V. The scattered field that results from interactions between plane waves and random media has an amplitude and phase that are also random functions of position. Such scattering media are called *random continua*.[21] The angle distribution factor Φ is now a random function because p_s is the coherent sum of waves scattered from random fluctuations in γ. Therefore the differential scattering cross-section per unit volume must be obtained by a statistical averaging process,

$$\sigma_d = \frac{1}{V} \langle \Phi(\mathbf{K})\Phi^*(\mathbf{K}) \rangle \qquad (53)$$

where $\langle \cdot \rangle$ represents the ensemble average of the quantity, i.e., an average over many different realizations of the process. Obviously, any single measurement will be determined by the detailed placement of inhomogeneities within that particular sample volume; however, the "true" value of σ_d is obtained by averaging over an infinite number of independent samples (ensemble average) or in practice by averaging a sufficient number of samples to attain the desired degree of precision.

It is often useful to view a random process ψ, e.g., the scatter field, as the sum of its average and fluctuating components. This is expressed mathematically as

$$\psi_1 = \langle \psi_1 \rangle + (\psi_1 - \langle \psi_1 \rangle) \equiv \overline{\psi}_1 + \psi_1'$$

$$\psi_2 = \langle \psi_2 \rangle + (\psi_2 - \langle \psi_2 \rangle) \equiv \overline{\psi}_2 + \psi_2' \tag{54}$$

where $\overline{\psi}_n$ and ψ_n' are, respectively, the average and fluctuating components and the subscripts identify the position (or time) of the realization ψ. Note that by definition $\langle \psi_1' \rangle = \langle \psi_2' \rangle = 0$. As with Equation 53, we are particularly interested in the autocorrelation of random processes, which is given by

$$\langle \psi_1 \psi_2^* \rangle = \langle (\overline{\psi}_1 + \psi_1')(\overline{\psi}_2^* + \psi_2'^*) \rangle$$

$$= \overline{\psi}_1 \overline{\psi}_2^* + \langle \psi_1' \psi_2'^* \rangle$$

$$= \langle \psi_1 \rangle \langle \psi_2^* \rangle + \langle (\psi_1 - \langle \psi_1 \rangle)(\psi_2^* - \langle \psi_2^* \rangle) \rangle \tag{55}$$

The first term on the right side of Equation 55 characterizes the *average* component, and for a stationary and homogeneous process is equal to $\langle \psi \rangle^2$. The second term characterizes the *fluctuating* component, and defines the autocovariance of the process. In the following section we will use Equation 55 to define coherent and incoherent components of the scattering intensity.

1. Coherent and Incoherent Scattering

The differential scattering cross-section per unit volume for a random continuum is found by substituting Equation 49 into Equation 53, to give

$$\sigma_d = \frac{k^4}{16\pi^2 V} \left\langle \int_V \gamma_1 e^{-i\mathbf{K} \cdot \mathbf{r}_1} \, dv_1 \int_V \gamma_2 e^{i\mathbf{K} \cdot \mathbf{r}_2} \, dv_2 \right\rangle \tag{56}$$

where $\mathbf{r}_1$ and $\mathbf{r}_2$ represent two positions in the medium, dv_1 and dv_2 are the corresponding volume elements, and for convenience we have written $\gamma_1 = \gamma(\mathbf{r}_1, \Theta)$ and $\gamma_2 = \gamma(\mathbf{r}_2, \Theta)$. We can expand Equation 56 into the sum of its average and fluctuating components using Equation 55 to find the expression

$$\sigma_d = \frac{k^4}{16\pi^2 V} \left(\left\langle \int_V \gamma_1 e^{-i\mathbf{K} \cdot \mathbf{r}_1} \, dv_1 \right\rangle \left\langle \int_V \gamma_2 e^{i\mathbf{K} \cdot \mathbf{r}_2} \, dv_2 \right\rangle \right.$$

$$\left. + \left\langle \int_V \int_V (\gamma_1 - \langle \gamma_1 \rangle)(\gamma_2 - \langle \gamma_2 \rangle) e^{-i\mathbf{K} \cdot (\mathbf{r}_1 - \mathbf{r}_2)} \, dv_1 \, dv_2 \right\rangle \right) \tag{57}$$

If we let $\Delta \mathbf{r} = \mathbf{r}_1 - \mathbf{r}_2$ and note that for statistically homogeneous media $\langle \psi_1 \rangle = \langle \psi_2 \rangle \equiv \langle \psi \rangle$, then Equation 57 becomes

$$\sigma_d = \frac{k^4}{16\pi^2 V} \left(\left| \int_V \langle \gamma(\mathbf{r}_o) \rangle e^{-i\mathbf{K} \cdot \mathbf{r}_o} \, dv_o \right|^2 \right.$$

$$\left. + \int_V \int_V \langle (\gamma_1 - \langle \gamma_1 \rangle)(\gamma_2 - \langle \gamma_2 \rangle) \rangle e^{-i\mathbf{K} \cdot \Delta \mathbf{r}} \, dv_1 \, dv_2 \right) \tag{58}$$

The first term in Equation 58 is the *coherent scattering* contribution to σ_d, which describes the *average* scatter field and correlations *among* inhomogeneities. For example, the boundary of a cloud of scatterers gives rise to a coherent scattering intensity. But inside the cloud, if the medium is random, the phases are such that the scattered waves destructively interfere, and there is no average scatter field, i.e., $\langle\gamma(\mathbf{r}_o)\rangle = 0$, and therefore no coherent term. There could also be a coherent contribution from scattering inside a cloud if the medium is not random, where, for example, the inhomogeneities are densely packed[17] or sparse but regularly positioned.[56] For the remainder of the discussion, however, we will assume that the media are random and that all coherent scattering contributions are negligible.

The second term in Equation 58 is the *incoherent scattering* contribution to σ_d, which describes the *fluctuating* portion of the scatter field and correlations *within* inhomogeneities. For example, we will show how the average scatterer size may be characterized by the incoherent scattering intensity. Since the incoherent term considers coherence only within the scatterers, it is not influenced by the position (phase) of scatterers in the medium, but sums the intensities of the scattered waves of each scatterer independently. For nonrandom media the autocorrelation function would contain contributions from correlations among the scatterers, and this term would no longer be properly labeled "incoherent".

Characteristics of the media are represented in the incoherent term by the autocovariance function for the medium, $\langle(\gamma_1 - \langle\gamma_1\rangle)(\gamma_2 - \langle\gamma_2\rangle)\rangle$, which considers the relationship between fluctuations in γ about the mean fluctuation at two points in the medium. Again, for statistically homogeneous random media, $\langle\gamma_1\rangle = \langle\gamma_2\rangle = 0$, and the autocovariance function reduces to[55] $\langle\gamma_1\gamma_2\rangle = \langle\gamma^2\rangle b_\gamma(\Delta\mathbf{r})$, where $\langle\gamma^2\rangle$ is the mean-square fluctuation in medium properties and $b_\gamma(\Delta\mathbf{r})$ is the correlation coefficient for the scattering medium as discussed below and in Appendix B. The differential scattering cross-section for incoherent scattering may be written as

$$\sigma_d = \frac{k^4}{16\pi^2 V} \int_V \int_V \langle\gamma^2\rangle b_\gamma(\Delta\mathbf{r}) e^{-i\mathbf{K}\cdot\Delta\mathbf{r}} \, dv_1 \, dv_2 \tag{59}$$

Next, to simplify Equation 59, we make two assumptions. First, we assume the medium is *isotropic*, so that we need consider only the magnitude and not the direction of $\Delta\mathbf{r}$ in the correlation coefficient, i.e., we can replace $b_\gamma(\Delta\mathbf{r})$ with $b_\gamma(\Delta r)$. Second, we assume that the size of the inhomogeneities, D, is much smaller than the size of V, so that the correlation distance in the medium is much smaller than the size of V. Where this is true, we can write the double integral in Equation 59 as the product of two single integrals, following a change of coordinates from $\mathbf{r}_1$ and $\mathbf{r}_2$ to $\Delta\mathbf{r} = \mathbf{r}_1 - \mathbf{r}_2$ and $\bar{\mathbf{r}} = (\mathbf{r}_1 + \mathbf{r}_2)/2$:

$$\sigma_d = \frac{k^4}{16\pi^2 V} \int_V \int_V \langle \gamma^2 \rangle b_\gamma(\Delta r) e^{-i\mathbf{K}\cdot\Delta\mathbf{r}} \, d\mathbf{r}_1 d\mathbf{r}_2$$

$$= \frac{k^4 \langle \gamma^2 \rangle}{16\pi^2 V} \int_V \int_V b_\gamma(\Delta r) e^{-i\mathbf{K}\cdot\Delta\mathbf{r}} \left| \frac{\partial(\mathbf{r}_1, \mathbf{r}_2)}{\partial(\Delta\mathbf{r}, \bar{\mathbf{r}})} \right| d\Delta\mathbf{r} \, d\bar{\mathbf{r}} \qquad (60)$$

where the Jacobian of the transformation,

$$\left| \frac{\partial(\mathbf{r}_1, \mathbf{r}_2)}{\partial(\Delta\mathbf{r}, \bar{\mathbf{r}})} \right| = \frac{\partial \mathbf{r}_1}{\partial \Delta\mathbf{r}} \frac{\partial \mathbf{r}_2}{\partial \bar{\mathbf{r}}} - \frac{\partial \mathbf{r}_2}{\partial \Delta\mathbf{r}} \frac{\partial \mathbf{r}_1}{\partial \bar{\mathbf{r}}} = 1 \qquad (61)$$

and we have used the notation $d\mathbf{r}_1 = dv_1$, and $d\mathbf{r}_2 = dv_2$. Therefore, we have

$$\sigma_d = \frac{k^4 \langle \gamma^2 \rangle}{16\pi^2 V} \left(\int_V d\bar{\mathbf{r}} \right) \left(\int_V b_\gamma(\Delta r) e^{-i\mathbf{K}\cdot\Delta\mathbf{r}} \, d\Delta\mathbf{r} \right)$$

$$= \frac{k^4 \langle \gamma^2 \rangle}{16\pi^2 V} V \left(\int_V b_\gamma(\Delta r) e^{-i\mathbf{K}\cdot\Delta\mathbf{r}} \, d\Delta\mathbf{r} \right)$$

$$= \frac{k^4 \langle \gamma^2 \rangle}{16\pi^2} \int_{-\infty}^{\infty} b_\gamma(\Delta r) e^{-i\mathbf{K}\cdot\Delta\mathbf{r}} \, dv_\Delta \qquad (62)$$

where $dv_\Delta = d\Delta\mathbf{r} = d(\Delta x) \, d(\Delta y) \, d(\Delta z)$ and the remaining integral is now extended to all space.

2. Small Discrete Inhomogeneities

Consider the special case of a medium containing discrete scatterers that are small in size as compared to the wavelength of sound. This is an important special case because closed-form solutions for σ_d are easily obtained, the results provide insight into the kinds of structural information available from scatter measurements, and because it is applicable to biological systems, e.g., acoustic scattering from low hematocrit blood in the low-MHz frequency range.[49] A complete discussion of this subject is given in the following chapter.

Following the discussion of Glotov,[17] we can simplify Equation 62 by assuming that the random medium contains discrete scatterers that are small compared to the wavelength ($kD \ll 1$, where D is the size of the scatterer). In that case, the Fourier transform of the correlation coefficient is approximately equal to the integral of b_γ (since $\exp(-i\mathbf{K}\cdot\Delta\mathbf{r}) \simeq 1$ for small $\Delta\mathbf{r}$), and, as shown in Appendix B, $\int b_\gamma(\Delta r) dv_\Delta$ defines the effective scatterer volume V_s, so that we obtain

$$\sigma_d \xrightarrow{kD \ll 1} \sigma_o = \frac{k^4 V_s^2}{16\pi^2} \bar{n} \langle \gamma_o^2 \rangle \qquad (63)$$

Note that Equations 75 and 64 are identical for spherical scatterers, as shown in the previous section. The form factor is the frequency-space analog of the correlation coefficient; the two are Fourier transform pairs. Analogous to the atomic form factor,[24] the acoustic form factor characterizes the shape and size of inhomogeneities and serves as a useful descriptor of scatterers with arbitrary shape. The advantage of studying F instead of σ_b is that the former depends only on the size of the scatterer D and the wavelength of sound $2\pi/k$. Therefore, if an appropriate correlation model for the medium can be found, an average scatterer size can be estimated.

Form factors corresponding to the correlation models above, Equations 67, 69, 71, and 73 are, respectively

$$F(2k, D) = \left(\frac{3}{2ka} j_1(2ka)\right)^2 \qquad (Fluid\ Sphere) \qquad (77)$$

$$= e^{-2k^2d^2} \qquad (Gaussian) \qquad (78)$$

$$= \frac{1}{(1 + 4k^2d^2)^2} \qquad (Exponential) \qquad (79)$$

$$= \left(\frac{\epsilon^4 d^4}{(1 + 4k^2d^2)^2} - \frac{1}{(1 + 4k^2/\epsilon^2)^2}\right)\left(\frac{1}{\epsilon^4 d^4 - 1}\right) \qquad (Modified\ Exp.) \qquad (80)$$

Examples of these functions are plotted in Figures 4 and 5.

Just as discrete isotropic media may be characterized by an average scatterer diameter $2a$, continuous isotropic media may be characterized by the correlation distance d. We can equate the two parameters by setting values of V_s for a continuum model equal to the volume of a sphere of radius a and computing an *effective diameter* $D = 2a$. In the case of the Gaussian model,

$$\frac{4\pi a^3}{3} = (2\pi d^2)^{3/2}$$

$$D_g \equiv 2a = (12\sqrt{2\pi})^{1/3}\ d \simeq 3.11\ d \qquad (81)$$

Similarly for the exponential model, $D_e = \sqrt[3]{48}\ d \simeq 3.63d$. These equations may be used to interpret characteristics of random continua in the more familiar terms of discrete scattering media. The three form factors plotted in Figure 4a as a function of the correlation distance, d, are replotted in Figure 4b as a function of the effective diameter, D, to show the similarity among the functions when D is used.

F. AN EXAMPLE: THE KIDNEY
1. Interpreting Backscatter Using an Isotropic Model

The kidney is an example of an anisotropic medium that may be considered isotropic in one plane, i.e., the plane perpendicular to the axis of symmetry

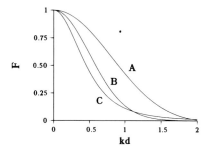

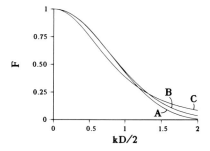

FIGURE 4. Form factors corresponding to the correlation coefficients in Figure 2 are plotted as a function of the correlation distance, d, in (top) and as a function of the effective diameter, D, in (bottom): (A) fluid sphere, Equation 77; (B) Gaussian, Equation 78; and (C) exponential, Equation 79.

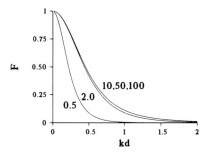

FIGURE 5. Form factors corresponding to the modified exponential correlation coefficients in Figure 3 are plotted (Equation 80). Numbers indicate the value of ϵd used in the plot.

determined by the orientation of the nephrons. Nephrons are the functional units of the kidney. A very simplistic description of microscopic renal anatomy pictures nephrons and blood vessels as tubular structures radiating out from the center of the kidney. Within the outer annulus of the organ (the cortex) at the ends of nephrons are spherical structures called glomeruli. Each of these structures contains a collagenous membrane with density and compressibility different from the surrounding tissue. Tubules in the adult human are approximately 50 μm in diameter and several millimeters in length; the average glomerular diameter is approximately 200 μm.

Broadband form factor measurements were obtained for planar regions within kidney cortex using experimental methods described elsewhere.[19,20] Measurements were fit to the form factor models described above to determine which isotropic model most closely represents the data, and to estimate the average scatterer size as determined by least-squares analysis. Backscatter data were recorded for several angles φ between the beam axis and the longitudinal axis of the dominant nephron orientation. For example, φ = 0° indicates the two axes are aligned and φ = 90° indicates that the beam is perpendicular to the nephron axis. Form factors measured for kidney cortex at these two angles are shown in Figures 6b and c.

The data are well represented by a *two-component* Gaussian correlation model at both scanning angles (Figure 6). At low frequencies, where F is a sharply decreasing function of frequency, scatterers with an average size $D_g = 220 \pm 15$ μm are estimated. At higher frequencies, where F gradually decreases with frequency, smaller scatterers with an average size $D_g = 55 \pm 10$ μm are estimated. These estimates closely correspond to the average glomerular diameter (216 ± 25 μm) and tubule diameter (41 ± 8 μm) measured for the same tissue samples using a calibrated light microscope.[20] Also, the scatterer size estimates were found to be independent of scanning angle (see Section IV.F.3).

The largest difference between the measurements at 0 and 90° is the increase in the magnitude of the high-frequency component for perpendicular incidence over that for parallel incidence (Figure 6a). The relative magnitude of the high-frequency component for oblique incidence falls between these two extremes and varies cyclically with angle (also discussed in Section IV.F.3).

In summary, backscattered ultrasound measurements indicate that there are two dominant structures in the renal cortex that scatter sound. Histological correlations reveal that these structures are blood vessels/tubules and glomeruli and that their relative contribution to the backscatter coefficient depends on the angle of incidence and the frequency of sound. In this example, isotropic scattering theory was used successfully to provide a framework for the description of anisotropic kidney tissue. In the next section we discuss a very simple anisotropic model that leads to a more complete understanding.

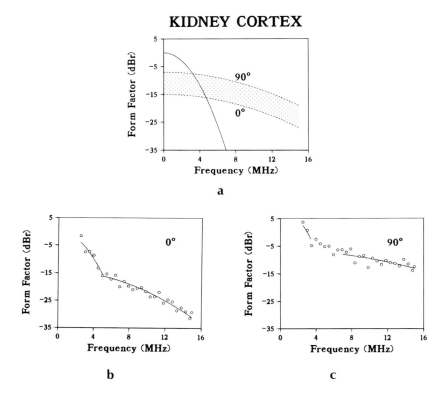

a

b   **c**

FIGURE 6. Modeled and measured form factors for renal cortex are plotted as a function of frequency. Plot (a) summarizes the two-component Gaussian model that is consistent with all the data described in Reference 20. The shape of the solid line indicates that scatterers with an average size of ~220 μm dominate backscatter at low diagnostic frequencies, and the shape of the dashed band indicates that scatterers with an average size of ~60 μm dominate at high frequencies. Although the magnitudes of both components vary cyclically with the angle between the acoustic beam and nephron axes (φ), the magnitude of the variation in *F* for the small-scatterer component is greater than that for the large-scatterer component. The magnitudes of both components are at a maximum for perpendicular incidence (90°) and at a minimum for parallel incidence (0°), as shown in Figure 9. The two acoustically determined average scatterer sizes correspond to the average diameter of glomeruli and renal tubules, respectively. Plots (b) and (c) are examples of *in vitro* acoustic measurement for the renal cortex of a dog kidney at two different scanning angles. Points indicate measured values and the solid lines indicate the best-fit isotropic Gaussian model, as in (a).

2. Backscatter Coefficient for Anisotropic Random Media

If the random medium is anisotropic, then the correlation coefficient depends on the relative orientation of the medium, so that Equation 62 becomes

$$\sigma_b = \frac{k^4 \langle \gamma^2 \rangle}{16\pi^2} \int_{-\infty}^{\infty} b_\gamma(\mathbf{\Delta r}) e^{-i\mathbf{K}\cdot\mathbf{\Delta r}} \, dv_\Delta \tag{82}$$

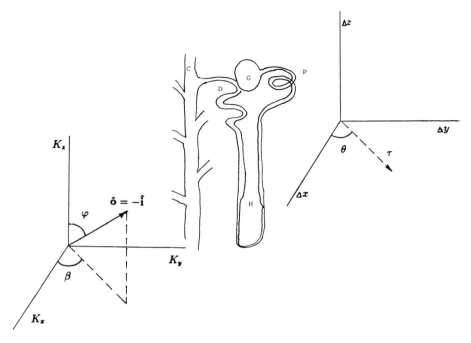

FIGURE 7. Scattering geometries in space (upper right) and spatial frequency (lower left) are shown relative to the orientation of the nephron (center). The correlation length in the (x,y) plane is much shorter than that along the z-axis. The diagram indicates several components of the nephron: collecting duct (C), glomerulus (G), loop of Henle (H), and proximal (P) and distal (D) convoluted tubules.

Consider a medium that is isotropic in the (x,y) plane, e.g., the kidney tissue described above. Further assume that we can write $b_\gamma(\mathbf{\Delta r}) = b_1(\Delta z)b_2(\Delta x, \Delta y)$. Then, with the help of Figure 7, we can define the following components:

$$\Delta x = \tau \cos \theta$$
$$\Delta y = \tau \sin \theta$$
$$\Delta z = \Delta z$$
$$K_x = 2k \cos \beta \sin \varphi$$
$$K_y = 2k \sin \beta \sin \varphi$$
$$K_z = 2k \cos \varphi \tag{83}$$

and write

$$
\sigma_b = \frac{k^4 \langle \gamma^2 \rangle}{16\pi^2} \int_{-\infty}^{\infty} b_1(\Delta z) e^{-iK_z \Delta z} \, d\Delta z \int_{-\infty}^{\infty} \int_{-\infty}^{\infty} b_2(\Delta x, \Delta y) e^{-i(K_x \Delta x + K_y \Delta y)} \, d\Delta x \, d\Delta y
$$

$$
= \frac{k^4 \langle \gamma^2 \rangle}{16\pi^2} \int_{-\infty}^{\infty} b_1(\Delta z) e^{-i2k\Delta z \cos\varphi} \, d\Delta z \int_0^{2\pi} \int_0^{\infty} b_2(\tau) e^{-i2k\tau \sin\varphi(\cos\theta\cos\beta + \sin\theta\sin\beta)} \tau \, d\tau \, d\theta
$$

$$
= 2\pi \, \frac{k^4 \langle \gamma^2 \rangle}{16\pi^2} \int_{-\infty}^{\infty} b_1(\Delta z) e^{-i2k\Delta z \cos\varphi} \, d\Delta z \int_0^{\infty} b_2(\tau) \, J_o(2k\tau \sin \varphi)\tau \, d\tau \tag{84}
$$

where

$$
J_o(\rho) = \int_0^{2\pi} e^{-i\rho\cos(\theta - \beta)} \, d\theta
$$

The first integral on the right side of Equation 84 is the one-dimensional Fourier transform of the correlation coefficient along the axis of symmetry. We have used the symmetry in the (x,y) plane to reduce the two remaining integrals to one, resulting in the Fourier-Bessel transform (Hankel transform)[18] of the correlation coefficient in the plane with symmetry. Since for backscatter the unit vectors defining the incident field and the observer are directed along the same line, φ is sufficient to define the scattering geometry.

Next, we assume the Gaussian model and let $b_1(\Delta z) = \exp(-\Delta z^2/2d_z^2)$ and $b_2(\tau) = \exp(-\tau^2/2d_\tau^2)$, where d_z is the correlation length along the axis of symmetry and d_τ is the correlation length along a radius in the (x,y) plane. Substituting these equations into Equation 84 and integrating, results in the following expression for the backscatter coefficient:

$$
\sigma_b = \left(\frac{k^4 V_s^2 \overline{n} \gamma_o^2}{16\pi^2} \right) e^{-2k^2(d_z^2\cos^2\varphi + d_\tau^2\sin^2\varphi)} \tag{85}
$$

where, as before, $\langle \gamma^2 \rangle = \overline{n} V_s \gamma_o^2$ and now $V_s = \int_{-\infty}^{\infty} b_\gamma(\Delta \mathbf{r}) dv_\Delta = (2\pi)^{3/2} d_z d_\tau^2$.

Equation 85 is very similar to that for the isotropic Gaussian model, Equation 69, as might be expected, except that it allows for two different correlation lengths, one characteristic of the z axis and the other characteristic of the (x,y) plane. In the case where $d_z = d_\tau$ (isotropic medium), Equation 85 is identical to Equation 69.

3. Interpreting Backscatter Using an Anisotropic Model

The form factor corresponding to Equation 85 is given by

$$
F_\varphi = e^{-2k^2(d_z^2\cos^2\varphi + d_\tau^2\sin^2\varphi)} \tag{86}
$$

and is now a function of the scanning angle φ. We have plotted F_φ in Figure 8 for the (arbitrarily selected) value $d_z/d_\tau = \sqrt{10}$. In this example the correlation distance along the axis of symmetry is approximately three times that

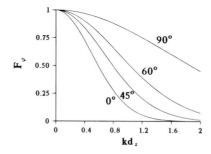

FIGURE 8. Form factors for the anisotropic Gaussian model, Equation 86, are plotted for four scanning angles. In this example $d_z/d_r = \sqrt{10}$.

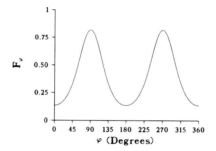

FIGURE 9. The form factor for the anisotropic Gaussian model, Equation 86, is plotted as a function of scanning angle (φ), where $kd_z = 1$ and $d_z/d_r = \sqrt{10}$. This plot shows that backscatter is greatest when the acoustic beam is oriented perpendicular to the axis of symmetry (z-axis) and minimum for parallel incidence.

in the (x,y) plane. In Figure 9 we have also plotted F_φ, but as a function of φ for the fixed value of $kd_z = 1$, to illustrate how the backscatter coefficient is expected to vary with scanning angle. We find experimentally[20] that σ_b does vary cyclically as predicted by the results of Figure 9. However, Figure 8 predicts that the scatterer size should also vary significantly with φ at high frequencies where blood vessels and/or tubules dominate scattering. Instead, measured values of D, the effective scatterer size, are nearly independent of φ at all frequencies.[20] The discrepancy between the model and the data is probably due to the fact that the model described by Equation 85 does not account for the variability in the alignment of structures that exists in real tissues. Nor does the model include the effects of scattering from *convoluted tubules* (Figure 7), which are not aligned and may be a large fraction of the total backscattered energy at high frequencies as compared to scattering from the aligned segments of tubules. It may be that the alignment of tissue structures more directly influences the magnitude of σ_b than the shape of σ_b vs. frequency. Testing these and other hypotheses is the goal of current research.

Nevertheless, the results demonstrate that simple scattering theory can provide insight into the structure of very complex media such as the kidney.

G. BACKSCATTER COEFFICIENT FOR DENSE MEDIA[†]

The final topic discussed in this chapter is backscatter from a densely populated, weakly scattering, isotropic, random medium. For this type of medium, the scatterers must be much smaller than a wavelength in order to be randomly positioned, and have a mean distance between scatterers less than one wavelength. Further, the random positioning ensures that there will be no coherent scattering except at the boundaries of the medium. Also, as with sparse media, we assume that the fluctuations in density and compressibility are small, that multiple scattering is negligible, and therefore that the Born approximation holds. The dense-medium model is important because it could apply to scattering from soft tissues, and yet the results are very different from that for sparse media, as we now show.

Recall that for incoherent scattering we are interested in the fluctuations in density and compressibility *about the mean values*. Therefore the large volume fraction of scatterers in the medium means that negative correlations become important. To see why this is true, consider a dense population of inhomogeneities, where each is more compressible than the surrounding fluid. As before, the compressibility of the inhomogeneities is greater than the average compressibility, but for the dense medium the compressibility of the surrounding fluid is significantly less than the average. Therefore the correlation between two points inside and outside an inhomogeneity will be negative.

Morse and Ingard[31] have suggested the following modified Gaussian correlation model to describe isotropic, dense random media:

$$b_\gamma(\Delta r) = \left(1 - \frac{\Delta r^2}{3d^2}\right)e^{-\Delta r^2/2d^2} \tag{87}$$

This function is negative for $\Delta r > \sqrt{3}d$ (Figure 10) and its integral is zero. Substituting Equation 87 into Equation 62 and integrating, we find the following backscatter coefficient for dense media:

$$\sigma_b = \frac{8\pi}{9}\, k^6 d^8 \bar{n} \gamma_o^2 e^{-2k^2d^2}$$

$$= \left(\frac{k^4 V_s^2 \bar{n} \gamma_o^2}{16\pi^2}\right)\left(\sqrt{8\pi}\left(\frac{e}{3}\right)^{3/2}k^2d^2\right)e^{-2k^2d^2} \tag{88}$$

[†] More precisely, this section describes scattering from media having an *intermediate* range of scatterer densities. For very high densities of extended particles, i.e., where the volume fraction of particles approaches one, we return to the sparse scattering models, where the spaces between particles scatter sound. This heuristic "hole" approach was described by Beard et al.[2] and later used by Shung et al.[49] to explain the relationship between the backscatter coefficients measured for blood and the number of red blood cells in the sample (hematocrit).

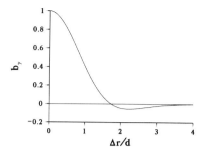

FIGURE 10. The correlation coefficient given by Equation 87 for isotropic, dense-packed, random media is plotted. Note how b_γ integrates to zero because, in the 3D integration, b_γ is weighted by a volume element proportional to Δr^2.

The last form was obtained by defining the scatterer volume as the integral over the region where b_γ is positive:

$$V_s = \int_0^{\sqrt{3}d} b_\gamma(\Delta r)(\Delta r)^2 \, d\Delta r$$

$$= \left(\frac{3}{e}\right)^{3/2} \frac{4\pi}{3} \, d^3 \tag{89}$$

Although somewhat arbitrary, this definition of V_s is also reasonable, since it is approximately equal to the volume of a sphere of radius d.

Equation 88 is similar to that for the Gaussian model, Equation 69, except for the additional and very important factor $(\sqrt{8\pi}(e/3)^{3/2}k^2d^2)$. As before, we can determine σ_b in the long-wavelength limit:

$$\sigma_b \xrightarrow{kd \ll 1} \left(\frac{k^4 V_s^2 \bar{n}\gamma_o^2}{16\pi^2}\right) \sqrt{8\pi} \left(\frac{e}{3}\right)^{3/2} k^2 d^2 \tag{90}$$

and therefore show that the form factor in this case is that for the Gaussian model, as given by Equation 78. That is, the ratio of σ_b for the finite-size scatterer, Equation 88, to that of the point scatterer, Equation 90, is simply $F(2k,D) = \exp(-2k^2d^2)$.

The most obvious distinction between dense and sparse random media, as modeled using a Gaussian correlation coefficient, occurs at low frequencies, where for dense media $\sigma_b \propto k^6d^8$ and for sparse media $\sigma_b \propto k^4d^6$. The difference in k dependence should make it easy to distinguish between the two conditions, but we find experimentally that this is not the case. There are no measurements of scattering cross-sections for tissues that show a frequency dependence greater than k^4. (Yuan and Shung[57] measured a backscatter frequency dependence as high as $k^{4.47}$ for low hematocrit flowing blood, although they explain the result could be a curve-fitting artifact.) This lack of definitive experimental evidence indicates that either this dense-medium model does not apply to biological tissues or the intrinsically low signal strength of the phenomenon makes it difficult to detect. There is, however, definitive experimental evidence showing that a frequency dependence greater than k^4 can be measured. Campbell and Waag,[9] for example, measured the total cross-section vs. frequency for a medium consisting of a dense concentration of Sephadex spheres in water and found it proportional to k^5. They suggested that the low signal-to-noise ratio of the measurement could reasonably account for the deviation from k^6 observed.

V. SUMMARY

Our intention in writing this chapter has been to review the basic equations of acoustic scattering theory in the context of ultrasound-tissue interactions. The framework for analysis follows from three easily understood principles, but the details of the analysis require us to make many simplifying assumptions in order to fully understand acoustic scattering in tissues, and therefore ultrasonic imaging. Future work in this area should aim to refine or eliminate as many of these assumptions as possible, which means that experiment and theory must continue to work together closely if we are to progress in our understanding of the basic interactions. Continued advancement in these basic areas of research promises an even brighter future for this already important diagnostic modality.

ACKNOWLEDGMENTS

The authors express their appreciation to Matthew R. Myers, Gerald R. Harris, and Robert F. Wagner for many helpful discussions and comments on the manuscript. This work was supported in part by a grant from the Whitaker Foundation.

APPENDIX A: VALIDITY OF THE FIRST ORDER APPROXIMATION

To adequately describe an acoustic system using first order equations, terms of higher order must be negligible by comparison. We will examine the terms discarded in obtaining the first order equations in order to determine the conditions required for discarding higher order terms. This treatment follows that of Pierce[44] who references Eckart.[13]

To begin, we *estimate* the derivatives in each expression by equating them to division by an appropriate length or time, i.e., the wavelength or period of the disturbance. In the equation of continuity, Equation 13, we have retained $\rho_o \nabla \cdot \mathbf{u}$ while discarding $\nabla \cdot (\delta\rho\mathbf{u}) = \delta\rho\nabla \cdot \mathbf{u} + \mathbf{u} \cdot \nabla(\delta\rho)$. Since both of the discarded terms are approximately $\delta\rho\mathbf{u}/L$, we must have

$$\frac{\delta\rho u}{L} \ll \frac{\rho_o u}{L}$$

$$\frac{\delta\rho}{\rho_o} \ll 1 \tag{91}$$

Similarly, in Euler's equation, Equation 15, we have discarded $\delta\rho \, \partial\mathbf{u}/\partial t$, $\delta\rho \, \mathbf{u} \cdot \nabla\mathbf{u}$, and $\rho_o\mathbf{u} \cdot \nabla\mathbf{u}$ in favor of $\rho_o\partial\mathbf{u}/\partial t$. Since $\delta\rho/\rho_o \ll 1$, we know that $\delta\rho\partial\mathbf{u}/\partial t \ll \rho_o\partial\mathbf{u}/\partial t$ and $\delta\rho \, \mathbf{u} \cdot \nabla\mathbf{u} \ll \rho_o\mathbf{u} \cdot \nabla\mathbf{u}$, but we must also require

$$\rho_o\mathbf{u} \cdot \nabla\mathbf{u} \ll \rho_o \frac{\partial\mathbf{u}}{\partial t}$$

$$\frac{u^2}{L} \ll \frac{u}{T}$$

$$u \ll L/T \tag{92}$$

Note that indirectly Euler's equation also provides us with a condition on p, since in order for it to be valid we must have $\rho_o u/T \approx p/L$ (see Equation 15), which when combined with Equation 92 yields the requirement

$$p \ll \rho_o(L/T)^2 \tag{93}$$

Finally, from the equation of state, Equation 10, we must have the first term in the expansion much larger than the second. This presents a difficulty only because the conventional measure of nonlinearity for the medium, called "B/A", is defined in terms of derivatives of p with respect to ρ rather than of ρ with respect to p. By solving, for example, the quadratic equation for δp ($\equiv p$) in the first two terms of the Taylor series expansion for $\delta\rho$, Equation 10, and comparing term by term with the Taylor series for δp, we verify the following relationships:

$$\frac{\partial \rho}{\partial p} = \left(\frac{\partial p}{\partial \rho}\right)^{-1} \tag{94}$$

$$\frac{\partial^2 \rho}{\partial p^2} = -\frac{\partial^2 p}{\partial \rho^2} \bigg/ \left(\frac{\partial p}{\partial \rho}\right)^3 \tag{95}$$

Therefore, for the second term in the Taylor series expansion for $\delta\rho$ to be negligible in comparison with the first, we must have

$$p \frac{\partial^2 \rho}{\partial p^2} \bigg/ 2 \frac{\partial \rho}{\partial p} \ll 1 \tag{96}$$

$$p \frac{\partial^2 p}{\partial \rho^2} \bigg/ 2\left(\frac{\partial p}{\partial \rho}\right)^2 \ll 1 \tag{97}$$

(Instead of beginning with the equation of state, this result may be obtained directly by requiring that the second term in the Taylor series expansion for $\delta p(\rho,s)$ be negligible in comparison to the first.) Noting that the dimensionless parameter of nonlinearity B/A is by definition[4]

$$\frac{B}{A} = \rho_o \frac{\partial^2 p}{\partial \rho^2} \bigg/ \frac{\partial p}{\partial \rho}$$

$$= \rho_o \frac{\partial^2 p}{\partial \rho^2} \bigg/ c^2 \tag{98}$$

where we have used Equation 94 and the relation $1/c^2 = \partial\rho/\partial p|_{s=0}$, where c is the speed of propagation of the acoustic disturbance, we arrive at the requirement

$$\left(\frac{B}{A}\right)\left(\frac{p}{2\rho_o c^2}\right) \ll 1 \tag{99}$$

Identifying p with $c^2\delta\rho$ and c with L/T, and using the plane wave relationship between particle velocity and pressure, $u = p/(\rho_o c)$, we find that Equations 91 through 93 are identical:

$$\frac{p}{\rho_o c^2} \ll 1 \tag{100}$$

and that Equation 99 differs only by the added factor $1/2(B/A)$.

For peak pressures of 100 atm, for B/A on the order of 7 for many soft biological tissues,[5] and taking $\rho_o c^2$ equal to 2.37×10^9 N/m^2, the ratios in Equations 100 and 99 are, respectively, 1:230 and 1:70. Consequently, the first order equations are valid under typical diagnostic imaging conditions. In addition, dissipative processes in real tissues lead to selective absorption of higher frequencies which further mitigates higher order effects.

in either arterial disease assessment or hematology. The general goal in the former is to use Doppler ultrasound to evaluate the severity of arterial stenoses, which usually entails spectral analysis of the backscattered Doppler signal from the artery.[8] It is generally recognized that in order to achieve this goal, a more complete understanding of the statistical nature of the Doppler signal is needed. In hematology, one of the main objectives is to develop ultrasonic techniques for measuring various rheological properties of blood such as hematocrit and the degree of RBC aggregation.[9,10] Towards this end, much theoretical and experimental work has been directed at understanding the relationship between the total backscattered signal power and hematocrit for different plasma protein concentrations and flow conditions.

While the basic laws of physics that govern the scattering of ultrasound in blood are essentially the same as in other soft tissues (Chapter 4), the problem of scattering in blood is different because the relative positions of the scatterers can vary significantly with time, depending primarily on the flow conditions. In fact, over the years much attention has been given to the effect of flow disturbances or turbulence on the ultrasonic backscatter. This is particularly important for Doppler quantitation of arterial disease since flow disturbances can occur in localized regions distal to a stenosis. The effects of flow disturbances are also important for potential hematological applications because the relationship between the total backscattered power and hematocrit can be significantly affected by the flow conditions.[11] As will be discussed later (Section IV), the explanations provided by different theoretical models regarding the physical causes of increased scattering under turbulent flow condition have not been consistent. Part of the problem, especially in the comprehensive studies that account for the effects of a nonuniform beam profile, is the use of a very formal approach based on complex signal representation, which often renders physical interpretation of the mathematical equations quite difficult.

With the above in mind, this chapter is organized as follows. A review of the fundamentals of acoustic theory with emphasis on ultrasound scattering by blood is given in the next section.* A semi-heuristic geometric ray approach is then used in Section III to introduce the Doppler effect and the basic wave interference concepts that are essential to understanding the ultrasonic backscatter from a dense and moving suspension of RBCs. In Section IV, two classical approaches to modeling the ultrasonic backscatter are described based on the mathematical framework established in Section II. In Section V, a recently proposed hybrid approach[12,13] that combines the advantages of the two classical approaches is used to model the backscattered Doppler signal. Section VI shows how this hybrid approach, which does not require the use of complex signal notation, can provide a better intuitive understanding of the effects of flow disturbances and RBC aggregation. In Section VII, the

---

* The reader can refer to the previous chapter for a more general and detailed treatment of the acoustic wave equation.

hybrid approach is used to derive the statistics of the Doppler power spectrum. The final section provides a summary of the theoretical results and a discussion of their practical implications.

II. LINEARIZED WAVE EQUATION AND RAYLEIGH SCATTERING

When a sound wave impinges upon an obstacle whose mechanical properties differ from those of the surrounding fluid, part of the wave is deflected from its original path. Morse and Ingard[14] defined the *scattered* wave as "the difference between the actual wave and the undisturbed wave, which would be present if the obstacle were not there". This is a very general definition which does *not* require the obstacle to be much smaller than a wavelength: if the obstacle is much larger than the wavelength, the scattered wave can be further divided into a *reflected* wave and an *interfering* wave that cancels with the original wave to create an acoustic shadow behind the obstacle.

A. LINEARIZED WAVE EQUATION

To model ultrasound propagation and scattering in blood, it is necessary to start with certain simplifying assumptions regarding the mechanical and thermodynamic properties of blood. Since most of these have been dealt with quite rigorously in the previous chapter, only a brief summary is given here. Suppose $p(\mathbf{r},t)$ and $\mathbf{u}(\mathbf{r},t)$* are, respectively, the pressure and fluid velocity vector of the longitudinal sound wave (compressions and rarefactions) at position $\mathbf{r}$ and time t. It is assumed that (1) the shear and bulk viscosities of blood are relatively small such that shear wave propagation and wave mode conversion effects at scattering sites in blood can be ignored; (2) attenuation of the incident and scattered ultrasound (2 to 8 MHz) due to viscous losses[15] is also negligible; (3) the insonified blood is quasi-stationary, i.e., any macroscopic movement is much slower than sound propagation in blood; (4) the heat conductivity of blood is zero; and (5) the sound-induced changes in the mass density of blood are small compared to the equilibrium density, such that the adiabatic blood compressibility can be considered as independent of $p(\mathbf{r},t)$. As pointed out by Insana and Brown (Chapter 4, Equation 15), the last assumption is weaker than the more commonly imposed condition of small pressure perturbations[14,16] which may be violated by pulse-echo imaging systems.

Under the above conditions, acoustic wave motion is governed by the following two equations:[14]

$$\rho \, \frac{\partial \mathbf{u}}{\partial t} = -\nabla p \qquad (1)$$

---

*   Boldface is used to denote vector quantities.

and

$$\kappa \frac{\partial p}{\partial t} = -\text{div } \mathbf{u} \tag{2}$$

where ρ and κ are the mass density and adiabatic compressibility of blood, respectively. The first equation is referred to as the *linear inviscid force equation* or the *linearized Euler's equation*. It basically states that a pressure gradient produces an acceleration of the fluid, in accord with Newton's second law (or momentum conservation). On the other hand, Equation 2 states that a velocity gradient produces a compression of the fluid, and this is derived from the continuity equation (or mass conservation) and an equation of state. Combining the divergence of Equation 1 and time derivative of Equation 2, and eliminating $\mathbf{u}$, yields the *linearized acoustic wave equation* (same as Equation 23 in Chapter 4) as follows:

$$\text{div}\left(\frac{1}{\rho}\nabla p\right) - \kappa \frac{\partial^2 p}{\partial t^2} = 0 \tag{3}$$

B. RAYLEIGH SCATTERING

To understand the nature of the scattered ultrasound from blood it is instructive to start with the *differential scattering cross-section* of a single RBC, defined as

$$\sigma_d(\gamma) = \frac{\text{power scattered/unit solid angle at } \gamma}{\text{incident intensity}} \tag{4}$$

where γ is the angle between the incident and the scattered wave vectors. As mentioned earlier, the RBC is a biconcave disc whose diameter is about 8 μm, which is much smaller than the wavelength (190 to 750 μm) of diagnostic ultrasound (2 to 8 MHz) in blood. In a classical paper[17] published in 1872, Rayleigh described two approaches that can be used to derive from Equation 3 the $\sigma_d(\gamma)$ of such a small obstacle. The first method assumes the obstacle is a perfect sphere of radius 'a', and expresses the solution as an infinite series of Bessel's functions whose coefficients are determined by matching boundary conditions. Suppose R is the distance between the observation point and the sphere, and $k_o = 2\pi/\lambda_o$ is the wavenumber, in which λ_o denotes the ultrasonic wavelength in the suspending medium. For $k_oR \gg 1$ and $k_oa \ll 1$, this boundary value method[14] yields

$$\sigma_d(\gamma) = \frac{V_c^2 \pi^2}{\lambda_o^4}\left[\frac{\kappa_e - \kappa_o}{\kappa_o} + \frac{3(\rho_e - \rho_o)}{2\rho_e + \rho_o}\cos\gamma\right]^2 \tag{5}$$

where $V_c = (4/3)\pi a^3$ is the volume of the RBC, $\kappa_e, \rho_e,$ and κ_o, ρ_o are the

TABLE 1
Acoustic Properties of Blood Constituents
at 20°C

Medium	Mass density, ρ (g/cm³)	Adiabatic compressibility, κ (10^{-12} cm/dyne)
RBCs	1.092[18]	34.1[19,20]
Plasma	1.021[18]	40.9[19,20]
0.9% saline	1.005[21]	44.3[19]
Distilled water	0.998[21]	46.1[19]

Note: The values reported by Urick[19] were based on measurements of sound velocity in fresh horse blood.

compressibility and mass density of the RBCs, and of the surrounding fluid, respectively.

The second method, known as the Green's function approach (Chapter 4), may appear at first glance to be more general since it can be applied to obstacles of *arbitrary shape*. However, in order to obtain an analytical solution using this approach, the obstacle is assumed to be a very weak scatterer so that the *Born approximation* is valid: the total wave can be approximated by the incident wave. This basically requires that the mismatches in density and compressibility between the obstacle and the surrounding fluid be fairly small. As shown in Table 1, this condition is satisfied for normal blood. With the above assumptions, the Green's function approach gives

$$\sigma_d(\gamma) = \frac{V_c^2 \pi^2}{\lambda_o^4} \left[\frac{\kappa_e - \kappa_o}{\kappa_o} + \frac{\rho_e - \rho_o}{\rho_e} \cos \gamma \right]^2 \tag{6}$$

The above equation will appear again in Section IV when we describe in detail the Green's function approach to solving Equation 3 for a dense suspension of RBCs. Therefore, it is worthwhile dwelling on the meaning of Equations 5 and 6 in the remainder of this subsection. First, it can be seen that $\sigma_d(\gamma)$ is generally comprised of two terms representing the fractional changes in compressibility and density. The compressibility term is independent of the angle γ, and is referred to as a *monopole* term since it corresponds to a pulsating point source. On the other hand, the density term is angle-dependent and is known as the *dipole* term because it arises from oscillatory motion like a dipole source.[14] Thus, one can see that the mismatches in different mechanical properties cause the obstacle to undergo different modes of vibrations. A mismatch in density causes the obstacle to oscillate back and forth about the undisturbed position, whereas a mismatch in compressibility causes the obstacle to pulsate (expand and contract). Note that Equation 5 reduces to Equation 6 as the density mismatch approaches zero. For a spherical particle Equation 5 is always more accurate than Equation 6, but for a deformable biconcave RBC suspended in plasma or saline, Equation 6 will suffice.

Equation 4 implies that the scattering particle can be viewed as a secondary source that emits sound waves of amplitude equal to $\sqrt{\sigma_d(\gamma)}$ times the incident sound pressure. From Equation 6 we see that in *Rayleigh scattering* for which the particle size must be much smaller* than λ_o, $\sqrt{\sigma_d(\gamma)}$ varies directly with V_c and with $1/\lambda_o^2$. In many textbooks these characteristics are explained only by a dimensionality analysis. However, in describing light scattering by particles suspended in an imaginary *aether*, Rayleigh[23] himself proposed a very helpful argument as follows. Suppose the particle differs only in density from its surrounding medium, and suppose the incident sound is a harmonic wave that causes the particle to undergo oscillatory motion in synchrony with the surrounding fluid. The particle displacement can then be expressed by $A\cos(\omega_c t)$, where A and ω_c are, respectively, the maximum amplitude and angular frequency of oscillation. Thus, the particle acceleration is given by $-A\,\omega_c^2\cos(\omega_c t)$. If the density difference between the particle and the surrounding fluid is $\Delta\rho = 0$, there will be no net force felt by the particle, and no scattering will occur. However, if $\Delta\rho \neq 0$, the net force that must be applied to the particle so that the acoustic field remains continuous and well behaved is $-(\Delta\rho V_c)A\,\omega_c^2\cos(\omega_c t)$. Further, by Newton's third law an equal but opposite force must be exerted on the surrounding fluid by the particle, and this explains the origin of the scattered wave. The important point to note is that the amplitude of the scattered wave is also proportional to $V_c\omega_c^2$, i.e., the scattered power is proportional to the square of the particle volume and to the fourth power of frequency. As is well known, Rayleigh showed[23] that the latter accounts for the blue color of the sky. For human RBCs suspended in saline, this fourth-order frequency dependence was confirmed by Shung et al.[24] over the diagnostic frequency range. The second-order dependence on RBC volume is also supported by measurements of the backscattered ultrasound power from suspensions of RBCs of various sizes obtained from different species.[11,25]

Shown in Figure 1 is a plot of the directivity pattern of the total scattered pressure wave and the *magnitude* of its monopole and dipole components, as computed from Equation 6 for a single human RBC suspended in plasma. It can be seen that the angle-dependent dipole term is smaller than the monopole term, and that the total scattered pressure shows a maximum variation of about 44% relative to the maximum amplitude at $\gamma = 180°$ (backward direction). This angular dependence was experimentally confirmed[26] for a span of $\gamma = 60°$ to $150°$ for human RBC suspensions. To obtain the *total scattering*

---

* As pointed out by Insana and Brown in Chapter 4, there is some experimental evidence[22] based on ultrasound scattering from various media, which suggests that in order to observe the fourth-order dependence on frequency, the particle size must be smaller than one tenth of a wavelength.

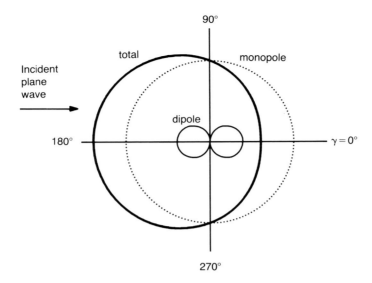

FIGURE 1. Plot of the normalized $|\sigma_d(\gamma)|^{1/2}$ for a human RBC suspended in plasma, showing the magnitude of the monopole and dipole components given in Equation 6 and using the values given in Table 1. Note that the incident plane wave originates from $\gamma = 180°$.

cross-section of an RBC, the σ_d must be integrated over all solid angles. However, from Equation 6 it can be easily shown that for $\lambda_o = 300 \ \mu m$, the maximum scattering cross-section given by 4π times the σ_d at $\gamma = 180°$, is about $6 \times 10^{-6} \ \mu m^2$, which is six orders of magnitude smaller than the geometric cross-sections of the RBC. This means that the presence of a RBC will hardly disturb the incident wave, which is consistent with the Born approximation.

III. DOPPLER EFFECT AND BASIC WAVE INTERFERENCE PHENOMENA

In this section we shall introduce in a progressive manner the basic Doppler and wave interference effects that may arise from scattering of a harmonic plane wave in blood. First of all, it will be noted that continuous wave (CW) Doppler ultrasound is assumed throughout this chapter, but the essential ideas embodied by the mathematical formalism should be equally applicable to pulsed Doppler ultrasound, which can simply be considered as a sampled version of the CW signal obtained from a selected region of blood. Second, to understand ultrasonic scattering in blood, it is important to remember that the longitudinal waves generated by the transducer crystal are *coherent* in time as well as in space, i.e., the amplitude of the pressure waves at different locations and at different times *vary in unison*. As we shall see in Section VII, the coherent insonification of blood accounts for the granular

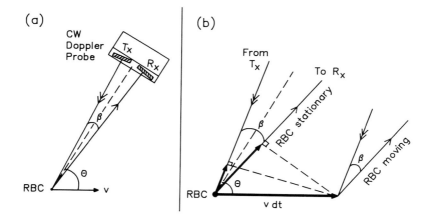

FIGURE 2. Geometry used for calculating the scattered pressure signal from a moving RBC. (a) Showing the path of an ultrasonic ray from the transmit crystal (T_x) to an RBC and then back to the receive crystal (R_x). (b) Exaggerated view showing the incremental displacement vector (bold arrow) of the RBC and its projections on the incident and returning ultrasonic rays.

structure of Doppler power spectrograms which is quite similar to *laser speckle*.

A. SCATTERING BY ONE RBC

Suppose the transmit crystal of a CW Doppler probe is emitting a sinusoidal wave represented by $A\cos\omega_c t$, where A is the amplitude and ω_c is the angular frequency. If an RBC is located at a large distance R from the transmit crystal, a geometric approach can be applied and the incident sound can be assumed to be a plane wave of amplitude A/R. If the sound propagation speed is c, then the time required by the wave to travel from the transducer to the RBC is $t_d = R/c$. Based on the above, the wave impinging on the RBC can be represented by $(A/R) \cos(\omega_c(t - t_d))$.

As illustrated in Figure 2a, suppose the receiving transducer crystal is oriented at a slight angle from the transmit crystal such that the incident and returning rays form a small angle β. Assuming the differential scattering cross-section $\sigma_d(\gamma)$ of the RBC is quite uniform over $\gamma = 180° \pm \beta$ (Figure 1), the RBC can be treated as a secondary source of amplitude $(A/R)\, \sigma_{bs}^{1/2}$, where $\sigma_{bs} = \sigma_d(180°)$ is the *backscattering cross-section* of the RBC. Further, the backscattered wave will be attenuated by a factor of 1/R and delayed by another t_d before it is received. Therefore, if the RBC is stationary, the received pressure signal can be written as follows:

$$\text{No RBC motion:} \qquad y(t) = (A/R^2)\sigma_{bs}^{1/2} \cos(\omega_c t + \phi_o) \qquad (7)$$

where $\phi_o = -2\omega_c t_d$ is the total phase shift determined by the round trip time delay.

Now suppose the RBC moves with a speed v at an angle θ with respect to the centerline between the incident and returning rays as shown in Figure 2b. As the RBC moves, the total length of the ray path changes; consequently, the phase shift of the received signal becomes a function of time as follows:

$$\text{With RBC motion:} \qquad y(t) = (A/R^2)\sigma_{bs}^{1/2} \cos(\omega_c t + \psi(t)) \qquad (8)$$

To determine the phase shift $\psi(t)$, we assume that $v \ll c$ such that over an incremental time interval dt during which the RBC will have moved by vdt, the phase of the impinging wavefront remains the same. From Figure 2b it can be seen that the lengths of the incident and returning ray paths will change by incremental amounts equal to the projections of the displacement vector on the two ray paths. Therefore, the change in phase shift given by $2\pi/\lambda_o$ times the change in total path length is

$$d\psi = (2\pi/\lambda_o)[v\ dt\ \cos(\theta - \beta/2) + v\ dt\ \cos(\theta + \beta/2)]$$

$$= (2\pi/\lambda_o)2v\ dt\ \cos\ \theta\ \cos(\beta/2) \qquad (9)$$

Integrating both sides yields

$$\psi(t) = 2\omega_c(v/c)\ \cos\ \theta\ [\cos(\beta/2)]t + \phi_o \qquad (10)$$

where ϕ_o represents $\psi(0)$. Consequently, by substituting Equation 10 into Equation 8, the received pressure signal is

$$y(t) = (A/R^2)\sigma_{bs}^{1/2} \cos((\omega_c + \omega_d)t + \phi_o) \qquad (11a)$$

where

$$\omega_d = 2\omega_c(v/c)\ \cos\ \theta\ \cos(\beta/2) \qquad (11b)$$

is the Doppler-shift frequency induced by the RBC motion. Equation 11 represents a coherent backscattered *wavelet* from a single moving RBC. In general, the value of ω_d is positive if the RBC is approaching the transducer, and negative if the RBC is receding. A positive ω_d can be interpreted as the result of a compression of the incident wavefront by an approaching RBC, similar to the increase in the sound pitch generated by an approaching train. Note that in most CW Doppler studies the factor $\cos(\beta/2)$ in Equation 11b should be very close to unity. For pulsed Doppler systems that use only a single transducer crystal for both transmission and reception, β is exactly equal to zero. Hence, from this point onwards the $\cos(\beta/2)$ factor will be ignored.

In Doppler studies, the phase shift ϕ_o of the backscattered wavelet from an RBC is often treated as a random variable, in which case y(t) can be considered as a *random process*. Now, from Equation 11 it can be easily shown that for any given ω_d and ϕ_o the average backscattered signal power *over time* is

$$\overline{y^2(t)} = \frac{1}{2} \sigma_{bs}(A^2/R^4) \tag{12}$$

In stochastic signal modeling, it is also useful to determine the mean or expected value of a signal attribute. As will be discussed later, in most of the existing scattering theories for blood, the insonified region is usually taken to be much larger than λ_o so that ϕ_o can be assumed uniformly distributed over the interval $[0,2\pi]$. In this case it can be easily shown that the expected value of $y^2(t)$ is identical to the time average given by Equation 12, and a random process with this property is said to be *ergodic*.[27]

B. SCATTERING BY TWO RBCS

In this subsection we will consider the effect of interference of the back-scattered wavelets from two RBCs that are moving at the same velocity. For the sake of convenience, we shall redefine $\sigma_{bs}^{1/2}(A/R^2)$ as A, so that the average backscattered power from a single RBC as given by Equation 12 is now simply $A^2/2$. There are two cases to be dealt with separately as follows.

1. Positions Uncorrelated

Suppose the phase shifts of the backscattered wavelets from two randomly positioned RBCs are ϕ_1 and ϕ_2, respectively. Ignoring the effects of multiple scattering between RBCs, the signal backscattered from two RBCs can be obtained by linear superposition as follows:

$$y(t) = A \cos(\omega t + \phi_1) + A \cos(\omega t + \phi_2) \tag{13}$$

where $\omega = \omega_c + \omega_d$. If ϕ_1 and ϕ_2 are uncorrelated and are both uniformly distributed over $[0,2\pi]$, it can be readily shown that the expected value of the total power $E[y^2] = A^2$, which is exactly twice the power from a single RBC. By induction it can be further shown that the total power received from a collection of N cells, whose positions are completely uncorrelated, is equal to the sum of the individual backscattered power. This is analogous to the case of *incoherent* illumination in optics for which intensities are additive.

2. Positions Perfectly Correlated

Now we consider the case when the positions of two RBCs are perfectly correlated, or more specifically, when two RBCs are separated by a known distance δ in the direction of the incident plane wave as illustrated in Figure 3. It is assumed that the transducer is located at a distance $R \gg \delta$ so that the difference between the round trip phase delays of the two RBCs is simply 2δ times $2\pi/\lambda_o$. The total backscattered wave can then be represented by

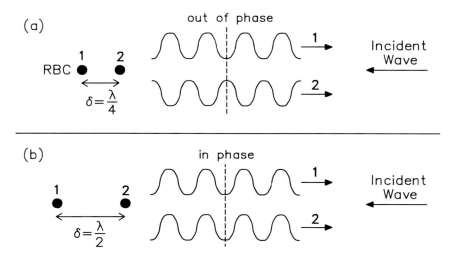

FIGURE 3.   Illustration of (a) perfect destructive and (b) perfect constructive interference of wavelets arising from two scattering RBCs that are separated by a distance $\delta = \lambda/4$ and $\lambda/2$, respectively.

$$y(t) = A \cos(\omega t + \phi_1) + A \cos(\omega t + \phi_1 + 2k_o\delta) \qquad (14)$$

in which $k_o = 2\pi/\lambda_o$. Using a trigonometric identity, this can be rewritten as

$$y(t) = 2A \cos(k_o\delta) \cos(\omega t + \phi_1 + k_o\delta) \qquad (15)$$

It can be seen that the amplitude and phase shift of the pressure wave have been modified by factors that are dependent on δ. The time or statistical average of $y^2(t)$ is $2A^2\cos^2(k_o\delta)$. This is analogous to *coherent* illumination in optics, and two important cases can be seen immediately. First, for $k_o\delta = \pi(2m + 1)/2$, where m is an integer, i.e., $\delta/\lambda_o = 1/4, 3/4, 5/4, \ldots$, the average backscattered power is zero. Physically, as shown in Figure 3a, this corresponds to the case of *perfect destructive interference* when the back-scattered wavelets from the two RBCs are exactly out of phase. Second, the average backscattered power reaches a maximum value of $2A^2$ when $k_o\delta = m\pi$ or $\delta/\lambda_o = m(1/2)$. This is the case of *perfect constructive interference* when the backscattered wavelets are exactly in phase with each other (Figure 3b). Note that the total power due to constructive wave interference is four times that of a single RBC, whereas the total power is only doubled for two position-uncorrelated RBCs.

An important implication of Equation 15 pertains to the resolution of a coherent ultrasonic transducer. It can be easily shown from Equation 15 that for $\delta \leq \lambda_o/20$, the amplitude of the total backscattered wave is almost the same

(a) Random Medium?   (b) Homogeneous

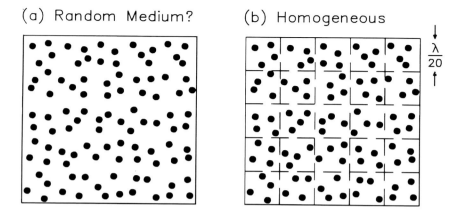

FIGURE 4. Two-dimensional illustrations showing (a) an apparently random distribution of particles; and (b) the same medium which now appears to be homogeneous because the number of particles in each resolution cell of dimension $\lambda/20$ is constant.

($>95\%$) as that for $\delta = 0$. In other words, the RBCs within an elemental volume of dimension $\lambda_o/20$ can be considered to act like an aggregate that is located at a single point in the insonified region. For a 5 MHz Doppler system, the value of λ_o for blood is about 300 μm; thus, if the hematocrit is 45% and the RBC volume is 90 μm^3, there will be an average of about 20 RBCs in a $(\lambda_o/20)^3$ volume which cannot be resolved by the transducer.

C. MODELING A DENSE SUSPENSION OF RBCS

Having dealt with the basic wave interference phenomena between two scatterers, we now consider the problem of how to represent blood which is a fairly dense suspension of RBCs. The most commonly used approach is to treat blood as a *random medium*, but this notion really needs to be examined carefully. First, the degree of randomness is a relative term that depends on the resolution of the observing system. For example, the medium shown in Figure 4a has a random appearance because the positions of the particles do not seem to follow any regular pattern. However, as concluded in the previous subsection, an ultrasonic transducer cannot resolve particles that are separated by less than $\lambda/20$, where λ is the ultrasonic wavelength in the medium. Therefore, as illustrated in Figure 4b, the same hypothetical medium which appears random to our eyes is actually highly homogeneous to an ultrasonic transducer because the number of particles in each elemental volume of dimension $\lambda/20$ is constant.

Second, the degree of randomness is dependent on the state of our knowledge about the medium. For RBCs suspended in plasma, if we have sufficient knowledge of the physical forces that act on the RBCs and of the position of every RBC at some time in the past, then in principle it is possible to predict the exact chaotic arrangement of the RBCs at the present time instant. However, if such a deterministic approach is used, the system of equations that must be solved are probably nonlinear and highly coupled (due to cell-cell collisions), and the resultant pattern of RBC positions can be regarded as a *fractal* image. Of course, in practice we do not have sufficient knowledge about the system and, therefore, it is necessary to resort to a stochastic approach and make certain assumptions about the probability distribution of variables such as the phase shift of a backscattered wavelet.

To summarize, we may say that any medium may exhibit some degree of randomness if we have enough resolving power, but none is truly random if we have sufficient knowledge. As we shall see in the next section, even if a stochastic approach is used to model blood, the problem is still quite difficult because the RBCs in normal human blood are so densely packed that the average separation between adjacent RBCs is less than 10% of the cell diameter. This means that the positions of any pair of RBCs are neither uncorrelated nor perfectly correlated; they are *partially correlated*. The equivalent problem which is also one of the most difficult in optics is known as *partial coherence*.

IV. THE PARTICLE VS. CONTINUUM APPROACH

As delineated in the previous section, the crux of the problem in developing a stochastic scattering model for blood is that the positions of the RBCs in normal hematocrit blood are partially correlated. In other words, the wavelets arising from different RBCs within the insonified region, whose dimensions are usually much larger than a wavelength, can interfere in a constructive as well as a destructive manner. From a deterministic viewpoint, the RBCs in blood are strongly interacting (colliding into, attracting, and deforming each other), and thus, they cannot be treated as independent scatterers. In the ensuing theoretical developments, we will attempt to provide both a deterministic and a stochastic viewpoint, in the hope of gaining deeper insights into the physical processes involved.

As shown in Table 2, over the last two decades numerous stochastic scattering models for blood have been proposed, and they may be described in terms of the approach taken and of the assumptions made regarding the RBC interaction, the flow field, and the insonifying beam profile. It can be seen that with the exception of a recently proposed hybrid approach[12,13] which is presented in Section V, the existing models can be classified as either a particle or a continuum approach. In this section, we will show that these two classical approaches are basically different ways of modeling the density and compressibility functions of the random medium, which lead to an inhomogeneous wave equation with different source terms. Specifically, the particle approach consists of summing the contributions from individual RBCs, whereas the continuum approach treats blood as a continuous medium in which local fluctuations in density and compressibility give rise to the scattered waves.

TABLE 2
Statistical Ultrasound Scattering Theories for Blood: A Historical Overview

Year	Investigators	Approach	Red cell interaction	Flow conditions	Insonifying beam profile
1971	Albright[28,30] (also, Albright and Harris[29])	Particle	None	Laminar and turbulent	Gaussian
1972	Brody[31] (also, Brody and Meindl[32])	Particle	None	General	General
1972	Ahuja[15] (see also Ahuja and Hendee[33])	Particle	None	Stationary	Plane wave
1973	Sigelmann and Reid[34]	Continuum	Dirac delta correlation	Stationary	Narrow uniform beam
1974	Atkinson and Berry[35] (also, Atkinson and Woodcock[36])	Particle	A constant excluded volume	Stationary	Finite and cylindrically symmetric beam
1976	Shung et al.[24] applying work of Beard et al.[37] and Twersky[38]	Particle	Heuristic "hole" model	Stationary	Plane wave
1979	Hanss and Boynard[39]	Particle	Spherical aggregates with trapped plasma	Stationary	Plane wave
1980	Yagi and Nakayama[40]	Continuum	Double-exponential correlation function	Stationary	Plane wave
1980	Angelsen[41]	Continuum	Dirac delta correlation	Laminar	Plane wave
1982	Shung[42] applying Twersky's work[43-45]	Particle	Percus-Yevick packing factor for hard spheres	Stationary	Plane wave
1986	Mo and Cobbold[46]	Particle	General packing factor	Laminar	Plane wave
1986	Fish[47]	Continuum	Dirac delta correlation	Laminar	General
1986	Williams[48]	Particle	None	Laminar	Plane wave
1986	Yang[49,50]	Particle	General	General	Plane wave
1987 1988	Routh et al.,[51] Gough et al.[52]	Particle (1-dimensional)	Hard slabs	Stationary	Plane wave
1987 and 1988	Twersky[53,54] (also, Berger et al.[55])	Particle	2-Parameter packing factor	General	Plane wave
1990	Mo[12] (also, Mo and Cobbold[13])	Hybrid	General	General	Plane wave

Of the models shown in Table 2, many[24,39,40,50,53-55] were aimed primarily at predicting the backscattering coefficient (BSC) of blood. This is defined as the average power (i.e., mean-square pressure) backscattered per steradian from a unit volume of blood when it is insonated by a plane wave of unit intensity. Some of the other models[28-32,47,48] were aimed at understanding the relationship between the Doppler power spectral density and the blood velocity profile; and a few[12,13,41,46] attempted to predict both the BSC and power spectral density.

The relationship between the BSC and hematocrit is particularly important because it is highly dependent on the RBC interaction in blood. Consequently, this section emphasizes the BSC prediction by various models, while the statistics of the Doppler spectrum are best deferred until a later section. It will be noted that both the particle and continuum approaches use the Green's function method (Chapter 4) to reduce an inhomogeneous wave equation into integral form, which is how the RBC scattering cross-section described in Section II.B is derived. The key steps in the two classical approaches, together with selected examples from Table 2, are discussed separately in the following subsections.

A. PARTICLE APPROACH

For blood both ρ and κ in Equation 3 are random functions of space that always vary with time due to Brownian motion of the RBCs and may additionally vary due to flow. Suppose ρ_e and κ_e, ρ_o, and κ_o represent the density and compressibility of the RBC and of the surrounding medium, respectively, then the functions ρ and κ can be expressed by

$$\rho(\mathbf{r}, t) = \rho_o + \rho_1(\mathbf{r}, t)$$

$$\kappa(\mathbf{r}, t) = \kappa_o + \kappa_1(\mathbf{r}, t) \tag{16}$$

such that

$$\rho_1(\mathbf{r}, t) = \begin{cases} \rho_e - \rho_o, & \text{inside a RBC} \\ 0, & \text{outside a RBC} \end{cases}$$

and similarly for $\kappa_1(\mathbf{r},t)$. Substituting into Equation 3 yields

$$\nabla^2 p - \frac{1}{c_o^2} \frac{\partial^2 p}{\partial t^2} = \frac{1}{c_o^2} \frac{\kappa_1}{\kappa_o} \frac{\partial^2 p}{\partial t^2} + \operatorname{div}\left(\frac{\rho_1}{\rho} \nabla p\right) \tag{17}$$

where $c_o = 1/\sqrt{\rho_o \kappa_o}$ is the wave propagation speed in the suspending medium. Compared with Equation 3, the right-hand side of this inhomogeneous equation can be interpreted as a source term that produces the scattered waves (whose energy is derived from the incident wave).

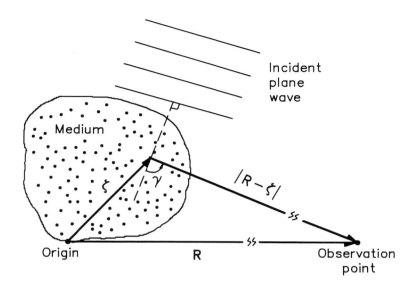

FIGURE 5. Coordinate system assumed for the plane wave insonation of a random medium.

As illustrated in Figure 5a, the ultrasound beam approaching the scattering medium is assumed* to be a harmonic plane wave (far-field) of unit amplitude and zero phase, as expressed by

$$p_o(\mathbf{r}, t) = \mathrm{Re}[P_o(\mathbf{r})e^{j\omega_c t}] \tag{18a}$$

where

$$P_o(\mathbf{r}) = e^{-j\mathbf{k_o} \cdot \mathbf{r}} \tag{18b}$$

is the complex envelope of p_o, and $\mathbf{k_o}$ is the incident wave vector such that $|\mathbf{k_o}| = \omega_c/c_o = 2\pi/\lambda_o$. In addition to the basic assumptions required in the derivation of Equation 3, it is also assumed that:

1.   The dimensions of the RBC are much smaller than λ_o.
2.   The acoustic mismatches between RBCs and the surrounding medium are so small that multiple scattering can be ignored and that the Born approximation is valid.
3.   The observation point $\mathbf{R}$ is sufficiently far from the scattering medium so that for any position vector ζ inside the medium, $|\mathbf{R} - \zeta| \cong R$.

* There is no loss of generality because, according to Equation 3, any nonplanar wave can be considered as a superposition of harmonic plane waves.

Equation 17 can then be solved using the Green's function method (Chapter 4). Noting that $\rho_1/\rho = \rho_1/\rho_e$ inside an RBC, the resultant complex envelope of the scattered pressure wave at the observation point is given by:

$$P_s(\mathbf{R}, t) = \frac{\pi}{\lambda_o^2} \frac{e^{-jk_oR}}{R} \int_\Omega \left\{ \frac{\kappa_1(\zeta, t)}{\kappa_o} + \frac{\rho_1(\zeta, t)}{\rho_e} \cos \gamma \right\} e^{j(\mathbf{k_s} - \mathbf{k_o}) \cdot \zeta} \, d^3\zeta \quad (19)$$

where Ω is the total insonified volume, and $\mathbf{k_s}$ is the scattered wave vector which makes an angle γ with respect to the incident wave vector. It can be seen from Equation 19 that the scattered pressure wave far from the medium is given by the spatial Fourier transform of the compressibility and density fluctuations within the medium. Note that in starting with Equation 3 the speed of RBC motion has been assumed to be much smaller than c_o such that any change in this quasi-stationary medium will have an instant effect at $\mathbf{R}$ as defined by Equation 19.

To obtain the total amplitude of the backscattered ($\gamma = \pi$) pressure wave, the integral in Equation 19 can be evaluated by summing over all RBCs as follows:

$$P_s(\mathbf{R}, t) = \frac{\pi}{\lambda_o^2} \frac{e^{-jk_oR}}{R} \left[\frac{\kappa_e - \kappa_o}{\kappa_o} - \frac{\rho_e - \rho_o}{\rho_e} \right] \sum_i \left\{ \int_{V_i} e^{j2\mathbf{k_o} \cdot \zeta} \, d^3\zeta \right\} \quad (20)$$

where V_i is the space occupied by the *i*th RBC. If the position of the *i*th RBC, is $\zeta_i = \mathbf{v}_i t + \zeta_i''$, in which $\mathbf{v}_i$ and ζ_i'' are the veolcity and initial position vectors, respectively, then from Equation 20 it is straightforward to show that the real part of $P_s(\mathbf{R},t)e^{j\omega_c t}$ is given by

$$p_s(\mathbf{R}, t) = (1/R) \sum_i \sigma_i^{1/2} \cos[(\omega_c + \omega_i)t + \phi_i] \quad (21a)$$

where

$$\sigma_i = \frac{\pi^2}{\lambda_o^4} \left[\frac{\kappa_e - \kappa_o}{\kappa_o} - \frac{\rho_e - \rho_o}{\rho_e} \right]^2 V_i^2 \quad (21b)$$

is the backscattering cross-section of the *i*th RBC, $\omega_i = 2\mathbf{k_o} \cdot \mathbf{v}_i$ is the Doppler frequency shift (cf. Equation 11b), and $\phi_i = (2\mathbf{k_o} \cdot \zeta_i'' - k_oR)$ is the phase shift corresponding to the round-trip distance between the exposed boundary of the medium (where the phase of the incident wave is zero) and the initial position of the RBC, plus the distance from the boundary to the observation point. The expression for σ_i which shows the characteristic fourth-order dependence on λ_o, is the same as Equation 6. Note that the above assumes that in a quasi-stationary medium, RBC motion may be significant enough to induce a phase shift, but any resultant change in local density is very slow compared to the speed of sound.

It should be noted that the starting points for most existing particle scattering theories are very similar, if not identical, to Equation 21. As summarized in Table 2, Albright[28] and Brody[31] should be credited* for developing two of the earliest particle models for Doppler ultrasound. In fact, both workers provided very comprehensive treatment of the effects of beam geometry and velocity profile on the Doppler power spectrum. However, both of their models assume that the RBCs are independent, point-sized particles which has long been proven invalid for normal blood.[11,24]

Mo and Cobbold[46] proposed a more general particle scattering model which treats blood as a suspension of RBC aggregates that are all much smaller than a wavelength. This model incorporates three parameters: the mean and variance of the aggregate size distribution, and a general packing factor W, which is a measure of the aggregate pair-position correlation. For the particular case in which all the RBCs in the insonified blood are assumed to be identical (mean scatterer volume = RBC volume, V_c), the model predicts that the backscattering coefficient of blood is given by

$$BSC = \sigma_{bs}(N_c/\Omega)W \tag{22a}$$

or

$$BSC = \sigma_{bs}(H/V_c)W \tag{22b}$$

where σ_{bs} is given by Equation 21b and N_c is the total number of RBCs in the insonified region of hematocrit H. As a matter of convenience, the insonified volume Ω can be assumed to be sufficiently large so that N_c is a constant. Physically, the packing factor W can be viewed as a measure of orderliness in the spatial arrangement of RBCs. At very low H when the RBC positions are completely random, W is equal to unity so that the BSC is simply proportional to H (sum of average power). As H increases, W decays gradually to zero since closer packing will invariably lead to greater order; the BSC will also eventually approach zero because in a densely packed medium that is much larger than a wavelength we can always find an RBC that will cancel or destructively interfere with the contribution from another RBC.

Most of the existing particle models can be considered as special cases of Equation 22b in which W is expressed as an explicit function of H. A packing factor which has found applications in various scattering problems[44,45] is based on the Percus-Yevick pair-correlation function for particles that are identical, hard, and radially symmetric in m-dimensional space (slabs, circles, and spheres for m = 1, 2 and 3, respectively). The Percus-Yevick packing

---

* Though Albright's work was probably motivated by Doppler measurement of urinary flow, his model is also applicable to blood flow since the key parameter values are very similar.

factor for hard spheres (m = 3) was first applied to blood by Shung,[42] and it can be expressed in terms of H as follows:

$$W = \frac{(1 - H)^4}{(1 + 2H)^2} \tag{23}$$

As will be seen in a later subsection, Equations 22 and 23 provide a fairly good fit to the observed BSC data only under laminar flow conditions. The larger discrepancies between this hard-sphere packing theory and the observations for turbulent or disturbed flow conditions have prompted workers to develop more sophisticated packing factors. In particular, Yang[50] attempted to develop a more general theory by including higher-ordered terms in the pair-correlation function. In this theory the backscattered signal is composed of a 'particle' and a 'turbulence' component, and Yang suggested that the latter is due to the "random fluctuation of the particle number density" which is comparable to that "due to the particles themselves".

Twersky[54] also proposed a generalized packing theory which is similar to that given by Mo and Cobbold[46] in that it contains three parameters: a 'particle population factor', an RBC shape parameter, and the variance of the particle size distribution. The main new feature was that the original Percus-Yevick packing factor W was modified by including the shape parameter and the variance of the particle size distribution (whereas in Mo and Cobbold's work, W was treated as an unknown function of H). These two parameters, whose values were estimated from Shung's data using nonlinear least-squares methods, provide a means for predicting a variation of BSC with flow conditions.[55] It was found that in order to fit the BSC data for saline suspensions of RBCs, the particle population factor needed was in the range from 66 to 82% of the known backscattering cross-section of RBCs. No physical explanation of this discrepancy was given. Furthermore, the shape parameter was originally intended as a measure of nonsphericity for hard, convex particles such as ovals or simple polyhedra averaged over all orientations. Hence, the present form of Twersky's packing factor is still somewhat empirical.

B. CONTINUUM APPROACH

In contrast to the particle approach which tracks the position of every individual RBC in the insonified region, continuum models basically sum the contributions from every point in the acoustic field. As shown in Table 2, it appears that Sigelmann and Reid[34] were first to report a continuum model which forms the theoretical basis for a substitution method for measuring the BSC of blood. However, in their model it was not made clear exactly how the strength of the backscattered wave from an elemental volume in space is related to the local RBC distribution. This was later determined by Angelsen[41] and by Yagi and Nakayama.[40]

In a continuum model, scattering is considered to arise from spatial fluctuations in density and compressibility created by the random distribution of

RBCs in the medium.* Thus, the initial steps in both Yagi and Nakayama's and Angelsen's work are very similar. The first key step is to split ρ and κ into a mean plus a fluctuation term as follows:

$$\rho(\mathbf{r}, t) = \rho_m(\mathbf{r}, t) + \Delta\rho(\mathbf{r}, t)$$

and

$$\kappa(\mathbf{r}, t) = \kappa_m(\mathbf{r}, t) + \Delta\kappa(\mathbf{r}, t) \tag{24}$$

Substituting the above into Equation 3 gives, after some manipulations,

$$\nabla^2 p - \frac{1}{c_m^2} \frac{\partial^2 p}{\partial t^2} = \frac{1}{c_m^2} \frac{\Delta\kappa}{\kappa_m} \frac{\partial^2 p}{\partial t^2} + \mathrm{div}\left(\frac{\Delta\rho}{\rho} \nabla p\right) \tag{25}$$

where $c_m = 1/\sqrt{\rho_m \kappa_m}$ is the *average* wave propagation speed in the random medium. Note that this has the same form as Equation 17 and thus it is also solved by the Green's function method provided that the same assumptions are made as in the particle approach. The resultant scattered wave at a large distance R is expressed by

$$P_s(\mathbf{R}, t) = \frac{\pi}{\lambda_m^2} \frac{e^{-jk_m R}}{R} \int_\Omega \left\{ \frac{\Delta\kappa(\boldsymbol{\zeta}, t)}{\kappa_m} + \frac{\Delta\rho(\boldsymbol{\zeta}, t)}{\rho_m} \cos\gamma \right\} e^{j(\mathbf{k}_s - \mathbf{k}_m)\cdot\boldsymbol{\zeta}} d^3\zeta \tag{26}$$

where $\lambda_m = c_m(2\pi/\omega_c)$ is the average ultrasonic wavelength in the random medium, and $\Delta\rho/\rho$ has been approximated by $\Delta\rho/\rho_m$. Although this equation has the same form as Equation 19 in the particle formulation, a rather subtle difference should be noted. The fluctuation terms $\Delta\kappa$ and $\Delta\rho$ in Equation 26 are random, whereas their counterparts κ_1 and ρ_1 in Equation 19 are functions that can only take on one of two values corresponding to the properties inside and outside an RBC.

Both Yagi and Nakayama's and Angelsen's models were derived from Equation 26; they were basically different approaches of solving the integral expression of $P_s(\mathbf{R},t)$. Specifically, Yagi and Nakayama assumed that the medium is stationary so that the fluctuation terms are functions of space only. The power of the scattered wave was obtained by taking the expected value of the product of $P_s(\mathbf{r})$ and its complex conjugate. This resulted in a double integral involving the spatial autocorrelation of $\Delta\rho/\rho_m$ and of $\Delta\kappa/\kappa_m$. But since

* It appears that the concept of fluctuation scattering originated from Morse and Ingard's work[14] on scattering from a cloud of particles. They started with a particle formulation similar to Equation 17, but instead of applying the principle of superposition and the ray approximation, the source terms in Equation 19 are separated into their mean values and fluctuations from the mean values. It was shown that the mean values cause refraction of the incident plane wave whereas the fluctuation terms give rise to incoherent scattering. The fluctuation terms were then expressed in terms of a 2-point spatial correlation as in the continuum approach.

the density and compressibility fluctuations arise from the presence or absence of RBCs, their autocorrelation functions must have the same form. In Yagi and Nakayama's model this spatial correlation function is represented by a sum of two exponential terms,* which was adapted from Debye et al.'s formulation[56] that describes the distribution of holes on the surface of inhomogeneous solids. The resultant expression for the backscattering coefficient is

$$
\text{BSC} = \frac{\pi^2}{\lambda_m^4}\,(N_c/\Omega)V_c^2\left[\frac{\kappa_e - \kappa_o}{\kappa_m} - \frac{\rho_e - \rho_o}{\rho_m}\right]^2
$$

$$
\times\ (1 - H)^4\left[\frac{0.37}{[1 + 44.4(a/\lambda_m)^2(1 - H)^2]^2}\right. \tag{27}
$$

$$
\left. + 0.63\exp\{-33.4(a/\lambda_m)^2(1 - H)^2\}\right]
$$

where the effective cell radius is 'a', such that $V_c = (4\pi a^3)/3$. A somewhat unusual feature of Equation 27 is that for Rayleigh scattering in which $a/\lambda_m \ll 1$, the spatial correlation function is weakly dependent on a/λ_m, so that the overall BSC is not exactly proportional to λ_m^{-4} and to V_c^2. This will be explained in Section VI.B.

Angelsen's formulation, on the other hand, made rather clever use of the fact that there are many RBCs in an elemental acoustic volume which cannot be resolved by the ultrasonic transducer. Thus, if the number of RBCs in an elemental volume Ω_e at position **r** and time t is expressed as $n(\mathbf{r},t) = \bar{n} + \xi(\mathbf{r},t)$, then the spatial autocorrelation of $\Delta\kappa$ and $\Delta\rho$, which are both functions of $\xi(\mathbf{r},t)$, can be assumed to be delta functions. Using this approach, the differential scattering cross-section of an elemental blood volume Ω_e was obtained (Equation 17 in Angelsen's paper[41]), which when normalized by Ω_e, yields

$$
\text{BSC} = \frac{\pi^2}{\lambda_m^4}\,V_c^2\left[\frac{\kappa_e - \kappa_o}{\kappa_m} + \frac{\rho_o - \rho_e}{\rho_m}\right]^2(1/\Omega_e)\overline{\text{var}(n)} \tag{28}
$$

where $\overline{\text{var}(n)}$ is the variance of n averaged over space and time. The advantage of using Angelsen's approach is that RBC motion is allowed. In fact, Angelsen's major contributions include a very general analysis of the Doppler receiver output for both continuous wave and pulsed systems in which near-field effects of the transducer are also taken into account.

C. COMPARISON OF RESULTS AND SUMMARY

In the preceding two subsections, the particle and continuum approaches have been described using a common mathematical framework. It was shown

---

* The exponential autocorrelation function is a natural choice if one thinks of the inhomogeneities as a random telegraph signal.[27]

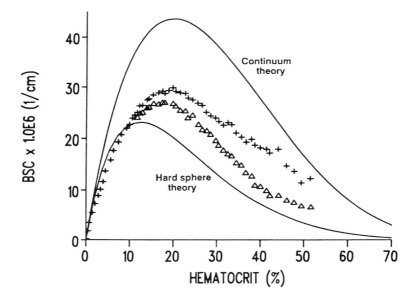

FIGURE 6. Plots of BSC vs. H for saline suspensions of human RBCs measured under stationary Δ, and stirred $+$, conditions. The data was obtained by Shung et al.[11] using a 7.5 MHz pulse-echo system. Solid curves represent the Percus-Yevick theory for the packing of hard spheres and Yagi and Nakayama's continuum theory.

that the fundamental difference between the two lies in the manner in which the random medium is modeled: particle models follow individual RBCs, whereas continuum models track local cell concentrations. In each case the linearized wave equation was solved by Green's function method which led to an integral expression for the backscattered wave in terms of density and compressibility changes. Now the existing particle theories generally start with an additional approximation: superposition of the backscattered wavelets from different RBCs. Since multiple scattering between RBCs can be ignored, this is a valid approximation; however, it does not account for the change in wavelength of the incident wave due to the presence of other cells in the medium. This explains why in Equation 21b the particle backscattering cross-section is inversely proportional to λ_o^4 (a constant), whereas the continuum theory shows a primary dependence on λ_m^4 (which varies with H). For the diagnostic frequency range the difference between the two is generally no more than a few percent, and λ_m can simply be used in place of λ_o.

For the purpose of comparison, a representative example from each approach can be considered; namely, the Percus-Yevick packing theory as defined by Equations 21 through 23 and Yagi and Nakayama's result in Equation 27. Figure 6 shows a plot of the BSC predicted by these two theories together with some experimental data on human RBC suspensions for stationary and stirred conditions. The experimental data was obtained by Shung et al.[11] using a substitution method and a 7.5 MHz pulse-echo system. The theoretical curve for the hard sphere model was calculated with λ_o replaced by λ_m to correct for the aforementioned refraction effect. It can be seen that the particle and continuum theories differ significantly over the whole range of H and that the experimental data generally falls between the two. These large discrepancies can be attributed to the different correlation functions being used.

Equation 28 shows that the backscattered power can also be expressed in terms of the average local variance in RBC concentration. This result may be ascribed to Angelsen's continuum model, but actually it was Twersky[44] who originally pointed out that for particles much smaller than a wavelength, $var(n)$ is equivalent to $\bar{n}$ times the packing factor W, so that $\overline{var(n)} = (N_c/\Omega)W$, and this unifies Equations 22 and 28. A rigorous proof of this together with a more detailed discussion of the packing factor can be found elsewhere.[57] The significance of this variance viewpoint is discussed in the next section.

In regard to the physical cause(s) of increased scattering in stirred and turbulent flow conditions, several conflicting viewpoints have been proposed. Shung et al.[11] suggested that the increased scattering is due to "macroscopic inhomogeneities" arising from large vortex structures, whereas Angelsen[41] proposed that scattering is caused soley by fluctuations in cell concentration and that increased scattering can be simply explained by increased fluctuations. Yang[50] contended that the backscattered signal generally consists of particle and turbulence components, of which only the latter arises from fluctuations in cell concentration. To further compound this problem, Yagi and Nakayama's result[40] suggests that the pair position-correlation function may vary with the ratio of the effective cell radius to the ultrasonic wavelength, but this has not been accounted for in other existing particle or continuum models.

V. THE HYBRID APPROACH

As discussed in the previous section, the main advantage of the particle approach is that it recognizes that the RBCs are such weak scatterers that geometric ray theory can be conveniently applied to sum their contributions. On the other hand, the continuum approach recognizes that within an elemental acoustic volume there may be tens or hundreds of RBCs which cannot be resolved by the transducer, and therefore, the medium can be treated as a continuum. In this section a recently proposed hybrid approach[12,13] that combines the strengths of these two classical approaches is presented.

A. BASIC MODEL FORMULATION

In the hybrid approach, the RBCs within an elemental acoustic volume are treated as a single scattering unit which moves with a single velocity. We shall refer to such an elemental volume as a *voxel*. It should be clear that the

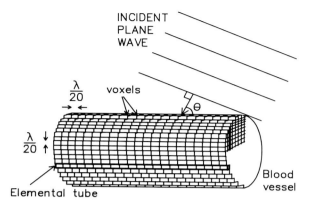

FIGURE 7. Illustrating the geometry of the insonation system with the blood vessel divided into elemental tubes which, in turn, are subdivided into elemental voxels of dimension λ/20.

voxel size must be chosen small enough that it approaches the transducer beam resolution or the smallest dimension of the flow field, whichever is smaller. As already pointed out in Section III.B, when two RBCs are closer than about λ/20, the total backscattered signal is almost indistinguishable from that of a single aggregate. For sound speed $c = 1540$ m/s and $f_c = \omega_c/(2\pi) = 5$ MHz, this distance corresponds to 15 μm. Thus, a 5 mm diameter artery such as the carotid will contain at least 300 voxels, which should be sufficient to provide an accurate representation of the flow velocity profile.

The assumptions required in the formulation of the hybrid scattering theory are essentially the same as those of the particle approach described in Section IV. As illustrated in Figure 7, a blood vessel is assumed to be insonified by a unit-amplitude, harmonic plane wave of wavelength much larger than the dimensions of an RBC; i.e., Rayleigh scattering holds. The receiving and transmitting crystals are assumed to be coincident and located at a distance from the vessel that is much greater than the vessel diameter. Beam divergence, attenuation, and multiple scattering effects are ignored. Further, the insonified blood is divided into elemental voxels of volume $\Omega_e = (\lambda/20)^3$. In the formulation of the basic Doppler signal model, no assumption about the flow field is required. However, in the next subsection, in order to obtain a closed form expression for the autocorrelation function of the Doppler signal, the flow is assumed to be *paraxial* (i.e., parallel to the vessel axis), though not necessarily axisymmetric; i.e., the insonified blood will be divided into elemental tubes flowing at various velocities, as shown in Figure 7.

With the above assumptions, the backscattered ultrasound can be modeled by first determining the contribution from a single scattering unit as in the particle approach, except the basic scattering unit is now a voxel containing many RBCs. Specifically, the amplitude of the pressure signal received by the transducer at time t from the k*th* voxel can be expressed by

$$y_k(t) = \sigma_{bs}^{1/2} n_k(t - t_k) \cos[(\omega_c + \omega_k)t + \phi_k] \tag{29a}$$

where

$$\omega_k(t) = 2(v_k(t)/c)\omega_c \cos \theta \tag{29b}$$

and

$$\phi_k = -(\omega_c t_k + (\omega_c + \omega_k)t_k) \tag{29c}$$

In the above, σ_{bs} is the backscattering cross-section of a single RBC, $n_k(t-t_k)$ is the number of RBCs in the k*th* voxel at time $t-t_k$, t_k is the time taken by the scattered wave to travel back to the transducer, $\omega_k(t)$ is the angular Doppler frequency shift corresponding to the flow velocity $v_k(t)$ of the k*th* voxel, and ϕ_k is the phase shift determined by the distance between the k*th* voxel and the transducer. In obtaining Equation 29c, the approximation $\omega_k(t-t_k)\cong\omega_k(t)$ has been made,[12] which requires that the stationarity interval of the blood velocity profile be much larger than t_k.

By summing the contributions from all the voxels, and then subjecting the total received signal to a frequency down-shift of ω_c, and removing any reflected component from vessel walls and other slow-moving interfaces, a general expression for the *demodulated* Doppler signal can be obtained as follows:

$$x(t) = \sigma_{bs}^{1/2} \sum_{k=1}^{N} n_k(t - t_k) \cos(\omega_k t + \phi_k) \tag{30}$$

where N is the total number of insonified voxels. In contrast to the particle formulation (Section IV.A), both the Doppler frequency and phase terms in Equation 30 are deterministic, and thus, the only random variables are the n_k's. If the RBCs are perfectly deformable, then the maximum number that can be placed in a voxel is given by $\aleph = \Omega_c/V_c$. Figure 8 shows a plot of $\aleph$ vs. ultrasound frequency, for which the average human RBC volume of $V_c = 90$ μm^3 is assumed. It can be seen that $\aleph$ may vary from over a thousand to less than ten over the diagnostic frequency range. For $f_c = 5$ MHz and 45% hematocrit, the average $n_k(t)$ is about 20, and as a result, N can be an order of magnitude smaller than the number of sinusoids to be summed in the particle approach.

As in Angelsen's continuum approach,[41] the next key step is to write $n_k(t) = \bar{n} + \xi_k(t)$, where $\bar{n}$ is the mean number of RBCs in a voxel and $\xi_k(t)$ is the deviation from $\bar{n}$ at time t. Substituting into Equation 30 gives the following basic model for the backscattered Doppler signal:

$$x(t) = \sigma_{bs}^{1/2}\bar{n} \sum_{k=1}^{N} \cos(\omega_k t + \phi_k) + \sigma_{bs}^{1/2} \sum_{k=1}^{N} \xi_k(t - t_k) \cos(\omega_k t + \phi_k) \tag{31}$$

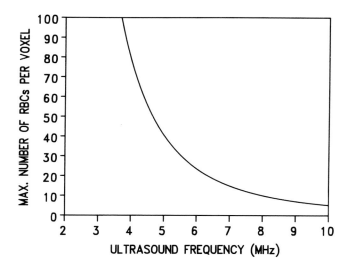

FIGURE 8. Plot of ℵ, the maximum number of 90 μm³ RBCs per (λ/20)³ voxel, vs. ultrasonic frequency. An average sound speed c = 1540 m/s for normal blood is assumed.

Since the insonified blood is divided uniformly into voxels, the first term represents the contributions from a crystalline structure and thus, can be referred to as the *crystallographic* term. For an insonified region that is much larger than a wavelength, the ϕ_k's as defined in Equation 29c should be uniformly distributed over $[0,2\pi]$, and thus, the crystallographic term should approach the average amplitude of a cosine wave which is zero. Physically, the backscattered wavelets from the crystalline structure interfere in a perfectly destructive manner so it may be considered as a homogeneous medium (as in Figure 4b) to the transducer.

The second term in Equation 31 arises from the random fluctuations in local H, and can be considered as the *fluctuation* term. Since the decorrelation distance of the inhomogeneities in the medium should be no more than two cell diameters long, the ξ_k's can be considered as independent random variables. This is in fact the basis for using the Dirac delta approximation[36,46] for the pair-correlation function in the particle approach, and for the spatial correlation of cell concentration in Angelsen's continuum approach.

In terms of the framework developed in the previous section, the hybrid approach has essentially reduced the integral of Equation 19 into a discrete summation over the insonified voxels; and in contrast to the classical approaches, the use of complex signal notation has been completely avoided, thereby enabling the mathematical developments to be physically interpreted in a straightforward manner. Figure 9 illustrates the essence of the hybrid approach in terms of random medium modeling. The previous approaches, especially the particle models, tend to attack directly the problem of modeling

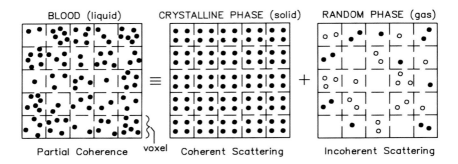

FIGURE 9. Two-dimensional representations that illustrate the essence of the hybrid approach. The insonified blood is divided into elemental voxels in which the filled circles represent RBCs. The partially coherent ultrasound backscatter from blood is modeled as a sum of two components: one arising from a crystallographic phase (which gives rise to coherent scattering) and the other from a random phase (incoherent scattering) in which unfilled circles signify negative RBC concentrations. Note that the number of RBCs in corresponding voxels of the crystalline and random phases add up to give the actual RBC concentrations in blood.

partially correlated RBCs in a dense medium, whereas in the hybrid approach, the total backscattered signal is decomposed into two much simpler components: one arising from a crystalline phase which gives no net contribution because of destructive wave interference; and the other representing contributions from independent fluctuations in RBC concentration. Note also the interesting analogy between blood and the three states of matter (except, unlike a gas, the RBC concentration in the random phase of blood can be negative). In fact, as pointed out in Section IV, the concept of a packing factor (W) and its relationship with var(n) were adapted from the statistical theory of liquids.[57]

It should be noted that in most of the existing particle and continuum theories, the crystallographic component has been tacitly ignored assuming that the dimensions of the insonified region are much greater than λ. This is why the Doppler signal can be considered to be caused by *fluctuation scattering*.[41] In the remainder of this chapter, we will also ignore the crystallographic term, though this may not be completely valid for pulsed Doppler or pulse-echo imaging systems in which the sample volume is not much larger than λ.

B. AUTOCORRELATION FUNCTION OF THE DOPPLER SIGNAL

By the Central Limit Theorem, the Doppler signal, which is given by a sum of independent and identically distributed random variables, is a *Gaussian random process*.[58] Further, since its mean is zero, the random process can be completely characterized by its autocorrelation function $R_x(t_1,t_2) \equiv E[x(t_1)x(t_2)]$; that is, all the information that governs the statistics of the scattering process is contained in $R_x(t_1,t_2)$. It is straightforward to show from Equation 31 that for *general flow conditions*, the autocorrelation function is given by:

$$R_x(t_1, t_2) = \sigma_{bs} \sum_{k=1}^{N} \sum_{i=1}^{N} E[\xi_k(t_1 - t_k)\xi_i(t_2 - t_i)]$$
$$\times \cos(\omega_k t_1 + \phi_k) \cos(\omega_i t_2 + \phi_i) \tag{32}$$

We will now reduce the above expression by assuming that the voxel velocities are paraxial and time-invariant over the period $t_2 - t_1$. Referring to Figure 7, this means that the RBCs in each voxel at t_1 will have simply moved into another voxel within the same elemental tube at t_2. Consequently, out of the N^2 terms in Equation 32, there must be N pairs such that $\xi_k(t_1 - t_k) = \xi_i(t_2 - t_i)$ and $\omega_k = \omega_i$ (which implies that $\phi_k = \phi_i$). Further, the remaining $N(N-1)$ terms should approach zero since the voxel contributions have zero mean and are statistically independent. Hence, Equation 32 becomes

$$R_x(t_1, t_2) = \sigma_{bs} \sum_{k=1}^{N} E[\xi_k^2(t_1 - t_k)] \cos(\omega_k t_1 + \phi_k) \cos(\omega_k t_2 + \phi_k) \tag{33}$$

Since $E[\xi_k^2] \equiv \mathrm{var}(n_k)$ which should not vary significantly over a period t_k, Equation 33 can be rewritten as

$$R_x(t_1, t_2) = \sigma_{bs} \sum_{k=1}^{N} \mathrm{var}(n_k) \cos(\omega_k t_1 + \phi_k) \cos(\omega_k t_2 + \phi_k) \tag{34}$$

If $\mathrm{var}(n_k)$ is further assumed to be constant along each elemental tube (Figure 7) that moves at velocity v_k, then the N voxels to be summed can be grouped as follows:

$$R_x(t_1, t_2) = \sigma_{bs} \sum_{k=1}^{N_T} \left\{ \mathrm{var}(n_k) \sum_{j=1}^{N_z} \cos(\omega_k t_1 + \phi_{kj}) \cos(\omega_k t_2 + \phi_{kj}) \right\} \tag{35}$$

where N_T is the total number of elemental tubes, N_z is the number of voxels in each element tube and ϕ_{kj} is the phase shift associated with the *jth* voxel of the *kth* tube. Using a simple trigonometric identity, this can be rewritten as

$$R_x(t_1, t_2) = \sigma_{bs} \sum_{k=1}^{N_T} \mathrm{var}(n_k) \left\{ \frac{N_z}{2} \cos(\omega_k(t_2 - t_1)) \right.$$
$$\left. + \frac{1}{2} \sum_{j=1}^{N_z} \cos(\omega_k t_1 + \omega_k t_2 + 2\phi_{kj}) \right\} \tag{36}$$

Since the width of the insonified region is assumed to be much greater than a wavelength, for every ω_k the inner summation in Equation 36 should approach zero, and as a result, the autocorrelation function for *paraxial flow* reduces to

$$R_x(\tau) = \frac{1}{2} \sigma_{bs} N_z \sum_{k=1}^{N_T} \text{var}(n_k) \cos(\omega_k \tau) \tag{37}$$

in which $\tau = t_2 - t_1$. The above is more general than those obtained by previous approaches in two respects. First, it does not require the velocity profile in the vessel to be axisymmetric. Second, since the autocorrelation of the backscattered signal from the kth elemental tube is proportional to $\cos(\omega_k \tau)$, Equation 37 shows that the total signal is given by the sum of contributions *weighted by the variance of the RBC concentration* in each tube.

Finally, by combining Equations 29b and 37, we obtain

$$R_x(\tau) = \frac{1}{2} \sigma_{bs} N_z \sum_{k=1}^{N_T} \text{var}(n_k) \cos[2\pi(\omega_c/c) v(r_k, \varphi_k) \cos \theta] \tag{38}$$

where $v(r_k, \varphi_k)$ is the velocity of the kth tube whose position is expressed in cylindrical coordinates. Note that for an axisymmetric velocity profile (v independent of φ_k) and for a constant value of $\text{var}(n_k)$, Equation 38 reduces to Angelsen's result[41] for infinite plane wave insonation and rectilinear flow field. Further, by equating $\text{var}(n)$ to $\bar{n}$ times the packing factor W (Section IV.C), Equation 38 can also be reduced to the particle theory of Mo and Cobbold.[46]

VI. THE BACKSCATTERING COEFFICIENT

A. DERIVATION USING THE HYBRID APPROACH

As stated in Section IV, the backscattering coefficient (BSC) is defined* as the average power backscattered per steradian from a unit volume of blood when it is impinged upon by a plane wave of unit intensity. Now the average backscattered signal power $E[x^2(t)]$ is the same as $R_x(t,t)$, which is a special case of the autocorrelation function defined by Equation 32 for general flow conditions. Since Equation 32 was derived for a sample volume Ω that is insonified by a plane wave whose mean-square amplitude is $1/2$, to calculate the BSC we need to multiply $R_x(t,t)$ by a factor of $2/\Omega$ as follows:

$$\text{BSC} = 2R_x(t, t)/\Omega \tag{39}$$

It can be easily shown that since no RBC motion can occur in zero time ($t_1 = t_2$), Equation 32 can be reduced to Equation 34 *without* imposing the paraxial flow condition. Thus, setting $t = t_1 = t_2$ in Equation 34, we obtain

$$R_x(t) = \sigma_{bs} \sum_{k=1}^{N} \text{var}(n_k) \cos^2(\omega_k t + \phi_k) \tag{40}$$

---

*   Equivalently, the BSC can be considered as the volumetric backscattering cross-section of blood.[34]

Suppose N_k is the number of voxels that are seen by the transducer to be moving at velocity v_k, and suppose M is the total number of velocity bins present, then Equation 40 can be rewritten as

$$R_x(t) = \sigma_{bs} \sum_{k=1}^{M} \sum_{j=1}^{N_k} \text{var}(n_{kj}) \cos^2(\omega_k t + \phi_{kj}) \tag{41}$$

If $\text{var}(n_{kj})$ is assumed to be a function of only the voxel velocity v_k, then $\text{var}(n_{kj}) = \text{var}(n_k)$ and this can be taken outside the inner summation in Equation 41. Further, if N_k is assumed to be sufficiently large for all k such that ϕ_{kj} is uniformly distributed, then Equation 41 reduces to

$$R_x(t) = (1/2)\sigma_{bs} \sum_{k=1}^{M} N_k \, \text{var}(n_k) \tag{42}$$

Substituting into Equation 39 and noting that $\Omega = N\Omega_e$, yields the BSC for *general flow conditions*:

$$\text{BSC} = (\sigma_{bs}/\Omega_e) \sum_{k=1}^{M} (N_k/N) \, \text{var}(n_k) = \sigma_{bs}\overline{\text{var}(n)}/\Omega_e \tag{43}$$

The above states that the BSC is given by the average variance in cell concentration over the whole insonified region. Alternatively, we can define $W_k = (1/\overline{n}) \, \text{var}(n_k)$ as the packing factor of the RBCs in the k*th* voxel, so that

$$\text{BSC} = \sigma_{bs}(\overline{n}/\Omega_e) \sum_{k=1}^{M} (N_k/N)W_k = \sigma_{bs}(\overline{n}/\Omega_e)W \tag{44}$$

That is, the BSC can also be considered to be determined by the average packing factor W. Note that Equations 43 and 44 are the same as Equations 28 and 22 of the continuum and particle approach, respectively.

B. UNDERSTANDING THE BSC VS. HEMATOCRIT RELATIONSHIP

As shown in Figure 6, use of the Percus-Yevick hard sphere packing factor as defined by Equation 23 predicts that the BSC vs. H curve should peak at a hematocrit of around H = 13%, which is in reasonable agreement with the observed data. However, the derivation of a packing factor involves some very complex statistical mechanics[44,45] and the Percus-Yevick approximation holds well up to only about H = 40%. Hence, the packing factor viewpoint does not easily provide clear physical insights.

An alternate viewpoint based on estimating the variance in local cell concentration appears to offer better insights into the shape of the BSC vs.

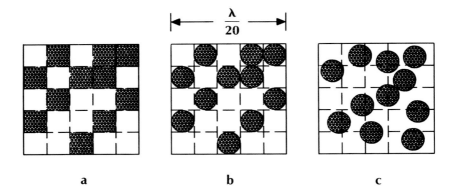

FIGURE 10. Two-dimensional representations of the random filling of an elemental voxel: (a) cube-shaped particles; (b) spherical particles fixed to a grid; and (c) freely distributed.

H curve. According to Equation 43, the backscattered power is proportional to the variance of the cell concentration in an elemental voxel. Figure 10a shows a *two-dimensional analog* of an elemental voxel that is divided into cubic sites each of volume equal to that of an RBC. If the RBC volume is V_c there will be an ideal total of $\aleph = \Omega_e/V_c$ sites within a voxel. From Figure 8 it can be seen that $\aleph > 10$ over the diagnostic frequency range of 2 to 8 MHz. Suppose RBCs are tossed randomly into these imaginary sites so that the probability of finding a particular site being occupied by an RBC is equal to the local average hematocrit H. As in a series of Bernoulli trials,[27] the probability of finding n RBCs in $\aleph$ sites is given by

$$p_\aleph(n) = Pr\{n \text{ of } \aleph \text{ sites occupied}\} = \binom{\aleph}{n}H^n(1 - H)^{\aleph-n} \qquad (45)$$

As H approaches zero, and for n of the order $\aleph H$, this can be approximated by the Poisson distribution[27]

$$p_\aleph(n) \simeq \frac{(\aleph H)^n}{n!} e^{-\aleph H} \qquad (46)$$

where $\aleph H$ is the mean as well as the variance of the distribution; i.e., var(n) = $H\Omega_e/V_c$. If var(n) is constant throughout the medium, then substitution into Equation 43 gives BSC = $\sigma_{bs}H/V_c$. Compared to Equation 44, it can be seen that the corresponding packing factor W must be unity. Physically, this implies that at very low H, the scatterer positions are completely uncorrelated (Section III.B) so that the BSC is a linear function of H. If H is not very small, Equation 46 will no longer apply. However, as long as the voxel is sufficiently large such that $\aleph > 10$, Equation 45 can be approximated by[27]

$$p_{\aleph}(n) \rightarrow \frac{1}{\sqrt{2\pi \aleph H(1 - H)}} \; e^{-(n - \aleph H)^2/[2\aleph H(1 - H)]} \tag{47}$$

which is a Gaussian distribution with $\bar{n} = \aleph H$ and $var(n) = H(1 - H)\Omega_e/V_c$. Substituting for the average variance in Equation 43 gives

$$BSC = (\sigma_{bs}/V_c)H(1 - H) \tag{48}$$

Again, compared to Equation 44, the effective packing factor $W = (1 - H)$ is that for impenetrable particles whose positions are independent of each other.[24,36] It should be noted that the accuracy of Equation 47 depends on $\aleph$, which is a function of the ratio a/λ. Thus, one can appreciate why a more refined packing theory like Yagi and Nakayama's[40] might predict a packing factor that varies with a/λ as shown in Equation 27. It should also be noted that Equation 48 implies that the maximum BSC peaks at $H = 50\%$, and this supports the notion that maximum scattering occurs when the nonuniformity of the medium is greatest.

Why then does the measured BSC peak at $H < 50\%$? This calls for a careful check of the implicit assumptions made in obtaining Equation 45. First, referring to Figure 10a again, it has been assumed that every site in the voxel can be fully filled by an RBC. This cannot occur in reality because the RBCs are not cube shaped but are asymmetrical and are generally randomly oriented. Thus, as illustrated in Figure 10b, a more realistic model is to approximate the RBCs as spheres of the same volume. In that case, it is not difficult to see that between adjacent spheres there would always be some dead space which has been referred to in the literature as *elbow room*[44] or *excluded volume*.[36] As H increases, the total fractional excluded volume approaches 100% and $var(n) \rightarrow 0$. It is interesting to note that the packing factor $W = \overline{var(n)}/(\bar{n})$ can then be interpreted as the fractional *free volume* in the medium, which approaches zero with increasing H. For hard spheres the highest H that can be achieved by hexagonal packing is about 74% and, thus, one might speculate that the maximum BSC should occur at around 37%. But this is not what actually happens because it is still unrealistic to assume that the two-dimensional voxels can be divided into squares.

To see what really happens, imagine that RBCs are tossed into the voxel one at a time. At very low H the model in Figure 10b may provide a good approximation to the actual spatial distribution. But as more and more cells are thrown in they can no longer be assumed to fall neatly into the squares, and consequently, as shown in Figure 10c, the packing factor or fractional free volume will diminish much more rapidly than the model in Figure 10b would predict. This explains why the BSC actually peaks at an H significantly below 37%.

Another important concept which can be seen from the more realistic model of Figure 10c is that as the number of RBCs within a voxel increases, they will soon overlap in space unless some rules are introduced to keep them from overlapping. Clearly, the exact form of the cell number distribution at high H (when n is no longer Poisson) is dependent on which packing rules are chosen. This means that in general there is no unique function which will describe the relationship between BSC and H. Given certain RBC characteristics such as size, shape, and deformability, these packing rules are determined by all the physical forces that act on the RBCs including, as discussed in the following two subsections, the shear forces of the flow field and the attractive forces between RBCs.

1. Effects of Flow Disturbances

Since there is no unique packing factor for RBCs, some discrepancies between the measured BSC and the Percus-Yevick theory should not be too surprising. The increased BSC due to stirring (Figure 6) can be explained by a broadening of the cell number distribution, or an increased var(n), due to the increased shear forces acting on the RBCs. The important point to note is that even a 50% increase in BSC can be explained by only a 22% increase in the standard deviation of the cell number distribution for each voxel; i.e., the inhomogeneities in the medium do not have to become macroscopic, as was previously suggested.[11]

A comment should also be made about Yang's model[50] which, as discussed in Section IV.A, consists of a particle and a turbulence component. The latter, which was said to be proportional to the variance of the particle concentration, appears to be equivalent to the fluctuation term in Equation 31. It seems that in Yang's conception of a system of particles, any deviation from a crystalline arrangement is attributed to 'turbulence' in the system. But since a medium can be random without *flow turbulence*, it is more appropriate to use the term fluctuation scattering.

2. Effects of RBC Aggregation

The effects of RBC aggregation on the ultrasonic backscatter can be readily understood from the standpoint of fluctuation scattering. Suppose all the RBCs within the medium appear in doublets, then it is not difficult to see why the var(n) would increase because each voxel would then either gain two cells or lose two cells, even though σ_{bs} and $\bar{n}$ remain unchanged. In fact, given a mean local H, the probability density function which has the largest variance is one that consists of two delta functions: one at n = 0 and the other at n = $\aleph$, which corresponds to the case in which the voxel is either empty or completely filled by a single RBC aggregate. However, it should be remembered that the hybrid theory is based on Rayleigh scattering, which assumes that the dimensions of the RBC aggregates are less than about $\lambda/10$:[22] if the cell aggregates become larger, the fourth-order frequency law will no longer apply.[59,60]

VII. DOPPLER SIGNAL STATISTICS AND SIMULATION

Due to the pulsatile nature of arterial blood flow, the frequency content of the backscattered Doppler signal is time varying, and is often displayed in the form of a gray-scale spectrogram of frequency vs. time. On many commercially available systems, the Doppler power spectrogram is computed in real time using conventional fast Fourier transform (FFT) methods. However, it has long been recognized that the granular structure, or speckle, seen in such power spectrograms may mask out any subtle changes caused by the presence of a minor arterial stenosis. Therefore, to improve sensitivity and accuracy of Doppler techniques, a proper understanding of the Doppler signal statistics is imperative.

The frequency domain statistics of any band-limited Gaussian random process like the Doppler signal can be determined from stochastic signal theory. In fact, a number of Doppler simulation models have been proposed (see Reference 61 for a concise review) that enable the synthesis of artificial Doppler signals for the testing of various signal processing systems.[62,63] However, in order to develop a full appreciation of the Doppler speckle, it is important to be able to derive its statistics directly from the fundamental scattering theory. In this section, we indicate how this has been recently achieved using the hybrid approach.

A. GAUSSIAN SIGNAL MODEL

Returning to the basic scattering model of Equation 31 and ignoring the crystallographic term, the Doppler signal x(t) is given by the total contribution from all the elemental voxels. The sum consists of different Doppler frequency components corresponding to the different velocities of the voxels. As in the previous section, N_k denotes the number of voxels seen by the transducer to be moving at velocity v_k, and M is the total number of velocity bins present. It should be clear that the Doppler signal can be expressed by

$$x(t) = \sum_{k=1}^{M} x_k(t) \tag{49a}$$

where

$$x_k(t) = \sigma_{bs}^{1/2} \sum_{j=1}^{N_k} \xi_{kj}(t - t_{kj}) \cos(\omega_k t + \phi_{kj}) \tag{49b}$$

represents the total contribution from the voxels that give rise to the Doppler frequency ω_k, and the subscripts of ξ_k, t_k, and ϕ_k in Equation 31 have been altered accordingly. Expanding the cosine term in Equation 49b yields

$$x_k(t) = A_k(t) \cos(\omega_k t) - B_k(t) \sin(\omega_k t) \tag{50a}$$

in which

$$A_k(t) = \sigma_{bs}^{1/2} \sum_{j=1}^{N_k} \xi_{kj}(t - t_{kj}) \cos \phi_{kj} \tag{50b}$$

and

$$B_k(t) = \sigma_{bs}^{1/2} \sum_{j=1}^{N_k} \xi_{kj}(t - t_{kj}) \sin \phi_{kj}$$

are the *in-phase* and *quadrature* components[58] of $x_k(t)$. Further, Equation 50a can be rewritten as

$$x_k(t) = a_k(t) \cos(\omega_k t + \psi_k) \tag{51a}$$

where

$$a_k(t) = \sqrt{A_k^2(t) + B_k^2(t)} \tag{51b}$$

and

$$\psi_k = \tan^{-1}(B_k/A_k) \tag{51c}$$

Note that A_k and B_k are Gaussian distributed random variables with zero mean because they are given by a weighted sum of many random variables ξ_{kj}'s that are zero mean and identically distributed. By definition, the sum of A_k^2 and B_k^2 is an exponential random variable (or chi-squared with two degrees of freedom). Consequently, the a_k's are Rayleigh distributed and the ψ_k's uniformly distributed.[58] By substituting Equation 50b into Equation 51b, the expected value of $a_k(t)$ can be shown[12] to be equal to the square root of $N_k \sigma_{bs} E[\xi_k^2]$ or $N_k \sigma_{bs} \text{var}(n_k)$, which is the total power backscattered from all the voxels associated with ω_k. Therefore, Equation 51b can be rewritten as

$$a_k = \sqrt{2S_x(f_k, t) \Delta f} \, \chi_k \tag{52}$$

in which χ_k is a standard (mean = 1) exponential random variable, and $S_x(f_k,t)$ is defined as the *power spectral density* of $x(t)$ such that the average signal power contained within an incremental frequency bin Δf around f_k is given by $2S_x(f_k,t)\Delta f$. Finally, substituting Equation 51a into Equation 49a and using Equation 52 yields the following Doppler signal model:[61]

$$x(t) = \sum_{k=1}^{M} \sqrt{2S_x(f_k, t) \Delta f} \, \chi_k \cos(\omega_k t + \psi_k) \tag{53}$$

B. DOPPLER POWER SPECTRUM

Let us denote the power spectrum or *periodogram*, of x(t) over a quasi-stationary time interval T (typically 10 ms) by $I_T(f) \equiv (1/T)|\mathscr{F}_T\{x(t)\}|^2$, where $\mathscr{F}_T\{\cdot\}$ is the Fourier transform operation performed over a time window T.

1. Power Spectrum Statistics

The total power of the *kth* sinusoidal component in Equation 53 is $a_k^2/2$, which should be equal to the amplitude of the periodogram at f_k, provided that the sinusoidal component can be observed over an indefinitely long time interval. Further, since the χ_k's are independent random variables, the covariance of any two sinusoidal components should approach zero if the observation period is long enough. In summary, we may conclude that from Equation 53,

$$\lim_{T \to \infty} I_T(f) = \frac{1}{2} \chi_2^2 S_x(f) \tag{54}$$

and

$$\lim_{T \to \infty} \text{cov}[I_T(f_m)I_T(f_n)] = 0, \qquad m \neq n \tag{55}$$

where $\text{cov}[\cdot,\cdot]$ denotes the covariance of two random variables, and χ_2^2 represents an exponential random variable. Since the mean and variance of χ_2^2 are 2 and 4, respectively,[27] Equation 54 implies that for finite T, $E[I_T(f)] \cong S_x(f)$, and $\text{var}[I_T(f)] \cong [S_x(f)]^2$. It can be shown[27] that the use of a finite window (or observation period) in computing the FFT of x(t) generally results in a convolution of $S_x(f)$ by a corresponding spectral window of effective width 1/T, which is usually much smaller than the Doppler signal bandwidth. Thus, in most practical situations where T is sufficiently large, the power spectral density $S_x(f)$ can be interpreted as the *ensemble-averaged* Doppler power spectrum.[64]

From the standpoint of statistical estimation,[65] the periodogram $I_T(f)$ which satisfies Equation 54 is said to be an *asymptotically unbiased* estimate of $S_x(f)$. However, $I_T(f)$ is also an *inconsistent* estimate because of its large variance, which is reflected in the large fluctuations of the sample power spectra. Equation 55 further implies that increasing the analysis time T will only decrease the covariance of the spectral components, which means that the spectrum will fluctuate more rapidly with frequency. If the speckle is considered as noise and $S_x(f)$ is the underlying signal, then the signal-to-noise ratio at a given frequency is $[S_x(f)]^2/\text{var}[I_T(f)]$, which is approximately equal to unity for a sufficiently large T. This is exactly analogous to the statistics of laser speckle which is also derived from scattered intensities. By contrast, the B-mode ultrasound speckle which represents fluctuations in the *amplitude* (Rayleigh-distributed) of the backscattered ultrasound from tissue, has a signal-to-noise ratio of 1.91.[66]

It should be noted that the signal model of Equation 53 has been derived directly from the basic scattering model of Equation 31, for which *no* assumption about the flow field was made (the restriction to paraxial flow was not introduced until after Equation 32). That is, the blood flow could follow a helical pattern as in curved vessels or may even be turbulent. This means that the power within a Doppler frequency bin Δf is always exponentially distributed *regardless* of the flow condition. With respect to the physical cause of these fluctuations, it should be remembered that the spectral component at any frequency f_k represents the net contribution from the elemental voxels whose velocities correspond to Doppler shifts within a frequency bin of width $\Delta f \cong 1/T$ centered at f_k. Since the backscattered wavelets from individual voxels can interfere constructively as well as destructively, the resultant amplitude a_k is a Rayleigh-distributed random variable and consequently, the spectral power is exponentially distributed with a mean equal to $S_x(f)$.

2. Relationship with Flow Velocity Distribution

How then is $S_x(f)$ related to the blood velocity profile? It has been stated that for paraxial flow the autocorrelation of the Doppler signal is $R_x(\tau)$ as defined by Equation 37, which is only a function of τ. In stochastic signal theory, such a random process is referred to as *wide-sense stationary*, and it is well known that under this condition $R_x(\tau)$ and $S_x(f)$ form a Fourier transform pair.[58] Accordingly, for a blood vessel segment that is uniformly insonified by a plane wave, it is not difficult to show from Equation 38 that the power spectral density is given by[31]

$$S_x(f) = K \; \sigma_{bs} \text{var}(n) \Lambda(f) \left[\frac{dv(r)}{dr} \right]^{-1} \bigg|_{r=\Lambda(f)} \tag{56}$$

where $r = \Lambda(f)$ is the solution to Doppler equation, $f - 2 f_c[v(r)/c] \cos\theta = 0$, $v(r)$ is the blood velocity profile which has been assumed to be axisymmetric and a *monotonic* function of radial position r, and K is a proportionality constant determined by the geometry of the insonation system. For example, it is well known that the ideal $S_x(f)$ for a parabolic velocity profile (as in steady laminar flow) is a constant up to a frequency corresponding to the maximum blood velocity. However, in practice the shape of the $S_x(f)$ is often significantly distorted due to a combination of factors including intrinsic spectral broadening,[67] non-uniform insonation,[68] and sound attenuation.[69] As a result, the relationship between $S_x(f)$ and $v(r)$ is usually so complicated that it can only be determined numerically.

The exponential statistics of the Doppler power spectrum for steady laminar blood flow has been verified[70] using an *in vitro* flow model and a 5 MHz CW Doppler system (Medasonics D9, Mountain View, CA). Specifically, a straight tube model with a 42% saline suspension of washed human RBCs

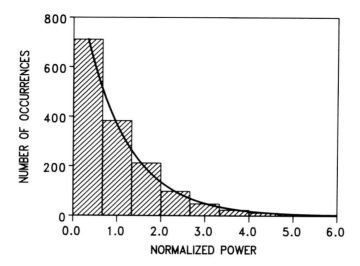

FIGURE 11. Histogram of the experimentally observed spectral power at 2 kHz, for a 42% saline suspension of human RBCs under laminar flow conditions.[70] Smooth curve represents an exponential distribution whose variance was estimated from the sample histogram.

was used. Figure 11 shows the histogram of spectral power at a frequency of 2 kHz, obtained from the FFT-based power spectra over 1500 successive 10 ms intervals. In the experiment a flow rate corresponding to a Reynolds number of 930 was used. It can be seen that the agreement is excellent between the histogram and an exponential function whose variance was estimated from the histogram. The agreement was also confirmed by the chi-squared goodness-of-fit test at the 5% significance level. However, it should be noted that in the same experiment when the Reynold's number was increased until the flow became turbulent, the power at specific frequencies in successive 10 ms spectra was found to deviate clearly from the exponential distribution. While this may appear to contradict the above theoretical finding, it should be remembered that in turbulent flow, the stationarity period of the velocity profile was probably less than 10 ms so that the underlying $S_x(f,t)$ between successive 10 ms spectra were no longer the same. Consequently, even though the fluctuations in each *individual* spectrum should always be governed by exponential distributions, the histogram of power over different 10 ms intervals might indicate much larger variability (as was observed).

C. DOPPLER SPECKLE SIMULATION

The signal model of Equation 53 was originally developed for the purpose of simulating Doppler ultrasound signals on a computer.[61] In fact, we have successfully used this model to synthesize both CW and pulsed Doppler spectrograms similar to those obtained from carotid arteries. Specifically, the time-varying $S_x(f,t)$ over a whole cardiac cycle was modeled by an empirical

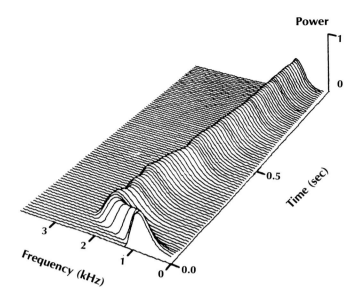

FIGURE 12.   Pseudo-3D plot of the empirical power spectral density function used in the carotid Doppler signal simulation model.[61]

function as shown in Figure 12, where the function parameters were estimated from the ensemble-averaged carotid Doppler spectrogram obtained from a 5 MHz CW system (Medasonics D10). The comparison between one realization of the model of Figure 12 and a real carotid Doppler signal is shown in Figure 13a. When the synthesized digital signal was converted into analog form and played back on a cassette tape recorder, the signal was found to be almost indistinguishable from the clinical recording on the basis of their audio quality. Further, when the synthesized and real signals (two cycles each) were analyzed on a commercial spectrum analyzer (Medasonics SP25A), the resultant gray-scale speckle patterns (display intensity proportional to power in decibels) shown in Figure 13b were also found to be very similar.

VIII. CONCLUDING DISCUSSION

In this chapter three approaches to modeling the backscattered Doppler ultrasound have been presented; namely, the particle, continuum, and hybrid approaches. It was shown that depending on which approach is taken, the RBC interaction in dense blood can be modeled by a particle pair-correlation function, or by a two-point spatial correlation function, or by the variance of local RBC concentration. The strength of the particle approach lies in the use of geometric ray theory and the principle of superposition, whereas the continuum approach recognizes that the random medium can be modeled simply

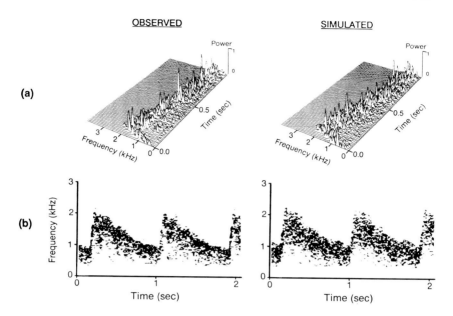

OBSERVED   SIMULATED

(a)

(b)

FIGURE 13. Frequency domain comparison between clinically recorded and simulated CW Doppler signals for normal carotid arteries. (a) Pseudo-3D plot of power spectrogram over one cardiac cycle. (b) Gray-scale display of speckle patterns of two cycles on commercial spectrum analyzer.

by tracking local RBC concentrations, without the need to follow every individual RBC. The recently proposed hybrid approach combines the strengths of the particle and continuum approaches; i.e., geometric ray theory is applied to sum contributions from elemental voxels.

Using the hybrid approach it was shown that the received Doppler signal from a dense medium is generally composed of a crystallographic and a fluctuation component, and this has led to the reconciliation of several existing theories. If the dimensions of the insonified region are much larger than a wavelength, the crystallographic component should be negligible compared to the fluctuation component.* Further, under the assumption of plane wave insonation, the autocorrelation function and the BSC were derived, and they were expressed in more general forms than those reported previously. In particular, it was found that the BSC is proportional to the variance rather than to the mean, of the local H.

Some practical implications of the theoretical findings should be noted. First, it will be recalled from Section I that considerable research in ultrasonic scattering was motivated by its potential applications in hematology, such as

---

* For pulse-echo imaging systems in which the sample volume size is not much larger than a wavelength, the effect of crystallographic scattering may not be negligible. This is an area which needs to be further investigated.

measurement of the degree of RBC aggregation in blood. Although we have argued in Section VI.B that a small degree of RBC aggregation will generally increase the BSC even at high H, it should be borne in mind that the BSC is not just determined by the mean scatterer size (or mean aggregate size), but also by the packing factor which generally varies with flow conditions. Thus, in practice, it would be quite difficult to quantify the degree of aggregation based on measurements of backscattered power alone.

Second, it should be noted that according to the experimental data shown in Figure 6, the relationship between the BSC and H will deviate significantly from a straight line when H >10%, presumably because n is no longer Poisson distributed. In fact, at an H of around 45%, the BSC can be 20 times lower than if the relationship remains linear. In other words, for a suspension of point-sized particles, the degree of fluctuation in local H relative to the mean H, can be 20 times greater than that for normal blood. Since the performance of Doppler signal processors are often evaluated by computer simulation[71] based on a point-sized particle scattering model, the effects of changing the voxel statistics should be investigated in future work.

It has also been widely assumed in the past that equal volumes of blood in the insonified vessel will give rise to the same backscattered power so that the Doppler spectrum can be considered as a measure of the flow velocity distribution. As pointed out in Section VII.B, the relationship between the two is given by Equation 56 if effects such as intrinsic spectral broadening and tissue attenuation are ignored. However, it should be remembered that the power in different frequency bins are determined by the variance of RBC concentrations in the corresponding acoustic voxels, which in turn are governed by the local shear rates or flow conditions. In other words, a uniform hematocrit distribution across the vessel lumen does not necessarily mean that the packing factor or echogenicity is also uniform.

Finally, in regard to the determination of the best Doppler spectral estimator for evaluation of arterial disease, an emerging trend is the development of efficient methods for reducing the speckle in Doppler power spectrograms. These include straightforward ensemble-averaging techniques[72] to more sophisticated methods such as parametric spectral modeling,[63,73-75] two-dimensional adaptive filters,[76] and Wigner distributions.[77] However, we have shown using the hybrid approach that the speckle in a sample spectrum is always governed by an exponential distribution whose standard deviation is as large as the mean. This implies that although the Doppler power spectral density may vary randomly and/or more rapidly with time due to disease-induced flow disturbances, it will be difficult for current speckle-reduction techniques to discriminate the resultant changes from the speckle in a sample spectrum.

ACKNOWLEDGMENTS

We are very grateful to Peter Bascom and Clement Fung for many constructive comments concerning this chapter. This work was financially sup-

Chapter 6

STANDARD SUBSTITUTION METHODS FOR MEASURING ULTRASONIC SCATTERING IN TISSUES

John M. Reid

TABLE OF CONTENTS

I. INTRODUCTION

The substitution method of measuring scattering was developed initially as a means of overcoming the problem of attaining acceptable accuracy when very many system parameters are involved in making a measurement. Since scattering measurements require both a transmitting and a receiving channel plus knowledge of the frequency dependent shape of the scattering region, more than twice as many parameters must be determined as for many other types of measurement. Since the errors of each separate determination accumulate, the total error may be very large if many measurements of normal accuracy are made. Just the measurement of output peak pulse power alone, for example, is seldom able to achieve high accuracy, yet this is only one quantity that must be known to measure scattering.

By substituting a reflector of known reflectivity for the sample being measured we can compare the signals scattered from this standard with those from the sample. By using the same measurement system in both cases we need to measure only the difference between the two scattered signals. The incident power, and transducer/receiver bandwidth and sensitivity, can be exactly the same for both signals, and do not have to be measured separately.

Standard substitution methods derived by using many simplifying assumptions are discussed here to clarify the physical principles involved in the measurement process. This presentation has advantages over the more mathematically rigorous methods that were developed later. For example, the nature of some fundamental limitations to scattering measurement will be made more clear. The method's simplicity allows easy understanding of the more exact measurement methods and thus serves as an introduction to them. In cases where accuracy is not possible, most often because of the presence of attenuation, the standard method is sufficiently accurate. The standard substitution methods require simpler equipment and less computation time than the other methods.

A particular problem in the quantitative study of scattering is the presence of attenuation in tissues. Since the magnitude of this loss of energy increases with frequency, it sets a limit to the shortness of wavelength that can be used, and hence to the resolution of both images and measurements. In imaging, to achieve best resolution, the frequency used often is raised to the limit set by attenuation. In this case all echo amplitudes represent the product of a local scattering strength and an appreciable attenuation loss factor. All true measurements of scattering from within tissues must thus be corrected for attenuation.

The presence of attenuation sets a limit to the accuracy of scattering measurements. If the total attenuation loss is 30 dB, for example, a 10% uncertainty in the total attenuation factor leads to a 3 dB, or 100%, error in the power scattering coefficient. This is not an extreme case since many measurements of attenuation vary by more than 10%. In clinical measurements the attenuation usually must be calculated from assumptions and may be of questionable accuracy. The variation of attenuation coefficient between individuals, particularly if disease is present, can be more than this. Therefore the approximate method is usually acceptable for clinical or *in vivo* estimates, with the advantage of being adaptable to real-time use. Besides being acceptably accurate in many cases, the standard substitution method is available to serve as a simple check on the others.

The other methods that are covered in this volume are that of Madsen in Chapter 7 and Waag in Chapter 8. These use more exact expressions than the approximations used here.

A. METHODOLOGY

Scattering measurement requires a basic theoretical framework which allows us to calculate the system gain factors that influence the scattered echo strength so that the tissue scattering function alone can be measured. We must first set up the equations describing the problem, then solve for the unknown, the scattering strength of the tissue. However, many problems exist in the exact solution of the partial differential equations that apply to elastic wave propagation, particularly when they are subject to both known (apparatus) and unknown (tissue) boundary conditions! Approximation methods must be used, and these require some justification. In particular, the assumptions about the scattered wave shape described just below limit the accuracy of all of the methods described in this volume.

Scattering measurements are limited in spatial resolution by the measuring system. Any measured scattering function is some average of the local values found within the system's resolution volume. The scattering structures within this volume will, in general, produce a scattered wave which has an unknown angular dependence of amplitude. This is the wave needed to meet the local boundary conditions at the scatterers. Specification of scattering ideally should include this angular distribution function, as well as the scattering strength. As a practical matter, we can seldom measure scattering right at the surface of the resolution volume to determine the true shape of the scattered wave. However, to make true measurements at a distance from the scatterer, we must know the loss in measured amplitude that occurs when this arbitrarily shaped scattered wave propagates back to the receiver.

The problem is usually handled by measuring sufficiently far from the scattering volume so that the scattered waves actually are of spherical shape. The propagation loss is then calculated as that due to spherical spreading (plus any attenuation). If the scattered wave is not spherical, then the propagation loss so calculated is inaccurate. This spherical wave assumption strictly is true only if the scattering volume is so small compared to the wavelength that it can be regarded as one very small scatterer. If this assumption is not true, the scattering function that is calculated by this procedure is an "equivalent" value. This value is the scattering strength of a scatterer or volume small enough, compared to the wavelength, to reradiate a spherical wave of amplitude that is equal to the amplitude measured at the measurement distance.

The concept of measuring scattering from an "equivalent scatterer" is as close as we can come to measurement of the true scattering without more detailed measurements of the scattered wave very close to the scattering volume. As we shall see later, the "equivalent" concept generally must include assumptions of weak, single scattering also. Since the minimum size of a resolution cell is about that of the wavelength, the equivalent scatterer measured in this case is the basic unit of measurement of scattering.

By separating the transmitting and receiving transducers, keeping their beam axes trained on the sample volume, the dependence of this equivalent scattering amplitude function on the angle between the transducer axes can be determined. If the scattered wave were truly a single spherical wave, there would be no angular dependence of the scattered amplitude. The existence of a measured angular scattering function could show, according to the above argument, that the true scattering amplitude has not been measured! Multipole scattering functions may be exempt from this restriction, since they are just appropriately phased sums of monopoles. Since multipole functions have angular symmetry they can be differentiated from the more irregularly shaped waves scattered by solid tissues.

To specify the scattering process we must include the angular dependence, and its measurement is important for determining mechanisms.

B. NOTES ON THE BASIC EQUATIONS

The equations to be derived and used here will be based on some approximations.[1] This is done for necessity in the solution of the equations for acoustic propagation, and for utility in solving these latter equations. Many large-distance and small wavelength approximations are not strictly necessary, except to obtain closed-form solutions that lead to physical insight into the scattering and echo-signal generation process. In many cases they are sufficiently accurate to lead to useful results. These restrictions are relaxed in the treatments of Madsen, Chapter 7 and of Waag, Chapter 8, which use exact expressions for field shapes and integrals.

Pulse operation will be assumed here so that the transmission and reception functions of the equipment can be considered separately, free of any standing waves. Continuous wave equations will be derived for convenience, and later extended to cover pulse operation through the Fourier transform or calculations using the band-center frequency, see Section III.A.

The acoustic waves are generated, usually, by conversion of an electrical signal to vibrations in the transducer. The waves travel to a region, v, diagramed in Figure 1, where they are scattered. The scattered waves are collected at another transducer (or the same transducer for measuring backscattering), then reconverted to an electrical signal that can be measured. The quantity we wish to measure is the angle-dependent scattering strength function within

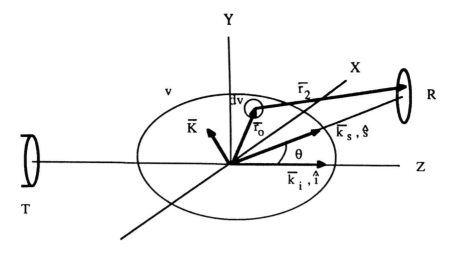

FIGURE 1. Directions and coordinates for a scattering experiment or calculation. The transmitting transducer is indicated by T and the receiver by R. The X-Y plane is defined by the plane containing the incident wave, $\hat{\imath}$, and the scattered wave, $\hat{s}$, unit vectors, and contains the scattering vector, $\overline{K}$. The position vector, $\bar{r}_0$, sweeps out the sample volume, v.

the volume, v (see Figure 1, which relates the incident to the scattered wave). The waves are related to the measurable electrical signals at the transducer terminals by three integrals: the first relates the motion generated by the transducer to the wave incident on the scatterers, the second the scattered wave to the scattering strength function within the scattering volume, v; and the third, the electrical echo signal to the scattered wave at the receiving transducer.

The calculation of the incident and scattered field integrals will use the elementary theory of lossless, isotropic fluids, with modifications. The wave-number, k', will be taken as complex, to include losses expressed as the pressure attenuation coefficient, α. The time-dependent variational pressure is written for a plane wave propagating a distance, x, with velocity, c, and angular frequency, ω, as:

$$p = p_o \epsilon^{j(\omega t - k'x)} \qquad \text{with} \quad k' = k - j\alpha \tag{1}$$

which reduces to

$$p = p_o \epsilon^{-\alpha x} \epsilon^{-jkx} \epsilon^{j\omega t} \tag{2}$$

where p_o is the pressure at $x = 0$. The exponential time factor will always be implied in equations for the acoustic pressure when it is not written. In some cases we will need the time average power density, the intensity, I, which is related to the pressure magnitude within a material of specific acoustic impedance, ρc, by;

$$pp^* = |p|^2 = |p_o \epsilon^{-\alpha x}|^2 = 2I\rho c \tag{3}$$

II. DERIVATIONS

A. SCATTERING MEASURES

We first will define the scattering function to be measured independently of any particular mechanism or model. It is a function having a local value which, when multiplied by the incident wave, allows the strength and angular distribution of the wave scattered by a particular region of tissue to be specified, assuming spherical spreading of the scattered wave, as discussed above.

The calculation of scattering functions from a knowledge of material properties is possible only in simple cases where the boundary conditions can be written in analytic form. It is also necessary that the scattering be so weak or the scatterers be so sparse, that we can assume that the wave incident at each scatterer is only that due to the transmitter (the first Born approximation). It is not necessary for our purposes, however, that this scattering coefficient be calculable from first principles. We will take the coefficient simply as a definition of the scattering strength. That is, the "equivalent" scattering cross-section, given the Born approximation and given spherical wave spreading of the scattered wave, which results in the same intensity at the receiver as is produced by the actual scatterer.

The scattered wave is specified not only by its amplitude but also by its spatial shape. In this sense the scattering measurement process is far more complicated than the attenuation measurement process where only a numerical coefficient must be determined. The shape of the scattered wave may differ radically from the shape of the incident wave. This wave appears to originate from the scattering volume (Figure 1) and its shape may be an important characterization of the scattering process. Present formulations make two consistent assumptions: a spherical scattered wave and measurement at a large distance from the scattering volume. Future work is directed toward removing this restriction through knowledge of the scattered waveshape. Spatially sampled scattered fields may be reconstructed by diffraction inversion methods, for example, to find not only the waveshape, but also values of the acoustic parameters in the medium.[2]

The general plan to be followed here is to derive expressions for the measurement process that relate measured quantities to the measuring system parameters and to a scattering coefficient characteristic of the tissues. By knowing the system parameters we can then solve for the unknown tissue coefficients. The coefficients can be related to the physical parameters of the tissues only if a model for the acoustical parameters responsible for the scattering process is adopted. Below we develop this relationship for a fluid model, which is assumed to be appropriate since the sound speed in soft tissues is nearly that of their major constituent, water. The development of the scattering equations for this model also clarifies the physics of the scattering process.

Several different scattering coefficients are used in the literature so it is useful to work out the relationships between them.[3] These have been written largely for power, which is very appropriate for the case of incoherent scattering where the mean power is additive. In tissues, such scattering certainly takes place from randomly located structures, but so does coherent scattering from structures whose spacing is not really random. Membranes exhibit phase interference effects from their boundary echoes, for example. Methods of integrating the scattering over space or frequency have been developed to handle these cases, and are covered in Section III.A.3.

The total scattering cross-section, σ_t, is an area that relates the scattered power, W_s, to the incident intensity, I_i;

$$W_s = I_i \sigma_t \tag{4}$$

Under the assumption of spherical scattered waves at a sufficient distance, r, from the scattering region, the scattered intensity, I_s, is written

$$I_s = W_s/4\pi r^2 \quad \text{or} \quad I_s = I_i \sigma_t/4\pi r^2 \tag{5}$$

where W_s is total scattered power.

This equation is also applied to the power scattered in a particular direction. For example, the σ_t in the second form of Equation 5 is called the backscattering cross-section, σ_b, in radar and remote sensing to describe scattering directly back to the transmitter.

Where the scattered wave shows a dependence on angle the scattering per unit solid angle is an appropriate measure. This differential scattering cross-section, σ_d, is defined for an isotropic scatterer as

$$I_i \sigma_d = W_s/4\pi \tag{6}$$

since the scattered power per unit solid angle is found by dividing the total scattered power, W_s, by 4π, the number of steradians in a unit sphere. For both isotropic and non-isotropic scatterers the scattered intensity, I_s, at the distance, r, is written;

$$I_s = I_i \sigma_d/r^2 \tag{7}$$

which is also taken as a definition of σ_d. Note that Equation 7 is based on the assumption that the scattered wave is spherical, but is applied to cases where it is not. This point was discussed in Section I.A. The inverse r dependence of pressure is used even when the scattered wave shows directivity, see Equations 16, 17, 21, and later.

We can obtain the relationship between the two types of scattering cross-section by equating the two expressions for scattered intensity, and obtain;

$$\sigma_t = \int_{4\pi} \sigma_d \, d\Omega \tag{8}$$

or $4\pi\sigma_d$ for an isotropic scatterer. Here, Ω is solid angle.

The above equations hold exactly if the scattering object is a single region which can be regarded as a point having the specified scattering cross-section. (Hence, they apply to the "equivalent" scatterer defined above.) It is more realistic to specify scattering in terms of a coefficient per unit volume, since measurements can only be made over a finite volume of tissue. If the scattering volume contained N discrete and non-interacting scatterers of cross-section, σ_n, we could write

$$\sigma = N\overline{\sigma} = \sum_{n=1}^{N} \sigma_n \tag{9}$$

by simply summing the scattered intensities to find the ensemble average for randomly located scatterers. The conditions under which this summation is allowable are discussed in Chapter 5. For a continuum, under the same conditions as above, we have:

$$\sigma = \int_V \rho_s \sigma_n \, dv \tag{10}$$

for a small volume, dv. Here σ_n is the scattering cross-section per scatterer and ρ_s is the scatterer density of the tissue. This expression defines the scattering cross-section per unit volume, η, or

$$\int_V \eta \, dv = \sigma \qquad \text{or} \qquad \int_V \eta_d \, dv = \sigma_d \tag{11}$$

The total scattering cross-section has the dimensions of an area. The scattering cross-section per unit volume, η, thus has the dimension of inverse length.

Both the total volumetric scattering cross-section and the pressure attenuation coefficient describe the loss of coherent energy from a plane propagating wave and have the same dimensions. Thus the attenuation affects a one-dimensional traveling wave as

$$I = I_0 \epsilon^{-(2\alpha + \eta)x} = I_0 \epsilon^{-2\alpha_t x} \tag{12}$$

The total loss, α_t, includes the scattering contribution and will be written simply as α in subsequent equations.

These cross-section measures of scattering based on intensity are related to the scattered pressures calculated from acoustic theory (see Chapter 4) through a scattering amplitude function. For an incident plane wave in the $\bar{r}$ direction of pressure, p_i, where p_o is the pressure at the origin of the coordinate system, of the form

$$p_i = p_o \epsilon^{-j\bar{k}_i \bar{r}} \tag{13}$$

where $\bar{k}_i = k \cdot \hat{i}$, and a scattered wave which at a large distance, r_2, is spherical, the scattered pressure, p_s, is of the form

$$p_s = p_i f(\hat{s}, \hat{i}) \epsilon^{-jkr_2}/r_2 \tag{14}$$

Here the amplitude function $f(\hat{s}, \hat{i})$ (which is dimensionally a length!) expresses the scattering strength independently of the spherical wave spreading factor. The directions of the $\hat{i}$ and $\hat{s}$ unit vectors and r_2 are specified in Figure 1. The relationship between the "equivalent" differential scattering cross-sections and the scattering amplitude is found by substituting the pressure expressions of Equations 13 and 14 into Equations 7 and 11, to obtain

$$\sigma_d = |f(\hat{s}, \hat{i})|^2 = \int_V \eta_d \, dv \tag{15}$$

The cross-sections are convenient for intensity expressions and the amplitudes for pressure functions.

Measurements of η or σ made under different conditions and for different values of the experimental parameters should give the same result, if the simple assumptions that we have made are true. If they are not true, then a different model for the scattering is called for. A specific example of this would be the case of measuring scattering from a flat, planar surface. Such a surface does not reradiate a spherical wave as will be seen below. The equivalent scattering cross-section measured as defined above will be a function of the measurement distance. Measurements made at different distances would give different values. However, if the scattering strength is expressed as the reflectivity of a planar interface instead, the measured reflectivity will not be a function of distance.

B. INCIDENT WAVE

The field incident upon the region of tissue to be investigated is best found by substitution methods that avoid the absolute calibration difficulty.[4] The incident field can, in principle, be directly measured using calibrated transducers.[5]

The simplest method is substitution, or comparison of the signal scattered from the region under test to the signal from a standard scattering target. Since the incident field can be kept constant during the comparison, it will divide out of the final equations and need not be measured separately, except to find the sample volume. The accuracy is improved since many difficult-to-measure factors are constant for both types of target and divide out in the final equations. This includes the frequency-dependent transfer functions that characterize the transducer and equipment. Because the transducer gain functions depend on the shapes of the scattered waves, and these may differ for the tissue and reflector waves, analytical expressions are needed for these functions so that suitable corrections can be made.[6]

One caveat is necessary here. Since a plane reflector is usually used for standardization, we must derive formulas that correct for the difference in the measured waveshapes of the sample and plane reflector, and approximations can introduce errors. If we had a standardizing phantom that also scattered spherical waves at the measurement distance, there would be no such errors, and the whole measuring procedure would be simplified, particularly the data reduction calculations. The waveshape scattered by the tissue will be spherical if the measurements are made sufficiently distant from the scattering volume, but internal resonances usually complicate the situation if balls are used as standard targets.

Incident wave equations are derived by considering the generation and diffraction properties of a transducer and the propagation properties of the medium. The incident pressure wave can be calculated from the Rayleigh equation, which is a form of the Helmholtz-Huygens integral that has been simplified by using Green's theorem and a special form of Green's function.[1] The pressure equation, including propagation losses, is

$$p_i = \frac{j\omega\rho}{2\pi} \int u_s \frac{\epsilon^{-jk'r}}{r} \, ds \tag{16}$$

The spatial distribution of the surface velocity, u_s, needed to evaluate the integral can be measured, or assumed to be uniform, which appears to be accurate for transducers constructed in the conventional way; that is, by plating electrodes on a ceramic element of uniform thickness, if the diameter is large compared to the wavelength. The field predicted by Equation 16 can be calculated for any region.

The calculated pressure in the far (or focused) field at a range, r_1, from a transducer with area, A, and uniform normal surface velocity, U_o, in attenuating tissue (see Equation 1), can be found from Equation 16, in terms of a normalized range-independent directivity factor, D_t, as

$$p(r_1) = \frac{j\omega\rho A}{2\pi r_1} U_o \epsilon^{-jk'r_1} D_t = B\epsilon^{-\alpha_1 r_1} D_t \epsilon^{-jkr_1}/r_1 \tag{17}$$

The equation defines B, and D_t has the familiar $2J_1(x)/x$ form for a round transducer.[1]

This incident field is usually considered to be a plane wave in most scattering theory, but in reality is a "beam" wave, which includes a directivity that is not a property of a plane wave. Plane waves are of constant amplitude over an infinite lateral extent and the beam wave is not. The beam wave does have an important property of a plane wave, it propagates in essentially one direction. In addition, it has the spherical-wave property of dependence upon the inverse propagation distance in the far field of the radiator.

Since any physically realizable beam wave can be expressed, via the two-dimensional Fourier transform, as an angular spectrum of plane waves (for each single-frequency component of the original wave) the scattering problem can be broken down into a large number of plane wave solutions which can then be added to obtain an accurate result. This procedure is included in the Waag approach covered in this volume (see Chapter 8).

Over a scattering region of limited axial and lateral extent, compared to the total distance from the transducer, the variation of the actual field from a plane wave is ignored here, and plane wave illumination is assumed. In a small volume centered at a distance r_1, where Equation 17 applies, the incident pressure is very nearly

$$p(\bar{r}_0) = |p(r_1)|\epsilon^{-j\bar{k}_i \cdot \bar{r}_0} \qquad (18)$$

where $\bar{r}_0$ is a vector specifying position within the scattering volume (see Figure 1). Note the change in the origin of coordinates.

C. SCATTERED WAVE

Calculation of the scattered wave proceeds in a manner that is similar to that above when considering the generation of the incident field. In this case, however, the scattering volume, Figure 1, acts as a three-dimensional source.[1]

In this case, the situation is much more complicated than for the incident wave. In the space being interrogated by an acoustic wave, we should find not only the wave originally radiated by the source transducer, but also the waves scattered by all inhomogeneities in the space. Generally we will have tissue overlying and attached to the sample volume being measured. The field at any point will be the incident field plus all of the waves scattered from other points. We have the quandary that we can find the field anywhere only if we know it everywhere else!

To allow the problem to be solved, we consider the scattering to arise from only one region, the sample volume. This approach also assumes that the scattering from the sample is not re-scattered by other regions or, if it is, that this secondary scattering is so weak that it may be neglected in comparison to the other scattering. This case is called single scattering.

We can look on scattering as a phenomenon that we wish to measure without concern for the physical details of the scattering process. Then the previous descriptions of the scattering strength, as given in Equations 4 through 15, can be used. The scattered pressure can be expressed in the form of Equation 14, and used with the other equations derived below in the measurement process. This will give numerical scattering measures in terms of an "equivalent" scattering function.

To relate the wave scattered from a region of inhomogeneities to the values of the material constants requires further derivation. The equations below do this for a simple fluid model. In either case the scattered field is written as an integral over the scattering volume, defined as the region in space in which the measurement system can detect scattered echoes. This volume is bounded by the borders of the incident beam wave, the sensitive region of the receiving transducer beam pattern, the transmitted pulse duration, and by any physical boundaries provided by the sample under test or its holder.

The variations in acoustic properties that lead to scattering have been described in several different ways. We use a simple fluid model in which the medium has variations in compressibility, κ, and density, ρ. The normalized variations are expressed by

$$\gamma_\kappa = \frac{\kappa_e - \kappa}{\kappa} \quad \text{and} \quad \gamma_\rho = \frac{\rho_e - \rho}{\rho_e} \tag{19}$$

where the subscripted values are those found in the scattering region, and no subscript in the surrounding medium.[7] Similar formulations can be written using elastic constants of isotropic solids.[8] If the above ratios are allowed to have non-zero means they can apply both to continuous and to discrete variations of properties inside the sample volume.[7,9] In the case of randomly distributed discrete scatterers or a random variation in medium properties the analysis must be extended to include statistical descriptions of the randomness. The scattering sources then can be formulated in terms of correlation functions.[9,10]

To find the scattered pressure we need a general solution of the Helmholtz equation, which involves knowing both the magnitude of the field and its spatial gradients at the scatterers.[1] Gradients can arise from the incident fields. Fields that are sharply focused in the axial direction, for example, have been suggested as being useful to emphasizing scattering, particularly from γ_ρ.[10] To date, however, smoothly varying fields have been used to make measurements. We will see that field gradients resulting from time gating and lateral beam diffraction are not sharp, even for rectangular gates. These gradual changes should prevent strong gradient waves.[12]

We will use the gradient operator for plane, continuous waves, which is simply a multiplication by $-jk$. The scattering function then becomes;[1,7]

$$\Gamma = \gamma_\kappa + \gamma_\rho \cos \theta \tag{20}$$

where θ is the scattering angle (see Figure 1). The pressure field, p_s, at a large distance, r_2, from the scattering region is found by integrating the contributions from within the sample volume.[1] The value of Γ is allowed to vary within the region. This yields

$$p_s = \frac{k^2 \epsilon^{-jkr_2}}{4\pi r_2} \int_V \Gamma(\bar{r}_0) p(\bar{r}_0) \epsilon^{j\bar{k}_s \cdot \bar{r}_0} \, dv \qquad (21)$$

where the vector, $\bar{r}_0$, sweeps out the sample volume, v, in Figure 1. The phase factor results from the varying distance between the differential volume, dv, and the observation point, calculated from the coordinate center in the direction of the receiving transducer. We see from Equations 20 and 21 that the simple fluid model predicts that the scattered wave will contain both a dipole term resulting from density changes and a monopole term from compressibility changes.

Under the Born approximation the pressure at $\bar{r}_0$ is assumed to be the incident pressure, and this is given by Equation 18 for the phase reference located at the origin of coordinates. Then, the general integral equation for the scattered field at a distant point, r_2, in tissue of attenuation coefficient α_2 is

$$p_s = \frac{k^2 B \epsilon^{-(\alpha_1 r_1 + \alpha_2 r_2)}}{4\pi r_1 r_2} \int_V D_t \Gamma(\bar{r}_0) \epsilon^{j(\bar{k}_s - \bar{k}_i) \cdot \bar{r}_0} \, dv \qquad (22)$$

The incident pressure is a function of the position vector, $\bar{r}_0$, inside the scattering volume through D_t.

The total effective scattering source function depends upon the incident field at each scatterer. Since this field includes the scattering from all other scatterers in the sample volume, we have used the Born approximation, and assumed that the incident field is only that due to the transmitting transducer in the absence of the scatterers. We also see that the volume integral will be a function both of the spatial extent of the scattering structures and of the incident field.

If Equation 22 applies to a scattering region so small that D_t is a constant within it, further simplification is possible. In this case, Equation 22 has the form of an inverse three-dimensional Fourier transform of the variations in material properties in the sample volume, evaluated in the $\overline{K} = \bar{k}_s - \bar{k}_i$ direction. This direction is halfway between the directions to the source and to the receiver, in the plane defined by the two transducer axes (Figure 1).

In the case of small sample volume, comparison of Equation 22 to Equations 14 and 15 shows that the various scattering measures are all related by

$$f(\hat{s}, \hat{i}) = \frac{k^2}{4\pi} \int_V \Gamma(\bar{r}_0) \, dv = \left[\int_V \eta_d \, dv \right]^{1/2} \qquad (23)$$

from which we can convert between the calculated Γ's and the measured scattering coefficients.

It appears that measurements need not be restricted to the far (or focused) field region of the incident wave, and we could use a field region closer to the transducer, since the general expression for such an incident field, Equation 16, can be evaluated. The resulting field could vary strongly over the region of the scatterers, even in the Born approximation of neglecting the scattered fields, and produce local amplitude and phase variations inside the volume integral. The resulting scattered field could be very complex, particularly if the receiving transducer is also close to the scattering volume. Of course, the field integration can be carried out, but only if the geometry is known. Accurate measurements would require that the scattered field be scanned point by point.

The gated received voltage, $V_r(\omega)$, is found by integrating Equation 22 over the area of the receiving transducer; this gives

$$V_s(\omega) = \int_S S_s p_s \, ds$$

$$= \frac{k^2 B \epsilon^{-(\alpha_1 r_1 + \alpha_2 r_2)}}{4\pi r_1} \int_S S_s \frac{\epsilon^{-jkr_2}}{r_2} \int_V D_t \Gamma(\bar{r}_0) G(\bar{r}_0) \epsilon^{j\bar{K}\cdot\bar{r}_0} \, (d\bar{r}_0)^3 \, ds \quad (24)$$

where S_s is the voltage per unit force sensitivity conversion factor of the transducer, (or of the whole receiver; in the substitution method there is no difference) and $G(\bar{r}_0)$ represents the effect of time gating inside the receiver. $G(\bar{r}_0)$ expresses the effects of time gating as a spatial function for inclusion in the frequency domain expressions and will be derived below.

If the scattering volume is in the far (or focused) field of the receiving aperture $G(\bar{r}_0)$ and $\bar{K}\cdot\bar{r}_0$ are independent of ds so the order of integration can be changed to find

$$V_s(\omega) = \frac{k^2 B \epsilon^{-(\alpha_1 r_1 + \alpha_2 r_2)}}{4\pi r_1 r_2} \int_V \left[D_t \Gamma(\bar{r}_0) G(\bar{r}_0) \epsilon^{j\bar{K}\cdot\bar{r}_0} \int_S S_s \epsilon^{-jkr_2} \, ds \right] (d\bar{r}_0)^3 \quad (25)$$

The surface integral can be found, by comparison to Equation 16, to be equal to $S_0 A_r D_r$, for uniform sensitivity, S_0, area, A_r, and normalized directivity function, D_r, all referring to the receiver transducer.

The mean square magnitude of the received voltage is needed for dealing with the probably random function, $\Gamma(\bar{r}_0)$. This is

$$\overline{V_s(\omega) V_s^*(\omega)} = \frac{k^4 S_0^2 A_r^2 B^2 \epsilon^{-2(\alpha_1 r_1 + \alpha_2 r_2)}}{16\pi^2 r_1^2 r_2^2} \int_V D_t^2 D_r^2 |\Gamma(\bar{r}_0)|^2 G^2(\bar{r}_0)(d\bar{r}_0)^3 \quad (26)$$

where the * indicates taking the complex conjugate. This result shows that $\Gamma(\bar{r}_0)$ cannot be directly measured since it appears in the integrand, accompanied by a number of spatial weighting factors.

However, the mean value can be found. Since we know that the intensities simply add for uncorrelated random scatterers, we can treat the problem as being linear in a power scattering coefficient, and find the mean coefficient. The value of Γ in terms of η_d is found from Equation 23. The mean value of η_d is then

$$\overline{\eta_d} = \frac{\int_V D_i^2 D_r^2 \eta_d(\bar{r}_0) G^2(\bar{r}_0)(d\bar{r}_0)^3}{\int_V D_i^2 D_r^2 G^2(\bar{r}_0)(d\bar{r}_0)^3} \tag{27}$$

which from Equation 26 is

$$\overline{\eta_d} = \frac{\overline{V_s(\omega)V_s^*(\omega)}\; r_1^2 r_2^2}{S_0^2 A_r^2 B^2 \epsilon^{-2(\alpha_1 r_1 + \alpha_2 r_2)} V'} \tag{28}$$

Here V', the weighted volume integral, is the denominator of Equation 27, which must be calculated from beam shape and other measurements. Equation 28 is the desired result for measuring scattering. The only terms in Equation 28 which are difficult to find are the absolute values of the incident pressure, B/r_1, and the receiving sensitivity S_0. These values can be most easily evaluated by a substitution measurement using a known reflecting target. This procedure and the calculation of the volume integral will be discussed in Section II.D.

There are pitfalls in assuming spherical symmetry of the scattered wave, which was implicit in the derivation of the equations above. Because of the transform relationship mentioned, such symmetry will follow only if the weighted scattering volume has spherical symmetry. It would be unusual to have spherical symmetry of V' since it is formed of the intersection of two radially shaded, but roughly cylindrical, beams and a time gate. This "source" of scattered waves may easily have a directivity function that can influence the magnitude of scattering.

The possible existence of such scattering directivity functions was initially not appreciated and the resulting influence of the receiver aperture size on the received scattered signal amplitude, in computer simulations, was interpreted as being due to a phase cancellation artifact caused by phase shifts between different field regions at a pressure averaging receiving transducer. Such a phase cancellation effect should be present, however, only for large phase sensitive apertures, such as piezoelectric transducers, that respond to the average pressure field. It should not be present for intensity-sensing transducers, except for the minor effect of inverse-square spreading. Results of computer simulation of scattering from cubical volumes show, however, that the average scattered field drops off for increasing receiver size with either type of transducer.[13] Phase cancellation does exist since the output of the pressure-averaging transducer per unit area does drop off more rapidly than the output of the intensity-sensing transducer. This effect should be small if the receiving aperture is no larger than that of the transmitter. It can be shown that this condition puts the scattering volume in the far field of the receiving aperture.

Some caution in future experiments is warranted to avoid or to assess coherent effects such as beam formation. That even randomly distributed sources can produce directive fields was found by Wolf.[14] This effect is the basis for the design of sparse and randomly filled apertures as directive radar antenna systems. There are two reasons why directive, coherent waves can be generated from a region of randomly positioned scatterers or randomly varying elastic properties:

1. Values of mean compressibility and density which differ from those of the medium in which the scatterers are embedded will give the scattering region the appearance of having a different acoustical impedance. The whole region then also will act as a single scatterer.
2. The scatterers which are next to a boundary will have a constant phase shift relative to each other. This scattering will be the same from ensemble to ensemble and will add coherently when determining the mean scattered power.

The above restrictions may be removed in future work, since it is possible to use a small probe to measure the scattered field on a point by point basis. Because of the loss of signal that results, such measurements are not always practical. These measurements may be made to obtain local amplitude and phase information for calculating the true scattered wave or for inverting the wave equation.[2]

A second consequence of non-spherical scattering volumes is that the region of tissue being interrogated in an angular scattering experiment would change as a function of angle. Changes in volume could change the angular dependence. If different tissue types were included as the volume changed, the mean scattering could also change. To avoid this effect the scattering volume can be made to have spherical symmetry.[15]

D. EVALUATION AND STANDARDIZATION

Each particular experiment to find a value of η_d from Equation 28 requires a choice of parameters based on the situation, facilities, and required accuracy. For each choice of parameters the terms of Equation 28 must be evaluated. The acoustical parameters will be discussed in this section, and the electronic signal processing in Section III.

1. Substitution Calibration

Although the unattenuated incident pressure amplitude given by B/r_2 in Equation 28 can be measured separately, it is better to use a method that uses the same transducer and circuits that were used to measure the scattered waves. Then the frequency response of the transducer, including that of its impedance and receiver input network, plus that of the rest of the receiver will affect both measurements in the same way. That is, these factors will divide out of Equation 28. In addition, any changes in amplitude and frequency content introduced by the properties of the detector or receiver circuits can also be canceled to a first order, if the two signal levels are kept comparable by using a calibrated attenuator, see Section III.

The formulas for the substitution method are derived by considering the same steps used in the previous section, for whatever reference reflector is used. Although small ball reflectors are convenient, the calibration of their absolute reflectivity is a major measurement problem. A planar interface has the advantage that its reflectivity can be calculated from knowledge of its acoustic impedance. The wave reflected from the plane interface is like the wave reflected from a mirror; it has been redirected and reduced by the pressure reflection coefficient, R, of the reflector. The pressure incident upon the receiving transducer after reflection from the plane interface is found from Equation 17 as

$$p_c = BD_r R \, \epsilon^{2jk(r_1 + r_2)}/(r_1 + r_2) \qquad (29)$$

Here we have assumed no attenuation in the calibrating liquid.

The output voltage from an unfocused transducer is found by integrating the pressure given by Equation 29 over the face of the receiving transducer following the procedure used to derive Equation 24

$$V_c(\omega) = \int_S S_s p_c \, ds \qquad (30)$$

This integral can be evaluated exactly or approximately as we did for Equation 29.

However, approximate methods of standardization of *focused* apertures by the substitution method apparently have not been rigorously justified, and differences are found if the equations above for unfocused transducers are used.[16]

If the reflecting plane used to calibrate focused transducers is placed at the same position as the sample volume, then the reflected field to be integrated over the focused receiver is the field at twice the focal length. Focused fields are very wide beyond the focal length, and the field structure over the receiving aperture may have large amplitude and phase variations. There are several ways to proceed. A diffusing reflector that scatters only spherical waves could be used at this position. A formalism appropriate to this case shows that the scattered field, which, in this case appears to originate from the reflector surface, is collected by the focused transducer without any phase cancellation.[12]

Computation of the calibration field by exact solution of Equation 16 for round, uniformly vibrating transducers is well known.[17] A similar calculation for focused apertures would be useful in allowing suitable approximations to be used with this method. Otherwise, Madsen's exact method, which carries out this calculation, must be used.[18] This is discussed in Chapter 7.

A useful approximation for unfocused transducers is to assume that S_s is constant in Equation 30. This is reasonable since the receiving transducer in this calibration experiment intercepts one fourth the area of the beam that was intercepted by the scattering volume when it was at the position of the reflector. Errors due to assuming uniform receiver transducer sensitivity will thus be less than errors from assuming constant incident beam pressure in evaluating the weighted volume integral.

The final result is now

$$\overline{\eta_d} = \frac{\overline{V_s V_s^*} \; r_1^2 r_2^2 R^2 A_r^2}{\overline{V_c V_c^*} \; (r_1 + r_2) A_t^2 V' \, \epsilon^{-2(\alpha_1 r_1 + \alpha_2 r_2)}} \tag{31}$$

since, from Equations 29 and 30

$$V_c = \frac{S_0 B R A_r}{r_1 + r_2} \tag{32}$$

if $D_t = 1$. We can assume that further equipment signal processing will not affect the ratio of the mean square transducer voltages used above if a calibrated attenuator is used to equalize the magnitudes of the two voltages at the input to the receiver.

To measure the scattering of a planar interface only Equation 32 is needed. The ratio of an unkown reflection coefficient to a known coefficient is simply the ratio of the reflected signal voltage from the unknown to that of the known. The comparison should be made at the same range so that any approximations to the integral of Equation 30 do not affect the result. Since a perfect reflector has $R = 1$, the reflectivity of interfaces is frequently reported as 20 times the logarithm of the above voltage ratio, corrected if the reference $R \neq 1$, as "decibels less than a perfect reflector". The scattering measures derived for scattering of spherical waves are not appropriate for a plane.

2. Weighted Volume Integral

This integral

$$V' = \int_V D_t^2 D_r^2 G^2(\overline{r}_0)(d\overline{r}_0)^3 \tag{33}$$

defines the effective scattering volume. The limits of integration are the boundaries of the physical region of scattering tissue, assuming $\eta_d = 0$ outside this region. *In vivo* this is not true, and the integral is over all space where the integrand is not zero. The integrand is the product of three dimensionless weighting factors, each of which can be taken to be effectively zero some distance from the center of the intersection of the transmitter and receiver beams. The volume may be visualized as the common volume of these two beams, a weighting factor that depends upon a time gate in the receiver, and the physical extent of the tissues. The value of Equation 33 can change as the gate delay is changed, or as the angle between the transducers is changed to measure angular scattering. The changes with angle can be minimized by making the shape of the volume roughly spherical, that is, by mutually centering the beam directivity patterns, gated volume, and sample. The time gate or transmitted pulse can be shaped to accomplish this.[15]

The exact value of Equation 33 can be calculated from analytical expressions for $G(\bar{r}_0)$ given by Campbell and Waag[19] (see Chapter 8). The first description of the gating effect given here will rely on physical arguments. The directivity functions are oscillatory for continuous waves, but are more nearly Gaussian for wideband pulses, and the appropriate functions must be considered for exact results. The simplest approximation is to take the acoustic beams as cylinders of constant intensity. The intersection volume of two equal diameter cylinders is inversely proportional to the sine of the angle between their axes.[20] For small scattering angles or for backscattering either the sample size or time gate will limit the volume.

The diameter of the constant intensity region that approximates an actual beam is commonly taken as the width at the -3 dB or the -6 dB points. It may be more justifiable to use the "equivalent width", the width of that beam of constant intensity, equal to the axial intensity, which transmits the same total power as the real beam. This width is the width between the 4.8 dB down points in the far, or focused, field of continuous-wave beam from a round transducer.[1]

A suitably delayed time gate is used in the receiver to define the tissue region being measured. It is necessary in backscattering for this reason, as well as to avoid measuring echoes from tissue boundaries such as the walls of sample containers. In angular scattering a time gate avoids the increase in volume as the transmitted and receiver beams become coaxial near the back-scattering condition.

The factor $G(\bar{r}_0)$, in Equation 33, was introduced to account for the effects of time gating and transmitted pulse length, which interact. The effects of these time functions must initially be considered in the time domain since the time delays are all the position information that the system has. The effects in the space domain must be deduced from the time domain behavior. We cannot simply assume, for example, that rectangular pulses and gates limit the integrand of Equation 33 to rectangular regions in the spatial domain. The angular scattering problem is simplified by the fact that the scattering vector axis, the $\bar{K}$-direction in Figure 1, is the axis of equal time delays for any

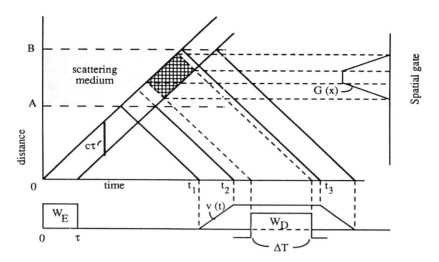

FIGURE 2. Distance-time graph of position of an incident pulse and the scattering from a region located between distances A and B. The received echo vs. time is shown at v(t). The spatial gate, G(x), results from the operation of a receiver time gate, W_D, on v(t).

scattering angle. That is, the phase factors in Equation 22 show that the phase shifts to any point specified by the vector $\bar{r}_0$, relative to the center of the coordinate system, are given by the projection of $\bar{r}_0$ on $\overline{K}$. Therefore, any angular scattering experiment can, in the time domain, be considered to be backscattering along the $\overline{K}$ vector axis. Note that this axis, if the transmitter and receiver beams are equal in size, runs through the center of the scattering volume and midway between the above two beams, making the problem nicely symmetrical.

The interaction of the transmitted pulse and the receiver gate in backscattering are straightforward to determine.[4,19] Consider the graph of Figure 2, which shows the range on the vertical axis and time on the horizontal. Straight lines with slope equal to the velocity show the range of any signal emitted at a time equal to the horizontal coordinate of the origin of the straight line. The transmitted pulse has envelope, w_E, of length, τ, and propagates outward through the medium, as shown starting at the origin. When the leading edge of the pulse reaches the scattering volume at range A, a portion of the incident wave is reflected back to the receiver. We assume that the space between A and B is uniformly filled with scatterers so that a scattered wave is returned continuously from this region.

The reflected pulse leading edge then follows the additional straight line of negative slope arriving at the receiver at time t_1. During this time the transmitted pulse continues outward, sending scattered waves back to the receiver from any part of the pulse that is between the distances marked A and B. At any one time the scattered waves are summed in the receiver from

ference effects. The resulting variations in scattering coefficient with frequency may obscure general trends. The interfering variations can be removed by integrating the data suitably. There are two methods that have been used to perform the calculation of the frequency domain integrated backscatter. In the first, the attenuation-corrected squared power scattering coefficient integrated over frequency is calculated by Rayleigh's theorem as the squared time domain value integrated over time. This is normalized as before by dividing by the integrated pulse from a plane reflector.[23] This is quite a simple and rapid procedure, but new time domain equations must be used to make the corrections outlined in this chapter.

In the second method the data are measured as a function of frequency and the averaging done in the frequency domain.[24] This approach allows direct comparison of the raw and integrated data, and these frequency domain correction equations can be used. The fluctuations can also be reduced by averaging the data in the spatial domain. The sample volume is viewed from many directions and the ensemble average of the scattering measures is calculated.[25] The result is a curve of average scattering coefficient vs. frequency that is smoother than the individual curves.

The ability of the measuring system to determine scattering data as a function of frequency accurately depends on the spectrum analyzer and on the time gate used in the receiver. These factors will be discussed separately. Analog spectrum analyzers have an analyzing bandwidth that requires a minimum analysis time per frequency that is of the order of the reciprocal of this bandwidth. They average the data over the same frequency range. To use these instruments effectively, the user must become familiar with the system and understand the instruction book procedures.

Fourier transform analyzers that use digital calculation of spectra use a sampling rate that similarly limits the operating speed in inverse proportion to the maximum desired frequency. When either type of spectrum analysis is used we need a record length sufficient to allow a spectrum analysis to be made. Hence, the receiver gate operates as a "window" and determines both the frequency and spatial resolution of the system.

The equations must be put into the proper general form to allow the frequency and time domain processing and signal formation properties of the system to be correctly expressed over the full bandwidth. This has been done for the backscattering case.[26] The modifications necessary can be outlined using the previous development. The frequency domain form of the transmitted signal given by Equation 38 is using capital letters for Fourier transforms of the time functions denoted by corresponding lower case letters, and * for convolution[22]

$$P(\omega) = W(\omega) * \delta(f \pm f_0) = B(\omega)/r_1 \qquad (39)$$

where δ is the impulse function. This expression can be inserted into Equation 25, which is a frequency domain transfer function, to give the system response after doing the necessary integrations to define the scattering volume. Since scattering is a linear process, and $V_s(\omega)$ is already a frequency domain quantity, the overall response can be put into the general frequency domain form

$$V_s = \overline{\eta_d}(\omega)P(\omega)F_1(\omega) \tag{40}$$

where $F_1(\omega)$ represents the system transfer function. The receiver time-gated output voltage is

$$v_o(t) = v_s(t)\omega_D(\tau) \tag{41}$$

which becomes, in the frequency domain

$$V_o(\omega) = [\overline{\eta_d}(\omega)P(\omega)F_1(\omega)] * W_D(\omega) \tag{42}$$

The convolution with the gating waveform shown above distorts the received spectrum, as was noted by Chivers.[27]

The normalization is carried out by dividing the output voltage spectrum by the transmitted spectrum obtained by analysis of the echo signal from the reference reflector discussed above.[12] By the same steps outlined above this signal is

$$V_c(\omega) = [RP(\omega)F_2(\omega)] * W_D(\omega) \tag{43}$$

Even if the system transfer functions F_1 and F_2 are known, using these $V(\omega)$s will give the true value of η_d only if η_d is a constant, independent of frequency. In general, to remove the convolution effect completely the echo signal must be deconvolved, or the effect can be included in the equations.[26]

Why then has the division by the reference spectrum given good results? One explanation is that the convolution in Equation 43 will not appear if a rectangular time gate only a little longer than the transmitted envelope is used. This gate does not affect the reference signal from a planar reflector at all. The gate in Equation 42 can similarly be arranged to have minimal effect by recognizing that the gating function is just a "window" in Fourier transform terms. Standard considerations for picking the window functions apply. This procedure has generally been followed.[12]

B. TRANSDUCERS

The transducers and their associated coupling networks must have adequate sensitivity so that weak echoes can be measured. System losses can be reduced considerably by placing matching networks at the transducer if the manufacturer does not supply the transducer already matched to a common cable impedance. If this is done then any length cable can be used for system interconnections provided that it has the proper characteristic impedance. Using a common impedance allows transmitter, receiver, and transducers to

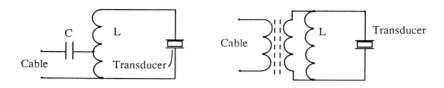

FIGURE 3. Impedance matching circuits for connecting transducers to low impedance cables (see text).

be interchanged at will to cover different situations. This also minimizes stray signal pickup since there are no voltage gradients or impedance unbalances along the length of the cables.

Two possible coupling circuits are shown in Figure 3. The inductance L is placed in parallel with the transducer to tune the static capacitance to parallel resonance at the band center frequency of the transducer. The transfer impedance of thickness resonant transducers is symmetrical if the mechanical resonance frequency is taken as the center frequency, as may be deduced from standard equivalent circuits.[28]

Since the tuned impedance of the transducer is unlikely to be at the same impedance level as available cables, to match impedances for maximum power transfer we must either tap down the impedance as shown at the top of Figure 3, if the transducer impedance is too large, or step up the impedance with a transformer, as shown on the bottom of Figure 3, if it is too small. It may be necessary to account for stray impedances. For example, the capacitance, C, shown in series with the cable at the top of Figure 3 is chosen to series resonate the leakage inductance of the tapped coil at the band center frequency.

The transformer shown at the bottom is a wide-band unit that uses a transmission-line winding on a toroidal core. Such transformers are available, usually, in a 4:1 impedance ratio and a variety of impedance levels. Frequency response is uniform between a few hundred kiloHertz to over 100 MHz. Such transformers can be used when the terminals are not exactly matched to the design impedance but the bandwidth will be less. This is readily checked with standard laboratory instruments. The value of L may have to be increased to compensate for the parallel-connected secondary inductance of the transformer.

C. ELECTRONICS

An overall block diagram of complete amplitude and phase measuring system is shown in Figure 4, with details of the important interconnections to the transducer in Figure 5. Either one or two transducers are used, depending on whether a backscattering or angular scattering measurement is to be made. The system (Figure 4) produces a gated sinusoid of width τ at a pulse repetition frequency (p.r.f.). The transducer used as a receiver is connected through a protected amplifier to a controlled gate which determines that part of the

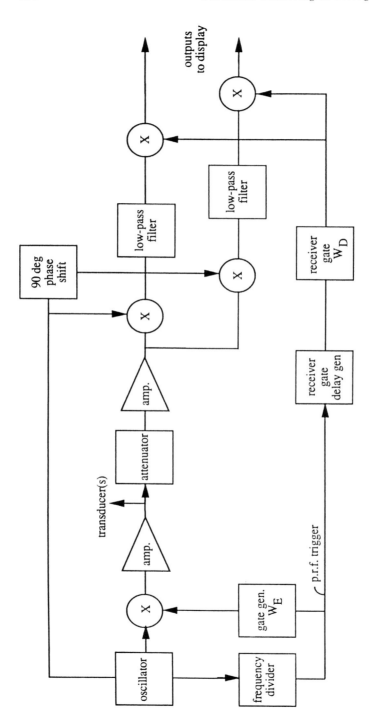

FIGURE 4. Block diagram of amplitude and phase measuring system for measuring scattering.

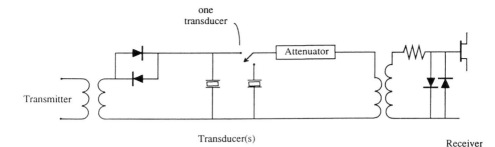

FIGURE 5. Detail of interconnection between transmitter output, on the left, the transducer(s) and receiver input circuits, right. The transmitter output is fed through disconnect diodes to eliminate low-level noise in the receiver. The receiver input has an attenuator terminated on one end (see text) for measuring ratios of scattered signals. Receiver protection diodes prevent damage to the input circuits when the transmitter and receiver are directly connected together in back-scattering measurements.

signal which is to be measured. Phase measuring features are described in Section III.C.3.

1. Transmitter

The transmitter should be fitted with a variable pulse repetition frequency generator to allow identification of second time around echoes, see Section III.D. The transmitter may generate either a sharp impulse by using avalanche, or similar, transistors driven by the p.r.f. trigger (Figure 4) or the conventional gated continuous-wave oscillator shown.

It is convenient to be able to vary the duration of the transmitted signal, a procedure that has been suggested as a means of separating the coherent and incoherent scattering contributions.[26] Since the coherent signal arises from a constant-width region near boundaries (if the field has no sharp gradients) its relative contribution to the total signal will decrease as the transmitter pulse duration is increased.

It is important that the transmitter be effectively disconnected from the receiver during the receive period. Otherwise, any random noise from the transmitter will be fed directly into the receiver input. Isolation can be accomplished either by designing the transmitter output stage to have no conduction current when not transmitting, or by the inclusion of parallel diodes in series with the transmitter output as shown in Figure 5. Silicon signal diodes are used because their conduction threshold is sufficiently high that they are an open circuit for echo signals, and their capacitance is low.

2. Receiver and Attenuator

To allow accurate measurement combined with high dynamic range the low-noise receiver radio frequency amplifier is preceded by an adjustable attenuator. This is necessary because of the wide dynamic range of signals encountered in most measurement situations. The most accurate reference reflectors are blocks of stainless steel or polished plate glass, so that the reflection coefficients can be accurately calculated from handbook values of wave velocity and density. Since the reference signals from these reflectors will be much larger than tissue echoes, it is necessary that the receiver not be desensitized by these strong signals. Any amplifying stages that precede the attenuator are subject to overload so they should be avoided if possible.

Note from Equation 31 that it is only the ratio of the mean square signal voltage from the tissue to the mean square voltage from the reflector that must be determined. The attenuator is adjusted for equal receiver output indications from the two signals, and the ratio read from the resulting change in attenuator readings. The simplest output indicator is a wideband true r.m.s. voltmeter; or, in a digital system, a program that calculates the same quantity. For the equations to be correct the time mean of the r.m.s. signals is required.

The requirements of efficient signal transfer to an input low-noise amplifier and the provision for the input attenuator are somewhat conflicting in design of a receiver front end, because of the low impedance of the usual switched Tee or Pi attenuators. The input circuit shown in Figure 5 illustrates a solution to the problem. The attenuator of the usual switched resistor type is accurate only if terminated in its characteristic impedance at both ends.[29] Fortunately, *changes* in attenuation will be read correctly if the attenuator is terminated on only one end. Because the transducer impedance is frequency sensitive, the termination resistance at the left end (Figure 5) may be inaccurate at the band edges. Therefore, it is wise to terminate the attenuator correctly, at the right end (Figure 5) by the receiver input impedance. With judicious design, including use of wideband transformers, this input impedance can have the usual 50 to 90 ohm impedance level of these attenuators.

Large-signal overload arises from the transmitted signal when using one transducer for backscattering and from large echo signals in other cases. Should the DC levels of any of the stages shift because of overload, the recovery time constant of most biasing and decoupling circuits is sufficiently long to make the receiver gain a function of time thereafter. The transformed output impedance of the attenuator in conjunction with a pair of low-threshold voltage (or Shottky) diodes can limit the transmitter voltage applied to the first stage of the receiver as shown in Figure 5. Because even this limited voltage is amplified in later stages, and the greatest voltage swing occurs at the receiver output, additional overload protection may be required all the way through the receiver. Alternatively, response down to zero frequency can reduce the recovery problem since the DC offset will not persist. In laboratory situations the water path delay may be sufficient to allow recovery from transmitter overload before the tissue signals arrive.

Protection from signals that are smaller than those discussed above but that can still overload receiver stages is also required. Tissues are heterogeneous and exhibit strong reflections from some smooth interfaces. Other sources of strong signals are reflections from the sample holders. These require a ''hard'' (abrupt transfer function) limiter in the receiver preceding the earliest receiver stage which overloads just above the full-scale output signal. It is always wise to check for overload effects by changing the p.r.f., which will change the receiver output voltage if the transmitter output remains constant.

3. Phase Measurement

To measure phase some form of coherence must be established between the transmitted and received signals. Accurate measurement of phase is essential if the small probe method of field mapping is to be used in connection with diffraction tomographic inversion methods of determining material properties of the tissue, or if Fourier transform techniques are to be used for wideband spectrum analysis.

The simplest system conceptually is a wide band analog to digital converter following the input circuits of Figure 5. Digital timing is facilitated if a delayed pulse from the PRF generator can be added to the input signal at some convenient time following the transmitted pulse and before reception of the tissue echoes. Digitization at a rate sufficient to sample the highest frequency in the transmitted spectrum, 5 to 10 times per cycle, will allow accurate measurements. A disadvantage of this radio-frequency system is the large amount of data that is generated by the high digitization rates. These data must generally be stored in a buffer memory to allow transfer to the main computer to be accomplished at a slower rate. Buffers in many available digitizers have an insufficient capacity for our use.

If the signal is heterodyned by mixing with the transmitter frequency and then low-pass filtered before digitization, a much lower frequency digitizer can be used, since only the modulation frequencies must be digitized. The data storage requirement therefore is less severe. The general scheme is shown in Figure 4. A reference signal at the center frequency of the transmitted pulse is fed in phase quadrature to two mixers and the low frequency difference signal is gated and then digitized in two channels. To perform spectrum analysis, the two channels are considered to be the real and imaginary parts of the input signal and used as input to a Fourier transform program.

The fully phase-coherent gated oscillator system shown in Figure 4 derives the receiver reference and the center frequency of the transmitted pulse from the same stable oscillator. This is the configuration used in pulse Doppler systems. The best spatial resolution and the maximum bandwidth for frequency analysis is obtained by impulse excitation of the transducer directly from the pulse repetition frequency trigger (Figure 4). Since the transmitted signal in this wideband system will have a spectrum of frequencies that are integer multiples of the p.r.f., one spectral component will have the same frequency as the receiver reference oscillator, and the system will still be coherent. The oscillator frequency must be adjusted to coincide with the center frequency of the transducer in either system.

D. TANK AND TISSUE SYSTEM

The scattering measurement equipment must include a holding tank for the water bath and transducer and sample mounting equipment. There is a great deal of freedom of choice in assembling this part of the equipment and it will not be discussed here in detail. A particular requirement when dealing with biological tissue is that of providing an environment which will not damage the tissue nor invalidate the experiments. Tissue mounting and de-mounting should be as convenient and flexible as possible, particularly if they are to be measured at or near body temperature (37°C) since tissues can deteriorate rapidly. Tissues should not be in contact with water but with a solution whose osmotic pressure is balanced to be close to that of the tissue, such as normal saline or Ringer's solution. These fluids can be contained in thin plastic bags and need only occupy the space close to the sample.

Contamination of the data must be avoided since the echo signals to be measured are generally extremely weak. The major sources of contamination by unwanted acoustic signals are reflections from the tank walls or water surface, the mounting apparatus, or multiple reflections between the sample and transducer faces. These signals can be reduced by the judicious application of absorbing material. Most low-impedance (those with minimal filling material) neoprene rubbers sold for other uses are convenient to use in the megaHertz frequency range and do not shed small fibers or particles into the tank. For lower frequencies the curled hair or plastic furnace-filter absorber is effective, particularly if filled with a viscous oil. To prevent oil leakage, enclosing the material within double plastic bags is required, as is eliminating air bubbles by heating the absorber before sealing the bags.

Checks must always be made for the presence of false or multiple echoes even if absorbers are installed. Any remaining highly reflective interfaces, such as the water surface, can sometimes produce pulses that come back several pulse repetition periods after their source pulse. These can occur at the same time as the echoes to be measured. The most reliable way of identifying such second-return echoes is to make a small change in the p.r.f. Any echoes due to these multiple-period returns will appear to shift their position in space if viewed on a time-base oscilloscope display, whereas true echoes will stay stationary. It is good practice to eliminate the multiple echoes rather than to select a sample and gate position free of them, as a protection against less wary workers in the future.

The apparatus should guard against contamination of the tank by corrosion of the apparatus which would also upset the smooth operation of mechanical components, particularly when doing angle scattering measurements. Many tap waters are corrosive and will attack everything except stainless steel or heavily anodized aluminum. The solutions recommended for use with tissues have a high salt and other electrolyte content and must be kept out of contact with the apparatus. It may be necessary to condition the water by adjusting the pH and filtering out particles, particularly if the source is ''hard''. A small

Another approximation common to current methods is that, corresponding to the echo signal at the time t, there exist two planar surfaces perpendicular to the axis of the transducer such that each scatterer lying between these planes contributes to the echo signal and all other scatterers (not lying between these planes) do not contribute. All previous data reduction schemes in the literature make this assumption. Consideration of scatterers far enough from the beam axis should convince the reader that this is not really true. Two such bounding surfaces will exist, but they are not planes; each surface likely has a shape lying between a plane and a spherical surface, the latter having its center at the center of the transducer face.

Because the gate function and the echo waveform are multiplied together, we can expect that the Fourier transform of the gated echo waveform will have contributions from a continuous band of frequencies involving both the pulse emitted and the gate function (the two functions are convolved). The presence of this convolution has been recognized by Chivers and Hill.[11] This situation is emphasized in importance, the broader the frequency bands of the pulse and gate.

In this chapter a previously published method of data reduction[12] is reviewed which avoids the restrictions and uncertainties of the above methods. It can be applied for any combination of focused or nonfocused and narrow or broadband transducers. In addition, the region interrogated is not limited to the far field or (in the case of a focused transducer) to the focal region of the transducer. The reason for the general applicability of the method is that the physics underlying the experimental process is faithfully modeled. Exhaustive experimental tests of the method, using tissue-mimicking media for which the backscatter coefficient can be independently estimated, are presented as well. The applicability to soft tissues, in particular, is addressed near the end of the chapter.

Correct values for backscatter coefficients presumes *a priori* knowledge of acoustic attenuation coefficients at the frequencies involved between the transducer and the position at which the backscatter coefficient is to be determined. However, if uniformity of backscatter and attenuation properties are assumed, such as might be reasonable for a single organ parenchyma, both parameters can be determined with the same data.

In the section dealing with use of broadband pulses, such as those generated by most current ultrasound imagers, a means for rapid estimation of the frequency dependence of the backscatter coefficient is described. Since this dependence can be related to mean scatterer size, the latter parameter could also be estimated.

II. THE BACKSCATTER EXPERIMENT AND NATURE OF THE RAW DATA

The experiment used for determining backscatter coefficients will be described using Figures 1 and 2. A single transducer is used to generate a

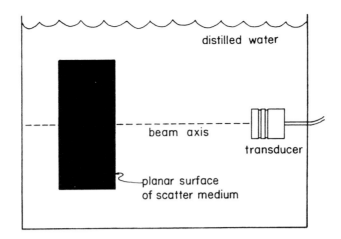

FIGURE 1.   Depiction of the experimental situation for collecting the echo signals backscattered from the scattering medium. The scattering medium is in the form of a cylinder with its plane parallel surfaces perpendicular to the beam axis. (From Madsen, E. L., Insana, M. F., and Zagzebski, J. A., *J. Acoust. Soc. Am.*, 76, 913–923, 1984. With permission.)

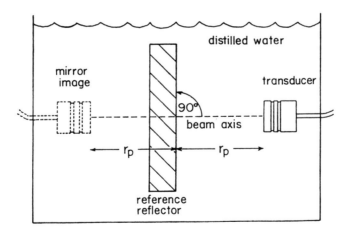

FIGURE 2.   Depiction of the experimental situation for collecting the reference echo signal from the planar reflector. The "mirror image" receiving transducer is shown on the left at a distance of $2r_p$ from the actual interrogating transducer. (From Madsen, E. L., Insana, M. F., and Zagzebski, J. A., *J. Acoust. Soc. Am.*, 76, 913–923, 1984. With permission.)

pulse and detect subsequent echoes from the scattering medium. The transducer involved can be focused or nonfocused and pulses created by the transducer are separated enough in time that all echo signals analyzed correspond to a single pulse. Also, all signal amplitudes are low enough — and the electronics are such — that all responses are linear.

In Figure 1 the transducer and block of scattering material are shown, the beam axis being perpendicular to the planar boundary of the sample's proximal surface. This boundary is chosen to be planar to simplify corrections for attenuation in the scattering medium.

The data collection consists of two parts. First, referring to Figure 1, a set of gated echo signals, due to scattering from within the scattering volume, are recorded. The onset and termination of the gate remains constant relative to the time of emission of the pulse. Between recordings of gated signals, however, the transducer is translated perpendicularly to the beam axis so that the positions of the scatterers contributing to any one gated signal can be considered to be uncorrelated spatially with the positions of those contributing to any of the other gated signals recorded. Also, the number of such gated echo signals recorded must be sufficiently large that mean values are significant; e.g., 25 such recorded signals would generally suffice.

The final piece of data consists of a recording of the echo signal from a reference reflector. An example of a reference reflector is shown in Figure 2. This consists of a planar specular reflector placed perpendicular to the beam axis. The reflector is assumed to be planar and specular in the method of data reduction described below. The method could, however, be modified to accommodate any reference reflector having a sufficiently well-known geometry. In the technique of Sigelmann and Reid a planar reflector is employed to estimate "the power reflected from a known interface".[1] In our technique the entire time-dependent echo signal is recorded and used in the determination of the backscatter coefficient.

III. THE METHOD OF DATA REDUCTION YIELDING THE BACKSCATTER COEFFICIENT

A. ASSUMPTIONS INVOLVING THE SCATTERING MEDIUM AND THE INTERROGATING INSTRUMENTATION

Five assumptions are made about the scattering medium:

1. The wave fronts of the scattered wave from each scatterer involved are approximated to be spherical in the region of the transducer face; this requires that the scatterer either be monopolar in nature or be sufficiently far from the transducer face. Thus, if the scatterers are not well-defined, as in tissues, the experiment should involve a sufficiently large distance between the transducer face and the volume of scatterers interrogated.

2. We assume that all scatterers are discrete and identical in this derivation. The method can easily be generalized, however, to apply to a finite number of sets of scatterers, each set containing scatterers identical to one another but unlike those in other sets.

3. It is assumed that the scatterers are randomly distributed in space and that the average number per unit volume is small enough that the only

(apparent) coherent scattering is related to the onset and termination of the gate. All other scattering is incoherent (see Appendix of Reference 12).

4.   The size of each scatterer is sufficiently small that the incident (primary) beam approximates a plane wave; i.e., the acoustic pressure can be expressed as a function of one rectilinear coordinate over the volume of the scatterer.

5.   The instrumentation parameters — characterized by the beam width at half maximum, pulse duration, and duration of the receiving time gate — assure that a sufficiently large number of scatterers contribute to the echo signal that the random positioning of the scatterers is fully represented in the echo signal at any instant of time.

B. DERIVATION OF THE METHOD AND ITS APPLICATION TO THE CASE OF NARROW-BAND PULSES

Define a pressure "wave packet" produced by the transducer with $p(\vec{r},t)$. This is the instantaneous pressure at the head of the position vector $\vec{r}$ at the time t. We can write this as the superposition of a complete set of continuous wave beams varying sinusoidally in time at any point in space.[13]

$$p(\vec{r},\, t) = \mathrm{Re} \int_0^\infty d\omega\, A(\vec{r},\, \omega) e^{-i\omega t} = \iint\limits_S ds'\, \mathrm{Re} \int_0^\infty d\omega\, A_{oo}(\omega)\, \frac{e^{ik|\vec{r}-\vec{r}'|-i\omega t}}{|\vec{r}-\vec{r}'|}$$

pressure at the field point $\vec{r}$
and time t due to pulsed monopole
radiators in the area element
ds' on the face of the transducer

where the beam is assumed to be formed from a superposition of pressure waves emitted in unison from monopole radiators uniformly distributed over the transducer face S. The quantity $A_{oo}(\omega)$ is a complex superposition coefficient, and Re means "real part of the quantity following". The complex wave number is denoted by $k = \omega/c(\omega) + i\alpha(\omega)$ where $c(\omega)$ is the speed of sound and $\alpha(\omega)$ is the attenuation coefficient at the angular frequency ω.

We can also write $p(\vec{r},t)$ in the form

$$p(\vec{r},\, t) = \mathrm{Re} \int_0^\infty d\omega\, A_{oo}(\omega) \iint\limits_S ds'\, \frac{e^{ik|\vec{r}-\vec{r}'|-i\omega t}}{|\vec{r}-\vec{r}'|}$$

$$= \mathrm{Re} \int_0^\infty d\omega\, A_{oo}(\omega) e^{-i\omega t} A_o(\vec{r},\, \omega)$$

where

$$A_o(\vec{r}, \omega) \equiv \iint_S ds' \frac{e^{ik|\vec{r} - \vec{r}'|}}{|\vec{r} - \vec{r}'|}$$

is proportional to the Rayleigh integral[14] for the case in which the normal component of the velocity at any instant of time is the same at all points on the radiating surface. Negative frequencies, though redundant, can be conveniently introduced. Let $c(\omega) = c(-\omega) = |c(\omega)|$ and $\alpha(\omega) = \alpha(-\omega) = |\alpha(\omega)|$. Then, since $e^{-i\omega t} A_o(\vec{r},\omega)$ is a solution of the wave equation, so is its complex conjugate; however, its complex conjugate also equals $e^{i\omega t} A_o(\vec{r}, -\omega)$. Requiring $A_{oo}(-\omega) = A^*_{oo}(\omega)$ where the asterisk denotes complex conjugation, we have

$$p(\vec{r}, t) = \frac{1}{2} \int_{-\infty}^{\infty} d\omega \, A_{oo}(\omega) e^{-i\omega t} A_o(\vec{r}, \omega)$$

Define $B_o(\omega) \equiv (1/2) A_{oo}(\omega)$. Then we have the simple form

$$p(\vec{r}, t) = \int_{-\infty}^{\infty} d\omega \, B_o(\omega) e^{-i\omega t} A_o(\vec{r}, \omega)$$

where $B_o(\omega)$ is a complex superposition coefficient. Notice that $A_o(\vec{r},\omega)$ depends only on ω, $c(\omega)$, $\alpha(\omega)$ and the shape and size of the transducer piezoelectric element.

Now suppose that a scatterer exists at the position $\vec{r}$. Assuming that the scatterer is small enough (e.g., less than a millimeter in diameter), we can approximate the incident wave packet as a superposition of plane waves each having a sinusoidal time dependence. In fact, we can deal only with the plane wave having frequency ω, i.e., that component having frequencies between ω and $\omega + d\omega$ where $d\omega$ is a differential. We have

$$dp_\omega(\vec{r}, t) \equiv d\omega \, B_o(\omega) e^{-i\omega t} A_o(\vec{r}, \omega) \tag{1}$$

The pressure wave scattered is given by

$$dp_{\omega s}(\vec{r}', t) = dp_\omega(\vec{r}, t) \frac{e^{ik|\vec{r} - \vec{r}'|}}{|\vec{r} - \vec{r}'|} \Phi(k, \cos \theta)$$

where we have assumed that $|\vec{r} - \vec{r}'| >> 1$ mm.[15] θ is the scattering angle between the direction of propagation of the incident plane wave and $\vec{r}' - \vec{r}$. $\vec{r}'$ is the point in space at which the scattered pressure is $dp_{\omega s}(\vec{r}',t)$ at the time t.

The force on the transducer face at time t is found by integrating over the area of the transducer face.

$$dF_s(\vec{r}, \omega, t) = dp_\omega(\vec{r}, t)\Phi(k) \iint_S ds' \frac{e^{ik|\vec{r}-\vec{r}'|}}{|\vec{r}-\vec{r}'|}$$

The area element ds' is at the head of $\vec{r}'$. We have invoked our assumption that the scatterer is monopolar in nature or that the scatterer is far enough from the transducer face that we can take $\Phi(k, \cos\theta) \approx \Phi(k, \cos 180°) \equiv \Phi(k)$; i.e., $\Phi(k)$ has no θ dependence. Notice that the area integral has already been defined as $A_o(\vec{r},\omega)$. Thus, we have

$$dF_s(\vec{r}, \omega, t) = d\omega\, B_o(\omega)e^{-i\omega t}[A_o(\vec{r}, \omega)]^2\Phi(k)$$

It can be shown[16] that

$$\Phi(k) \equiv \psi(\omega) = \psi^*(-\omega) \qquad (2)$$

Also, $\|\Phi(k)\|^2 = \|\psi(\omega)\|^2$ is the differential scattering cross-section of the scatterer at $180°$.

The above discussion applies to a single scatterer at position $\vec{r}$. The scattering medium will possess many scatterers, however, and we can define $N(\vec{r})$ to be the number of scatterers per unit volume at $\vec{r}$. $N(\vec{r})$ could, for example, be a sum of delta functions

$$N(\vec{r}) = \sum_{i=1}^{M} \delta(\vec{r}_i - \vec{r})$$

where the ith scatterer is at position $\vec{r}_i$. (Note that $\vec{r}_i$ is the position vector specifying the position of some reference point in the ith scatterer; $\vec{r}_i$ *and the* δ-*function do not imply anything about the size of the scatterer.*) Then the force at time t on the transducer face due to frequencies between ω and $\omega + d\omega$ can be written

$$dF_s(\omega, t) = d\omega\, B_o(\omega)e^{-i\omega t}\psi(\omega) \iiint_\Omega d\vec{r}\, N(\vec{r})[A_o(\vec{r}, \omega)]^2$$

where Ω is a volume containing all scatterers in the field.

The total force at time t on the transducer face due to all scatterers and including all frequencies is

$$F_s(t) = \int_{-\infty}^{\infty} d\omega\, B_o(\omega)e^{-i\omega t}\psi(\omega) \iiint_\Omega d\vec{r}\, N(\vec{r})[A_o(\vec{r}, \omega)]^2$$

Because the force is computed at some specific time t, not all scatterers contribute; e.g., if t is short enough, $F_s(t) = 0$ because the scattering volume is assumed not to be in contact with the transducer.

Let $T(\omega)$ be the (frequency dependent) complex receiving transfer function of the transducer. Then the echo signal voltage has the form

$$V_s(t) = \int_{-\infty}^{\infty} d\omega \ T(\omega)B_o(\omega)e^{-i\omega t}\psi(\omega) \iiint_{\Omega} d\vec{r} \ N(\vec{r})[A_o(\vec{r}, \omega)]^2 \quad (3)$$

$V_s(t)$ is the echo signal voltage which will be time gated and then Fourier analyzed. Before doing that, however, let us consider the echo voltage due to the presence of a reference reflector, the scattering medium being removed. Knowledge of the latter allows determination of the factor $T(\omega)B_o(\omega)$ in Equation 3.

Recall from Section II that the pulses created for the scattering situation are identical to those for the reference reflector situation. Thus, at some point $\vec{r}$, we have the pressure at time t due to frequencies between ω and $\omega + d\omega$, viz.,

$$dp_\omega(\vec{r}, t) = d\omega \ B_o(\omega)e^{-i\omega t}A_o(\vec{r}, \omega)$$

For simplicity of discussion, let the reflector be planar and perpendicular to the beam axis (as in Figure 2). If the distance from the center of the transducer face to the plane is r_p, then the force on the transducer face will be given by

$$dF_r(r_p, \omega, t) \equiv d\omega \ B_o(\omega)e^{-i\omega t}R \iint_{S_{mir}} ds \ A_o(\vec{r}, \omega)$$

The integral is over the area S_{mir} of a ''receiving'' transducer face which corresponds to the mirror image of the transmitting transducer in the reflecting plane. The distance between source and receiver is $2r_p$. R is the amplitude reflection coefficient for the planar reflecting surface. The differential area element, ds, lies at the head of $\vec{r}$.

Introducing the complex receiving transfer function and integrating over all ω, we have the echo signal voltage at time t due to the presence of the reference reflector:

$$V_r(t) \equiv \int_{-\infty}^{\infty} d\omega' \ T(\omega')B_o(\omega')e^{-i\omega' t}R \iint_{S_{mir}} ds \ A_o(\vec{r}, \omega')$$

For didactic convenience, we have changed the dummy integration variable from ω to ω'. Multiplying by $(1/2\pi) \ e^{i\omega t}$ and integrating over all time, we have the Fourier transform of the echo voltage:

$$\tilde{V}_r(\omega) \equiv \frac{1}{2\pi} \int_{-\infty}^{\infty} dt \, V_r(t) e^{i\omega t} = T(\omega) B_o(\omega) R \iint_{S_{mir}} ds \, A_o(\vec{r}, \omega) \qquad (4)$$

Notice that R should be calculable if the materials have known acoustic properties, $V_r(t)$ is recorded directly in the experiment, and the quantity

$$\iint_{S_{mir}} ds \, A_o(\vec{r}, \omega) \qquad \text{can be calculated, e.g., using a single}$$

integration for focused or nonfocused transducers.[13] Therefore, Equation 4 can be solved for $T(\omega)B_o(\omega)$ for any angular frequency ω. Thus, returning to Equation 4, $T(\omega)B_o(\omega)$ can be considered known and is given by

$$T(\omega)B_o(\omega) = \tilde{V}_r(\omega) \left[R \iint_{S_{mir}} ds \, A_o(\vec{r}, \omega) \right]^{-1}$$

The next step in our data reduction scheme is to take the Fourier transform of the set of gated echo voltages. Assuming a simple rectangular gate function of the form

$$G(t) = \begin{cases} 1 & \text{if } T_1 \leq t \leq T_2 , \\ 0 & \text{for all other } t \end{cases}$$

then

$$\tilde{V}_s(\omega) = \frac{1}{2\pi} \int_{-\infty}^{\infty} dt \, G(t) e^{i\omega t} \int_{-\infty}^{\infty} d\omega' \, T(\omega')B_o(\omega')e^{-i\omega' t} \, \psi(\omega') \iiint_{\Omega} d\vec{r}'$$

$$\times \, N(\vec{r}')[A_o(\vec{r}', \omega')]^2$$

$$= \frac{1}{2\pi} \int_{T_1}^{T_2} dt \, e^{i\omega t} \int_{-\infty}^{\infty} d\omega' \, T(\omega')B_o(\omega')e^{-i\omega' t} \, \psi(\omega') \iiint_{\Omega} d\vec{r}'$$

$$\times \, N(\vec{r}')[A_o(\vec{r}', \omega')]^2$$

and

$$\overline{\tilde{V}_s^*(\omega)\tilde{V}_s(\omega)} = (\tau/2\pi)\int_{-\infty}^{\infty} d\omega'\ T^*(\omega')B_o^*(\omega')\psi^*(\omega')\ \mathrm{sinc}[(\omega - \omega')(\tau/2\pi)]$$

$$\times \int_{-\infty}^{\infty} d\omega''\ T(\omega'')B_o(\omega'')\psi(\omega'')\ \mathrm{sinc}[(\omega - \omega'')(\tau/2\pi)]$$

$$\times \left[\overline{N}\iiint_{\Omega} d\vec{r}'\ [A_o^*(\vec{r}',\ \omega')]^2[A_o(\vec{r}',\ \omega'')]^2 \right.$$

$$\left. + \overline{N}^2 \iiint_{\Omega} d\vec{r}'\ [A_o^*(\vec{r}',\ \omega')]^2 \iiint_{\Omega} d\vec{r}''\ [A_o(\vec{r}'',\ \omega'')]^2 \right]$$

This expression can be further simplified by rearranging orders of factors and orders of integration.

$$\overline{\tilde{V}_s^*\tilde{V}_s(\omega)} = (\tau/2\pi)^2 \left[\overline{N}\iiint_{\Omega} d\vec{r}' \int_{-\infty}^{\infty} d\omega'\ T^*(\omega')B_o^*(\omega')\psi^*(\omega') \right.$$

$$\times \ \mathrm{sinc}[(\omega - \omega')(\tau/2\pi)]\,[A_o^*(\vec{r}',\ \omega')]^2$$

$$\times \int_{-\infty}^{\infty} d\omega''\ T(\omega'')B_o(\omega'')\psi(\omega'')\ \mathrm{sinc}[(\omega - \omega'')(\tau/2\pi)]\,[A_o(\vec{r}',\ \omega'')]^2$$

$$+ \ \overline{N}^2 \int_{-\infty}^{\infty} d\omega'\ T^*(\omega')B_o^*(\omega')\psi^*(\omega')\ \mathrm{sinc}[(\omega - \omega')(\tau/2\pi)]$$

$$\iiint_{\Omega} d\vec{r}'\ [A_o^*(\vec{r}',\ \omega')]^2$$

$$\times \int_{-\infty}^{\infty} d\omega''\ T(\omega'')B_o(\omega'')\psi(\omega'')\ \mathrm{sinc}[(\omega - \omega'')(\tau/2\pi)]$$

$$\left. \iiint_{\Omega} d\vec{r}''\ [A_o(\vec{r}'',\ \omega'')]^2 \right]$$

$$= (\tau/2\pi)^2 \left[\overline{N}\iiint_{\Omega} d\vec{r}'\ \|J_\omega(\vec{r}')\|^2 + \overline{N}^2\|M_\omega|^2 \right] \tag{12}$$

where

$$J_\omega(\vec{r}') \equiv \int_{-\infty}^{\infty} d\omega'\ T(\omega')B_o(\omega')\psi(\omega')\ \mathrm{sinc}[(\omega - \omega')(\tau/2\pi)]$$

$$\times \ [A_o(\vec{r}',\ \omega')]^2 \tag{13}$$

and

$$M_\omega \equiv \int_{-\infty}^{\infty} d\omega' \; T(\omega') B_o(\omega') \psi(\omega') \; \text{sinc}[(\omega - \omega')(\tau/2\pi)]$$

$$\iiint_\Omega d\vec{r}' \; [A_o(\vec{r}', \omega')]^2 \tag{14}$$

To separate out the differential scattering cross-section per scatterer at 180°, $\|\psi(\omega')\|^2$, we are faced with the problem that, in $J_\omega(\vec{r}')$ and M_ω, $\psi(\omega')$ appears in the integrand where ω' is a dummy variable of integration. If the pulse and gate durations are long enough, however, there should be a very strong peaking of $J_\omega(\vec{r}')$ and M_ω at ω_o, the carrier frequency of the pulse; thus, one method of dealing with the problem is to assure that we use narrowband pulses and sufficiently large gate durations. In general, the frequency dependence of $\psi(\omega')$ will then be small enough that, for $\omega = \omega_o$, $\psi(\omega') \approx \psi(\omega_o)$, and we have

$$\|J_{\omega_o}(\vec{r}')\|^2 \approx \|\psi(\omega_o)\|^2 \; \|J'_{\omega_o}(\vec{r}')\|^2 \tag{15}$$

$$\|M_{\omega_o}\|^2 \approx \|\psi(\omega_o)\|^2 \; \|M'_{\omega_o}\|^2 \tag{16}$$

where $J'_{\omega_o}(\vec{r}')$ and M'_{ω_o} are the same as $J_{\omega_o}(\vec{r}')$ and M_{ω_o}, respectively, with the $\psi(\omega')$ factors missing from the integrands.

The backscatter coefficient, or differential scattering cross section per unit volume at frequency ω_o and 180° scattering angle, is given by $\eta(\omega_o) = \overline{N}\|\psi(\omega_o)\|^2$. Substituting the approximations 15 and 16 into Equation 12, we get

$$\eta(\omega_o) \approx \frac{\overline{\tilde{V}_s^*(\omega_o) \tilde{V}_s(\omega_o)}}{(\tau/2\pi)^2 \left\{ \iiint_\Omega d\vec{r}' \; \|J'_{\omega_o}(\vec{r}')\|^2 + \overline{N}\|M'_{\omega_o}\|^2 \right\}} \tag{17}$$

Expression 17 is suitable for determining $\eta(\omega_o)$ only if $\overline{N}$ is known or if the term in the denominator containing $\overline{N}$ is negligible. A requirement that $\overline{N}$ be known is inconvenient, particularly for making measurements on tissues. As noted earlier, however, for sufficiently long gate durations, τ,

$$\frac{\overline{N}\|M'_{\omega_o}\|^2}{\iiint_\Omega d\vec{r}' \; \|J'_{\omega_o}(\vec{r}')\|^2} \xrightarrow{\tau \text{ increasing}} 0 \tag{18}$$

or

$$\zeta(\omega_o) \equiv \frac{\overline{\tilde{V}_s^*(\omega_o)\tilde{V}_s(\omega_o)}}{(\tau/2\pi)^2 \iiint\limits_{\Omega} d\vec{r}' \; \|J'_{\omega_o}(\vec{r}')\|^2} \xrightarrow[\tau \text{ increasing}]{} \eta(\omega_o) \qquad (19)$$

Thus, if one plots $\zeta(\omega_o)$ vs. τ, this ratio should decrease to the asymptotic constant value, $\eta(\omega_o)$, making knowledge of $\overline{N}$ unnecessary.

Once $\eta(\omega_o)$ has been determined in this fashion, an estimate of $\overline{N}$ can itself be obtained by solving Equation 17 for $\overline{N}$:

$$\overline{N} = \|M'_{\omega_o}\|^{-2} \left[(2\pi/\tau)^2 \eta^{-1}(\omega_o) \overline{\tilde{V}_s^*(\omega_o)\tilde{V}_s(\omega_o)} - \iiint\limits_{\Omega} d\vec{r}' \; \|J'_{\omega_o}(\vec{r}')\|^2 \right]$$

However, the above equation is useful only if $\overline{N}$ is large enough and if the expectation value is obtained for sufficiently small values of τ that the coherent effects are significant.

C. APPLICATION TO BROAD-BAND PULSES

Some measurements of backscatter coefficients, $\eta(\omega)$, reported in the literature, involve use of broad-band pulses.[5,6,8] In these cases it is assumed that in the data reduction one needs to be concerned only with the one specific frequency component in any one analysis. Section III.B above makes it clear that, for pulses with very broad frequency bands, such an assumption deserves scrutiny and that, to get a value of $\eta(\omega)$ which is truly independent of other frequencies, one needs to properly account for the influence of other frequencies on the data. Particular attention needs to be paid to points on spectral distribution curves where the magnitude of the slope is large.

The method of data reduction in Section III.B may find its greatest use in evaluating $\eta(\omega)$ for a range of frequencies when broad-band pulses are employed. Very importantly, when analyzed properly, data acquired using broad-band pulses can yield values of $\eta(\omega)$ over a range of frequencies whereas, for the same data acquisition time, the backscatter coefficient at only one frequency could reasonably be obtained using a narrow-band pulse.

For data analysis in which broad-band pulses are employed, use of an iterative method is proposed. Equation 12 will be used as a starting point in developing this method for determining $\eta(\omega)$, viz.,

$$\overline{\tilde{V}_s^*(\omega)\tilde{V}_s(\omega)} = (\tau/2\pi)^2 \left\{ \overline{N} \iiint\limits_{\Omega} d\vec{r}' \; \|J_\omega(\vec{r}')\|^2 + \overline{N^2}\|M_\omega\|^2 \right\} \qquad (12)$$

where $J_\omega(\vec{r}')$ and M_ω are given in Equations 13 and 14. $\psi(\omega')$ appears in the integrands of $J_\omega(\vec{r}')$ and M_ω where ω' is a dummy variable. It is proposed that a guess be made at the frequency dependence of $\psi(\omega')$ and that this frequency dependence be separated out as a factor, $g(\omega')$: $\psi(\omega') \equiv \psi_o(\omega')g(\omega')$. The quantity $\psi_o(\omega')$ will be a more slowly varying function of ω' than $\psi(\omega')$ if $g(\omega')$ was judiciously chosen; e.g., for scatterers which are thought to be small compared to the wavelength (Rayleigh-like), one might try $g(\omega') = \omega'^2$. (For true Rayleigh scatterers, $\psi_o(\omega')$ is then independent of ω', of course.) In our development we ignore any frequency dependent phase factor of ψ and pursue only the frequency dependence of $\|\psi\|$.

Now suppose that τ is made large enough that the coherent ($\overline{N}^2$) term is negligible in Equation 12. (This can be determined experimentally in the fashion described in Section III.B.) Also, we take $\psi_o(\omega')$ to be nearly constant in the neighborhood of ω and move it outside the integral sign in $J_\omega(\vec{r}')$; i.e., $J_\omega(\vec{r}') \approx J_{\omega,L}(\vec{r}') \equiv \psi_o(\omega)K_{\omega,L}(\vec{r}')$ where

$$K_{\omega,L}(\vec{r}') \equiv \int_{-\infty}^{\infty} d\omega' \ T(\omega')B_o(\omega')g(\omega') \ \text{sinc}[(\omega - \omega')(\tau_L/2\pi)] [A_o(\vec{r}', \omega')]^2$$

The subscript L refers to a long gate duration. ψ_o is designated above as a function of ω due to the facts that there will in general be residual frequency dependence in ψ_o and that there is some peaking in the integrand of $J_\omega(\vec{r}')$ at ω due to the sinc function. Thus, Equation 12 becomes

$$\overline{V_s^*(\omega)\,\overline{V}_s(\omega)}\Big|_L \approx (\tau_L/2\pi)^2 \overline{N}\|\psi_o(\omega)\|^2 \iiint\limits_\Omega d\vec{r}' \ \|K_{\omega,L}(\vec{r}')\|^2$$

or

$$\overline{N}\|\psi_o(\omega)\|^2 \approx \frac{\overline{V_s^*(\omega)\,\overline{V}_s(\omega)}\Big|_L}{(\tau_L/2\pi)^2 \iiint\limits_\Omega d\vec{r}' \ \|K_{\omega,L}(\vec{r}')\|^2}$$

A plot of $(\overline{N})^{1/2}\|\psi_o(\omega)\|$ vs. ω (a set of ωs must be chosen from the frequency band, of course) then yields, via curve fitting, a frequency dependence of $\|\psi_o(\omega)\|$; i.e., we can determine a real function f such that $\psi_o(\omega') = \psi_{oo} f(\omega')$ where ψ_{oo} likely is nearly independent of frequency and f is a real function of frequency. This completes the first iteration with $\psi(\omega) \propto f(\omega)g(\omega)$.

The second iteration is accomplished by repeating the above process, yielding

$$\overline{N}\|\psi_{oo}(\omega)\|^2 \approx \frac{\overline{V_s^*(\omega)\,\overline{V}_s(\omega)}\Big|_L}{(\tau_L/2\pi)^2 \iiint\limits_\Omega d\vec{r}' \ \|K'_{\omega,L}(\vec{r}')\|^2}$$

where

$$K'_{\omega,L}(\vec{r}') \equiv \int_{-\infty}^{\infty} d\omega' \; T(\omega')B_o(\omega')f(\omega')g(\omega') \; \text{sinc}[(\omega - \omega')$$

$$(\tau_L/2\pi)] \, [A_o(\vec{r}', \omega')]^2$$

It may result (and seems likely) that $\psi_{oo}(\omega)$ will be almost frequency independent. If so, we can use the results of the first iteration to express $\eta(\omega)$, viz.,

$$\eta(\omega) \approx \overline{N}\|\psi_{oo}\|^2 f^2(\omega)g^2(\omega) = \frac{\overline{\tilde{V}_s^*(\omega)\,\tilde{V}_s(\omega)}\big|_L \, f^2(\omega)\,g^2(\omega)}{(\tau_L/2\pi)^2 \displaystyle\iiint_\Omega d\vec{r}' \; \|K'_{\omega,L}(\vec{r}')\|^2} \tag{20}$$

Otherwise, another iteration could be performed. Note that the backscatter coefficient, including its frequency dependence, is obtained over approximately the bandwidth of the interrogating pulse.

For sufficiently large values, the mean scatterer concentration, $\overline{N}$, can also be determined in a way analogous to that described in Section III.B. Let us assume for this discussion that ψ_{oo} can be considered constant and, therefore, that Equation 20 represents a good approximation for $\eta(\omega)$. Then, with this knowledge of $\eta(\omega)$, a determination of $\overline{\tilde{V}_s^*(\omega)\tilde{V}_s(\omega)}$ for a short enough gate duration, τ_S, will cause the $\overline{N}^2$ term to be significant in Equation 12 and $\overline{N}$ can be determined using the expression

$$\overline{N} = \|N'_{\omega,S}\|^{-2}\left[(2\pi/\tau_S)^2\,\eta^{-1}(\omega)\,\overline{\tilde{V}_s^*(\omega)\,\tilde{V}_s(\omega)}\big|_S\, f^2(\omega)\,g^2(\omega) \right.$$
$$\left. - \iiint_\Omega d\vec{r}' \; \|K'_{\omega,S}(\vec{r}')\|^2 \right] \tag{21}$$

which is obtained from Equation 12 using the relation $\eta(\omega) = \overline{N}\|\psi_{oo}\|^2 f^2(\omega)g^2(\omega)$ and the definitions

$$K'_{\omega,S}(\vec{r}') \equiv \int_{-\infty}^{\infty} d\omega' \; T(\omega')B_o(\omega')f(\omega')g(\omega') \; \text{sinc}[(\omega - \omega')(\tau_S/2\pi)]$$
$$[A_o(\vec{r}', \omega')]^2$$

and

$$N'_{\omega,S} \equiv \int_{-\infty}^{\infty} d\omega' \; T(\omega')B_o(\omega')f(\omega')g(\omega') \; \text{sinc}\left[(\omega - \omega')(\tau_S/2\pi) \right]$$
$$\iiint_\Omega d\vec{r}' \; [A_o(\vec{r}', \omega')]^2$$

where the subscripts "S" refer to a "short" gate duration.

IV. EXPERIMENTAL TESTS OF THE ACCURACY OF THE METHOD

Extensive experimental tests of the accuracy of the method have been carried out using phantom materials in which the scatterers are spatially randomly distributed glass beads with diameters less than 100 μm. Knowledge of diameter distributions and bulk properties of the glass — including shear wave properties — allowed independent computation of backscatter coefficients using the theory of Faran[23] for comparison. Consistency of the two methods is taken as conclusive evidence that the method of Section III is accurate. Accuracy has been tested and verified over a broad range of parameters, including the transducer nominal frequency, whether it is focused or nonfocused, reception time gate duration, transducer-to-scattering-volume distance, and the attenuation coefficient of the medium.

The material in this section is based on previously published work by Insana et al.,[24] Hall et al.,[25] and Hall.[26]

A. NONFOCUSED TRANSDUCERS IN LOW ATTENUATION MEDIA USING THE NARROW-BAND METHOD

The accuracy of the method was tested for narrow-band pulses as a function of carrier frequency and of transducer-to-scattering-volume distance. In addition, the dependence of $\xi(\omega_o)$, defined in relation 19, on echo reception gate duration τ was evaluated to determine how large τ must be in order that $\xi(\omega_o) = \eta(\omega_o)$.

1. Materials

Two samples consisting of glass bead scatterers randomly distributed in agar were used to test the accuracy of the data reduction analysis. In sample 1, the glass spheres exhibit a strong peak in their diameter distribution at 40 μm. The mean concentration of scatterers is $\overline{N} = 46$ scatterers per mm³. The scatterers in sample 2 are spheres of the same type of glass, also having narrow diameter distribution with a 59 μm mean diameter. The mean scatterer concentration for sample 2 is $\overline{N} = 7.7$ scatterers per mm³. Measurements of the diameter distribution were done using an optical microscope with a calibrated ocular micrometer. Histograms for the diameter distributions are shown in Figure 3 (sample 1) and Figure 4 (sample 2). Mean concentrations of scatterers were found by counting the number of beads in a thin slab of the material, the mass of the slab and its density also having been determined.

Measurements of speed of sound and attenuation coefficients were made using a through-transmission technique[27] at 20°C at five discrete frequencies between 1 and 7 MHz. Curve fitting was done assuming that the attenuation coefficient $\alpha(f)$ is proportional to a power of the frequency f; i.e., $\alpha(f) = \alpha_o f^n$, where α_o and n are constants. For sample 1, $\alpha_o = 0.052$ dB cm^{-1} MHz$^{-1.5}$ and n = 1.5. For sample 2, $\alpha_o = 0.055$ dB cm^{-1} MHz$^{-1.4}$ and n = 1.4.

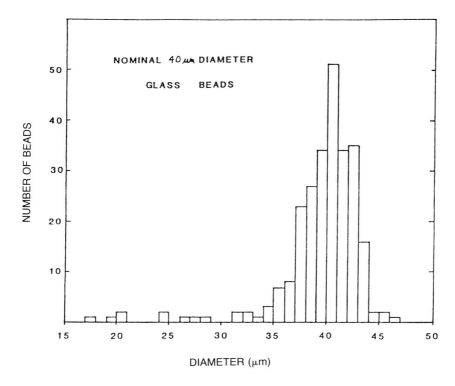

FIGURE 3. Histogram of the diameter distribution of the glass bead scatterers in sample 1. Measurements were done using an optical microscope with a calibrated ocular micrometer. (From Insana, M. F., Madsen, E. L., Hall, T. J., and Zagzebski, J. A., *J. Acoust. Soc. Am.,* 79, 1230–1236, 1986. With permission.)

Backscatter coefficients for these samples were computed independently using the theory of Faran.[23] Required parameters for the glass beads were the density (2.4 g/cm³), Poisson's ratio (0.21), and the longitudinal speed of sound (5570 m/s). The agar was taken to be water-like with the density of 1.00 g/cm³ and a speed of sound of 1525 m/s (our measured values). Faran's theory accounts for both longitudinal and transverse waves and has been found to agree with experiments for direct measurements of differential scattering cross-section over a broad range of scattering angles and frequencies.[28] Assuming incoherent scattering only, the calculated, or ''theoretical,'' backscatter coefficient (defined as the differential scattering cross-section per unit volume for a 180° scattering angle) is given by $\eta_{theory}(\omega_o) = \overline{N}d\sigma/d\Omega|_{180°}$. Because the backscattered intensity for small scatterers is a strong function of scatterer diameter, $\eta_{theory}(\omega_o)$ for each sample was calculated as a weighted average using the histograms in Figures 3 and 4; i.e.,

$$\eta_{theory}(\omega_o) = \sum_{i=1}^{L} f_i \eta_i(\omega_o) \qquad (22)$$

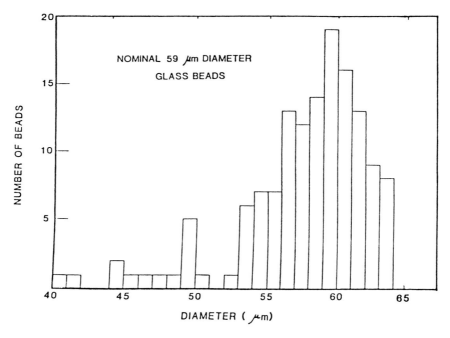

FIGURE 4. Histogram of the diameter distribution of the glass bead scatterers in sample 2. Measurements were done using an optical microscope with a calibrated ocular micrometer. (From Insana, M. F., Madsen, E. L., Hall, T. J., and Zagzebski, J. A., *J. Acoust. Soc. Am.*, 79, 1230–1236, 1986. With permission.)

where f_i is the number fraction of scatterers (particles) in the ith bin and $\eta_i(\omega_o)$ is the calculated backscatter coefficient corresponding to the diameter of the scatterers in the ith bin.

2. Experimental Procedure

A block diagram of the equipment used to obtain scattered echo signals from the sample is shown in Figure 5. The sample was placed in water at a specific distance from the transducer. Nonfocused 13 mm diameter transducers were used to transmit narrow-band pulses and receive the backscattered echoes. The pulse repetition frequency was adjusted so that echoes from the entire sample were received before the next pulse was transmitted. From these echo signal waveforms, axial regions of interest were selected via time gate positioning. Using the Biomation 810° transient recorder, 25 time-gated echo signal waveforms were then recorded, each corresponding to a different position of the transducer beam axis through the sample. These different positionings were accomplished by translating the sample perpendicularly to the beam axis in a raster fashion. The Fourier transforms of each of these 25 waveforms were then computed yielding values of $V_s(\omega_o)$. The mean value of the square of the moduli of these Fourier transforms then yields $\|V_s(\omega_o)\|^2$, the numerator of $\zeta(\omega_o)$.

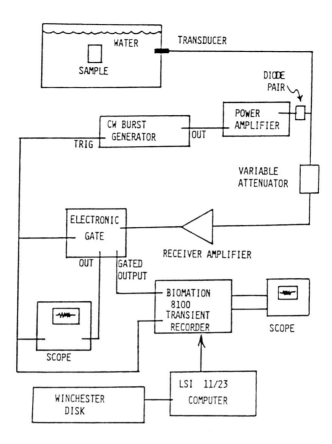

FIGURE 5. Block diagram of the experimental apparatus for acquiring and storing data. (From Insana, M. F., Madsen, E. L., Hall, T. J., and Zagzebski, J. A., *J. Acoust. Soc. Am.*, 79, 1230–1236, 1986. With permission.)

To account for pulser-receiver characteristics for each set of experimental parameters, a recording was made of the echo signal waveform due to reflection from a planar Lucite*-to-water interface placed at half the transducer-to-sample distance. This reflector position was chosen so that the echo pressure wave received most closely represents that incident on the involved scatterers.** This echo waveform allows the determination of the (complex) function $B_o(\omega)T(\omega)$. Care was taken to maintain the same transmission and reception conditions for recording the plane reflector signal as those for recording the scatter signals from the sample.

* Lucite is a trade name for polymethyl acrylate.

** In theory, the specific choice of the distance between the transducer and the plane reflector is not important; this flexibility depends on the accuracy of the beam model, however, and remains to be tested.

All data acquisition and storage in these measurements were under control via an LSI 11/23 microcomputer.* The data sampling rate of the transient recorder was 20 MHz. Further details regarding experimental techniques can be found in Reference 29.

3. Numerical Methods

All data reduction was done on a PDP 11/23-PLUS computer configured with an array processor.** Discrete Fourier transforms of echo signal waveforms were computed according to Bracewell.[30]

The quantity $\zeta(\omega_o)$ was determined for each measurement. Recall that $\zeta(\omega_o) \to \eta(\omega_o)$ for sufficiently large gate duration τ.

Numerical integrations over volume and frequency were done to determine values of the denominator of $\zeta(\omega_o)$, viz.,

$$\iiint_\Omega d\vec{r} \; \left\| \int_\infty^\infty d\omega' \; T(\omega') B_o(\omega') \; \text{sinc}[(\omega - \omega')(\tau/2\pi)] A_o(\vec{r}, \omega') \right\|^2$$

Computing time was minimized without compromising accuracy by judicious choice of integration limits and intervals. It was found that the frequency integral converges when the lower limit of integration is less than $\omega_o - 6\pi/\tau$ and the upper limit is greater than $\omega_o + 6\pi/\tau$, where τ is the gate duration (time). This range of frequencies includes the five central lobes of the sinc function. Thirty-one frequency values were necessary for convergence of the frequency integral. Considerable savings in computer time resulted from introducing a Taylor series expansion of the expression for $A_o(\vec{r},w)$ about ω_o.

The pressure field has axial symmetry so that the volume integral in Equation 7 was reduced to two dimensions: radial and axial. The volume integration limits extend over the entire sample volume Ω. However, a smaller integration volume, Ω' having lateral margins extending through the first sidelobe of the pressure field and containing at least 80 points over that radial distance, was used with no loss of accuracy. Axially, Ω' must have a length of at least $L = c_s(\tau+T)/2$, where c_s is the speed of sound in the scattering medium and T is the pulse duration. The distance d between the transducer face and the proximal end of Ω' must be no greater than

$$d = d_w + (c_w/2)(t_{on} - T - 2d_w/c_w)$$

where d_w is the water path distance from the transducer to the sample, c_w is the speed of sound in water, and t_{on} is the onset time of the electronic gate relative to a zero time corresponding to the beginning of the emission of the

* Trademark of the Digital Equipment Corporation, Maynard, MA.

** SKYMNK-Q, Sky Computers, Inc., Lowell, MA.

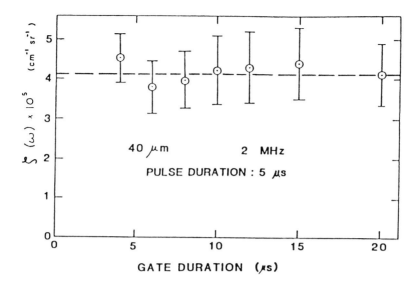

FIGURE 6. Plotted values of $\zeta(\omega_o)$, determined with the data reduction method on sample 1 at $\omega_o = 2$ MHz and at various gate durations. The pulse duration was maintained at 5 μs. The horizontal dashed line corresponds to the theoretical value.[23] The transducer-to-scattering-volume distance was 20 cm. Error bars correspond to standard deviations of the means. (From Insana, M. F., Madsen, E. L., Hall, T. J., and Zagzebski, J. A., *J. Acoust. Soc. Am.*, 79, 1230–1236, 1986. With permission.)

pulse from the transducer. The distance between the distal end of Ω' and the transducer face must be no less than d + L. In fact, Ω' should be extended over a somewhat larger axial extent than that defined here to account for the fact that the boundaries separating scatterers contributing from those not contributing are not simply planes perpendicular to the axis of symmetry of the beam.[12] An axial integration increment of less than $2\lambda_o$ (where $\lambda_o = 2\pi c_s/\omega_o$) was found to be sufficient to maintain accuracy.

4. Results and Discussion

One of the major objectives was to gain insight into the dependence of $\zeta(\omega_o)$ on the gate duration and frequency for some reasonable pulse duration. The results of the gate duration studies shown in Figures 6 and 7 (2 and 5 MHz, respectively) indicate that, for 5 μs pulse durations and the ranges of gate durations employed, there is negligible dependence of $\zeta(\omega_o)$ on gate durations. Thus a pulse duration of 5 μs and gate duration of 6 μs or higher are adequate to ensure that $\zeta(\omega_o)$ is a good approximation to the backscatter coefficient $\eta(\omega_o)$. This criterion was used in setting $\zeta(\omega_o)$ equal to $\eta(\omega_o)$ in the results displayed in Figures 8 and 9 and Tables 1 and 2.

In Tables 1 and 2, good agreement between theory and the results of our data reduction method is shown to exist for both samples over the frequency range from 1 to 6 MHz.

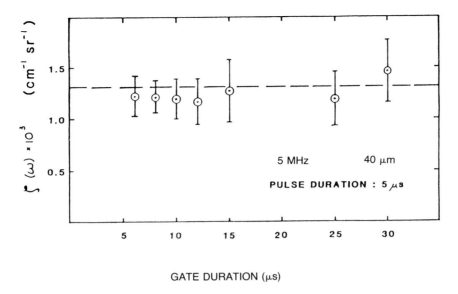

GATE DURATION (μs)

FIGURE 7.   Plotter values of $\zeta(\omega_o)$ determined with the method of data reduction on sample 1 at $\omega_o = 5$ MHz and at various gate durations. The pulse duration was maintained at 5 μs. The horizontal dashed line corresponds to the theoretical value.[23] The transducer-to-scattering-volume distance was 20 cm. Error bars correspond to standard deviations of the means. (From Insana, M. F., Madsen, E. L., Hall, T. J., and Zagzebski, J. A., *J. Acoust. Soc. Am.,* 79, 1230–1236, 1986. With permission.)

Perhaps the most important results of the present work have been the demonstration that backscatter coefficients determined are nearly independent of transducer-to-scattering-volume distance. The results at 2 MHz for five distances from 5 to 25 cm agree with one another and with the theoretical value. The mean experimental value, shown as the solid horizontal line, is about 6% below the theoretical value, shown as the dashed horizontal line. In Figure 9, the results for 4 MHz also show reasonably good agreement over the same range of transducer-to-scattering-volume distances, their mean value (solid horizontal line) being about 13% below the theoretical value (dashed horizontal line).

B. FOCUSED TRANSDUCERS USING THE NARROW-BAND METHOD
The same experimental technique was used in data acquisition as that described in Section IV.A.2. In addition, the focused transducers were characterized regarding the radius of curvature of the radiating element and its projected diameter.

1. Transducer Parameters
To calculate the function $A_o(\vec{r},\omega)$ for a focused transducer, it is necessary to know the projected diameter and radius of curvature of its radiating element.

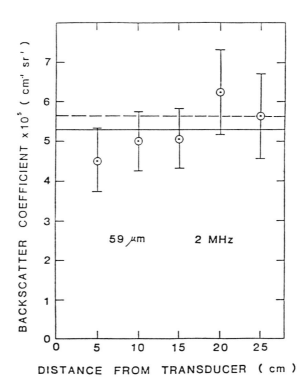

FIGURE 8. Backscatter coefficients for sample 2 at 2 MHz for various distances between the transducer face and scattering volume interrogated. The pulse and gate durations were maintained at 5 and 10 μs, respectively. The horizontal solid line corresponds to the mean value of these experimental determinations and the horizontal dashed line corresponds to the theoretical value based on Faran.[23] Error bars correspond to standard deviations of the means. (From Insana, M. F., Madsen, E. L., Hall, T. J., and Zagzebski, J. A., *J. Acoust. Soc. Am.*, 79, 1230–1236, 1986. With permission.)

As O'Neil has shown,[31] the directivity function[32] for a planar circular disk accurately describes the relative pressure amplitude in the "focal plane" of a weakly focused transducer driven by a sinusoidal voltage. The focal plane is defined as a plane perpendicular to the axis of symmetry of the radiating element and passing through the center of curvature of the radiating element. Agreement with the directivity function is unique to this plane, and this fact can be used to determine, experimentally, the radius of curvature and projected radius as described in Reference 13.

The approximate radius of curvature of each transducer employed was given by the manufacturer. Scanning the pressure field in the appropriate region with a 1 mm diameter hydrophone, we found the plane that yielded the best approximation to the directivity function of a disk transducer. The axial distance from the center of the transducer to this plane was taken to be the transducer's radius of curvature. The uncertainty in this distance is less

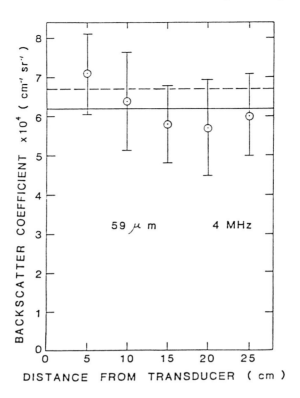

FIGURE 9.   Backscatter coefficients for sample 2 at 4 MHz for various distances between the transducer face and scattering volume interrogated. The pulse and gate durations were maintained at 5 and 10 μs, respectively. The horizontal solid line corresponds to the mean value of these experimental determinations and the horizontal dashed line corresponds to the theoretical value based on Faran.[23] (From Insana, M. F., Madsen, E. L., Hall, T. J., and Zagzebski, J. A., *J. Acoust. Soc. Am.*, 79, 1230–1236, 1986. With permission.)

than 1% based upon establishment of least minima between side lobes. The first off-axis minima defining the main lobe of the beam were then used to estimate the projected diameter. Propagating uncertainties in the positions of off-axis minima and radius of curvature yield an uncertainty in the effective transducer aperture of about 0.1 to 0.2 mm for the transducers used in this work. An example of the experimentally measured relative pressure is shown in Figure 10 along with a directivity function fitting the experimental data. The directivity function and the measured relative pressure have been normalized to their respective peaks. The nominal transducer frequency as well as the carrier frequency of the pulse was 5.0 MHz. The measured profile follows the directivity function quite well. The absence of completely zero values between side lobes is likely the result of electronic noise or the fact that the hydrophone perturbs the pressure field.

TABLE 1
Theoretical and Experimental Backscatter Coefficients at Eight Frequencies for Test Sample 1 in Which the Nominal Scatterer Diameter is 40 μm

Frequency (MHz)	Backscatter coefficient (sr^{-1} cm^{-1})	
	Theory	Experiment
1.0	2.64×10^{-6}	$(3.35 \pm 0.64) \times 10^{-6}$
1.2	5.45×10^{-6}	$(5.85 \pm 1.15) \times 10^{-6}$
2.0	4.10×10^{-5}	$(4.45 \pm 0.88) \times 10^{-5}$
2.5	9.79×10^{-5}	$(9.98 \pm 1.93) \times 10^{-5}$
3.0	1.98×10^{-4}	$(1.50 \pm 0.26) \times 10^{-4}$
4.0	5.84×10^{-4}	$(5.48 \pm 1.02) \times 10^{-4}$
5.0	1.31×10^{-3}	$(1.09 \pm 0.22) \times 10^{-3}$
6.0	2.44×10^{-3}	$(1.80 \pm 0.36) \times 10^{-3}$

Note: The theoretical values were computed using the theory by Faran.[23]

From Insana, M. F., Madsen, E. L., Hall, T. J., and Zagzebski, J. A., *J. Acoust. Soc. Am.,* 79, 1230–1236, 1986. With permission.

TABLE 2
Theoretical and Experimental Backscatter Coefficients at Eight Frequencies for Test Sample 2 in Which the Nominal Scatterer Diameter is 59 μm

Frequency (MHz)	Backscatter coefficient (sr^{-1} cm^{-1})	
	Theory	Experiment
1.0	3.74×10^{-6}	$(3.32 \pm 0.54) \times 10^{-6}$
1.2	7.69×10^{-6}	$(7.84 \pm 1.40) \times 10^{-6}$
2.0	5.63×10^{-5}	$(5.31 \pm 1.21) \times 10^{-5}$
2.5	1.32×10^{-4}	$(1.30 \pm 0.25) \times 10^{-4}$
3.0	2.58×10^{-4}	$(2.92 \pm 0.58) \times 10^{-4}$
4.0	7.08×10^{-4}	$(6.20 \pm 1.12) \times 10^{-4}$
5.0	1.44×10^{-3}	$(1.45 \pm 0.24) \times 10^{-3}$
6.0	2.37×10^{-3}	$(2.03 \pm 0.33) \times 10^{-3}$

Note: The theoretical values were computed using the theory by Faran.[23]

From Insana, M. F., Madsen, E. L., Hall, T. J., and Zagzebski, J. A., *J. Acoust. Soc. Am.,* 79, 1230–1236, 1986. With permission.

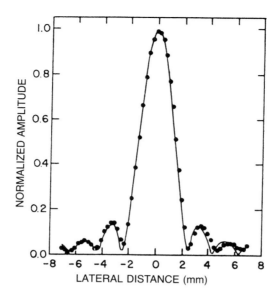

FIGURE 10.   Relative acoustic pressure amplitude values, shown as small circles, defining a lateral beam profile at an axial distance of 8.5 cm from a 5 MHz transducer excited by an approximately 40-wavelength, 5.0 MHz pulse. The solid line represents the directivity function chosen as a best fit to experimental measurements. The directivity function and measured values have been normalized to their respective peaks. These data allow determination of the radius of curvature and projected diameter of the transducer. (From Hall, T. J., Madsen, E. L., Zagzebski, J. A., and Boote, E. J., *J. Acoust. Soc. Am.*, 85, 2410–2416, 1989. With permission.)

This procedure was used to determine the radius of curvature and effective aperture for three transducers employed in this work. Table 3 shows the nominal resonant frequency, radius of curvature, and effective transducer aperture for each transducer. In each case, the carrier frequency of the narrow-band pulses used to determine the projected diameter and radius of curvature was chosen to equal the nominal frequency specified by the manufacturer.

2. Materials

Two phantoms were used in this study, one a cylinder (13 cm diameter, 6 cm thick), called phantom 1, and the other a rectangular parallelepiped (6 × 13 × 16 cm³), called phantom 2. Both have parallel 50 μm-thick Saran Wrap℗ acoustic windows. The scattering medium in both phantoms consists of glass bead scatterers randomly positioned in an agar matrix such the mean number per unit volume is uniform. Phantom 1 is identical to "Sample 2" described in Section IV.A.

Phantom 2 is very similar to phantom 1. The mean diameter of the glass bead scatterers is about the same as that in phantom 1. In addition, very finely powdered graphite is uniformly distributed throughout the phantom, raising the frequency-dependent attenuation to one typical of most soft tissues.[27] The graphite particles are small enough that their contribution to the backscatter

TABLE 3
Values of Radii of Curvature and Projected Diameter for the Three Transducers Employed in This Work

Nominal resonant frequency (MHz)	Radius of curvature (cm)	Effective aperture (mm)
2.25	13.7 ± 0.7	18.3 ± 0.1
3.5	9.65 ± 0.07	19.2 ± 0.2
5.0	8.50 ± 0.06	18.6 ± 0.02

Note: These parameters were determined from lateral beam profile measurements as in Figure 10. In each case, the carrier frequency of the narrow-band pulses employed was chosen to equal the nominal frequency.

From Hall, T. J., Madsen, E. L., Zagzebski, J. A., and Boote, E. J., *J. Acoust. Soc. Am.,* 85, 2410–2416, 1989. With permission.

coefficient is negligible compared to that of the glass beads. This fact was verified using a sample containing graphite powder, but no glass beads.

Ultrasonic speeds and frequency-dependent attenuation coefficients were measured using a through transmission-technique[27] at 20°C. Ultrasonic speeds were measured at discrete frequencies between 1 and 7 MHz. Attenuation coefficients were then curve fitted assuming the functional form $\alpha(\omega) = \alpha_1\omega + \alpha_2\omega^2$ where ω is frequency and α_1 and α_2 are constants. The small dependence of ultrasonic speed on frequency was determined using the Kramers-Kronig relation given by Ginzberg[33] and was used in computing the complex wavenumber. Ultrasonic speeds, attenuation coefficients, and densities for both phantoms are given in Table 4.

Backscatter coefficients for these phantoms were computed independently using the theory of Faran. Required parameters for the glass beads were the density (2.4 g/cm³), Poisson's ratio (0.21), and the longitudinal speed of sound (5570 m/s).* The density and ultrasonic speed of the media surrounding the beads are given in Table 2. Backscatter coefficients were computed for each particle size in the distribution (such as shown in Figure 4), and these results were combined in a weighted sum using Equation 22.

Use of phantom 2 involved a somewhat different experimental arrangement, shown in Figure 11, in which phantom 2 was in contact with the transducer face and tissue-like attenuation was present over the full acoustic path. This setup mimics typical *in vivo* studies and is another important step toward clinical application of backscatter coefficient measurements.

---

* Glass bead density was measured in our lab using water displacement in a graduated cylinder with a large mass of glass beads. Poisson's ratio and the longitudinal speed of sound were calculated from parameters supplied by the manufacturer (Potter's Industries).

TABLE 4
Density and Ultrasonic Properties of the Two Phantoms Used in This Work

| | Attenuation coefficient | | | |
Phantom	α_1 (dB cm^{-1} MHz^{-1})	α_2 (dB cm^{-1} MHz^{-2})	Speed of sound (m s^{-1})	Density (g cm^{-3})
1 (Low attenuation)	0.046	0.013	1532 ± 3 (2.00 MHz)	1.00 ± 0.005
2 (Tissue-like attenuation)	0.48	0.018	1525 ± 3 (2.50 MHz)	1.04 ± 0.005

Note: The attenuation coefficient is described in terms of two constants, α_1 and α_2, corresponding to a curve fitting to experimental values of the form $\alpha(\omega) = \alpha_1\omega + \alpha_2\omega^2$.

From Hall, T. J., Madsen, E. L., Zagzebski, J. A., and Boote, E. J., *J. Acoust. Soc. Am.*, 85, 2410–2416, 1989. With permission.

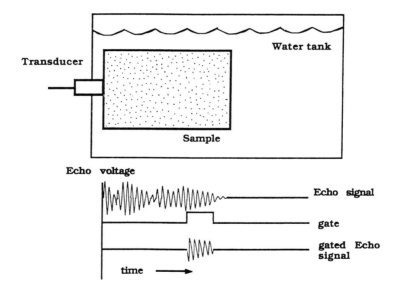

FIGURE 11.   Illustration of the experimental arrangement in which the phantom is maintained in direct contact with the transducer. Echo signal voltages are depicted along with a rectangular gate applied to the detected signal. Gated echo signals are sent to ta transient recorder to be digitized and stored for later analysis. (From Hall, T. J., Madsen, E. L., Zagzebski, J. A., and Boote, E. J., *J. Acoust. Soc. Am.*, 85, 2410–2416, 1989. With permission.)

3. Results and Discussion

To test the accuracy of the method over a broad range of frequencies, backscatter coefficients were determined at ten frequencies in the range from 2 to 6 MHz using phantom 1. All three focused transducers were employed. In each case, the volume interrogated was in the focal region of the transducer.

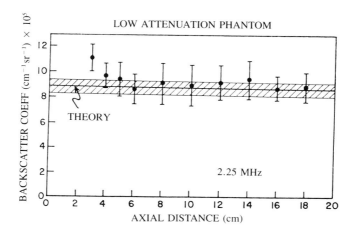

FIGURE 12. Experimental values of the backscatter coefficient $\eta(\omega_o)$ for phantom 1 (low attenuation) at various transducer-to-scattering-volume distances using the nominal 2.25 fMHz transducer at $\omega_o = 2.25$ MHz. The error bars correspond to our estimate of the uncertainty. The pulse and gate durations were 5 and 10 μs, respectively. The horizontal cross-hatched band corresponds to the predicted value based on Faran's theory[23] $\pm$ our estimate of the uncertainty as determined by uncertainties in the physical properties of the glass beads and agar medium. (From Hall, T. J., Madsen, E. L., Zagzebski, J. A., and Boote, E. J., *J. Acoust. Soc. Am.*, 85, 2410–2416, 1989. With permission.)

The 2.25 MHz transducer was used at frequencies of 2.0, 2.25, 2.5, and 3.0 MHz; the 3.5 MHz transducer was used at frequencies of 3.5 and 4.0 MHz; and the 5 MHz transducer was used at frequencies of 4.5, 5.0, 5.5, and 6.0 MHz.

Backscatter coefficients η obtained for various experimental situations are shown in Figures 12 through 16. In Figures 12 through 15, the frequency at which η was determined was either 2.25 or 5.0 MHz employing the transducers with nominal frequencies of 2.25 and 5.0 MHz, respectively. In each figure, error bars correspond to our estimate of the total uncertainty (accuracy, as opposed to precision) in the measured value. This total uncertainty includes the standard deviation of the mean[34] of $\|V_s(\omega_o)\|^2$ combined, using the standard method for propagation of errors,[34] with instrumental uncertainties in measurements of medium attenuation coefficients and transducer diameters and radii of curvature. The uncertainty in $\|V_s(\omega_o)\|^2$ dominates in most cases, that due to the attenuation coefficient becoming comparable only for the most distal values of η (on the right) in Figure 15. The horizontal cross-hatched bar in each figure corresponds to the backscatter coefficient computed using the theory of Faran as described in Reference 24. (The computed value is at the center of the bar and the height of the bar corresponds to the estimated uncertainty in that value; a discussion of the basis for this accuracy estimate is given below.)

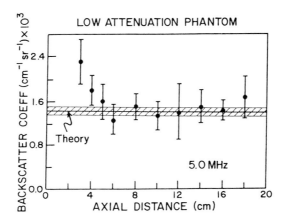

FIGURE 13.   Experimental values of $\eta(\omega_o)$ for phantom 1 (low attenuation) at various transducer-to-scattering-volume distances using the nominal 5 MHz transducer at $\omega_o = 5.0$ MHz. All other parameters are the same as described for Figure 12. (From Hall, T. J., Madsen, E. L., Zagzebski, J. A., and Boote, E. J., *J. Acoust. Soc. Am.*, 85, 2410–2416, 1989. With permission.)

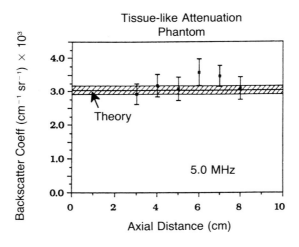

FIGURE 14.   Experimental values of $\eta(\omega_o)$ for phantom 2 (tissuelike attenuation) at various transducer-to-scattering-volume distances using the nominal 5 MHz transducer. All other parameters are the same as given for Fig. 13. (From Hall, T. J., Madsen, E. L., Zagzebski, J. A., and Boote, E. J., *J. Acoust. Soc. Am.*, 85, 2410–2416, 1989. With permission.)

Figures 12 and 13 show backscatter coefficients measured at 2.25 and 5.0 MHz, respectively, for phantom 1. Here, η is plotted for a broad range of transducer-to-scattering-volume distances, viz., 3 to 18 cm. In both cases, the volume interrogated was a few centimeters beyond the proximal scanning window, as depicted in Figure 1.

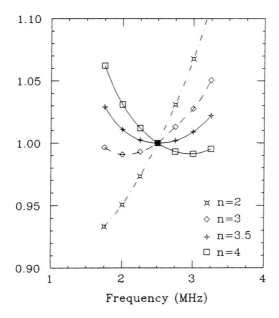

FIGURE 17. Backscatter ratios for data acquired with the 2.25 MHz transducer excited by broad-band pulses. Power laws, $g^2_{trial}(\omega) = \omega^n$, were assumed for several values of the exponent n. Values were normalized to that at 2.5 MHz. (From Hall, T. J., Ph.D. thesis, University of Wisconsin, Madison, 1988. With permission.)

low attenuation compared to soft tissues. ω_o is the frequency at the peak of $\|T(\omega)B_o(\omega)\|$. None of the power laws results in zero slope.

Analysis of Figure 17 prompted the modified method for estimating the correct $g^2(\omega)$. The power law giving the zero slope point in Figure 17 at the higher frequency (ω^3) was used to compute η and $d\eta/d\omega$ at the higher frequency. η and $d\eta/d\omega$ were also computed at ω_o using the power law giving zero slope in Figure 17 at $\omega_o(g^2(\omega) = \omega^{3.5})$, and the same procedure was used for the lower frequency. Thus, six equations result. These six equations were then used to determine the six coefficients in the following truncated power series representation of $\eta(\omega)$:

$$\eta(\omega) = a_0 + a_1\omega + a_2\omega^2 + a_3\omega^3 + a_4\omega^4 + a_5\omega^5$$

The resulting function was used to define a new $g^2(\omega)$ which was then used in the right side of Equation 24 to yield the final frequency dependent backscatter coefficient.

In Figure 18 are shown backscatter coefficients determined with this modified (polynomial) method using a nominal 2.25 MHz focused transducer and a nominal focused 5 MHz transducer. The solid curve is that computed using the theory of Faran.[23] One polynomial for each case yielded the results.

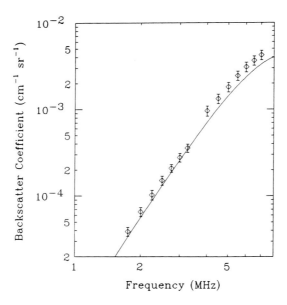

FIGURE 18.   Backscatter coefficients for a low attenuation phantom with glass bead scatterers for data acquired with focused transducers excited with broad-band pulses and analyzed using deduced polynomials for $g^2(\omega)$. Values below 3.5 MHz were determined using a nominal 2.25 MHz transducer and those above 3.5 MHz using a nominal 5 MHz transducer. The solid line corresponds to values predicted with the Faran theory.[23] (From Hall, T. J., Ph.D. thesis, University of Wisconsin, Madison, 1988. With permission.)

The phantom consisted of glass beads of mean diameter 59 μm distributed in agar gel.

In Figure 19 are shown results using a nominal 1.6 MHz focused transducer and two different phantoms having the glass bead scatterers with the same diameter distribution and concentrations, but one has a low attenuation (phantom 3) and the other a higher tissue-mimicking attenuation (phantom 4). The best fit to the Faran curve occurs for the low attenuation phantom, although in magnitude, good agreement is demonstrated for both phantoms.

V. APPLICABILITY OF THE METHOD TO SOFT TISSUES

The variations in compressibility and density in soft tissues cannot generally be described as a set of discrete scatterers isolated from one another. For example, the connective tissue matrix binding parenchymal cells together involves a complicated continuum aspect and might be described as a continuous random medium.

Such tissue, however, could be modeled by a set of distinct scatterers which overlap one another. For example, consider a set of scatterers consisting

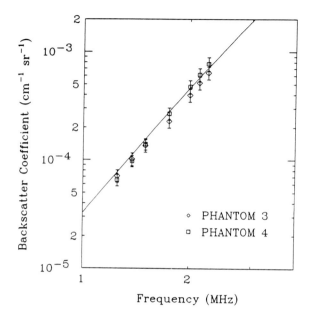

FIGURE 19. Backscatter coefficients for phantoms 3 (low attenuation) and 4 (tissue-like attenuation) using a nominal 1.6 MHz transducer and broad-band analysis. Results were determined with deduced polynomials for $g^2(\omega)$ for each phantom. The solid line corresponds to Faran values.[23] (From Hall, T. J., Ph.D. thesis, University of Wisconsin, Madison, 1988. With permission.)

of M subsets, the ℓth subset consisting N_ℓ identical scatterers randomly positioned in space within some volume Ω_o. A scatterer of the ℓth subset might be a three-dimensional Gaussian distribution in density and compressibility variations characterized by values for full-widths-at-half-maximum and peak values in density and compressibility variations. If the scatterer is not spherically symmetric, then the orientation of all scatterers would be the same, meaning that the scattered waves from each would be identical for the same incident wave. Some point can be identified with the "position" of this scatterer; if the peaks in density and compressibility occur at the same point, this point would reasonably be chosen to represent the position of the scatterer. Since the positions of the scatterers in each subset are random, the scatterers can overlap to any extent, resulting in a continuum aspect of the subset itself. If the concentration of scatterers (scatterer positions) and the size of the scatterers in the subset satisfy the conditions of Section III.A., then a backscatter coefficient, η_ℓ, will be defined in Section III.B. where $N = N_\ell$ and $\vec{r}_i$ is the position vector for the ith scatterer in the subset.

Likewise, there will be a backscatter coefficient defined for each subset and we have a total backscatter coefficient

$$\eta(\omega) = \sum_{\ell=1}^{M} \eta_\ell(\omega) \tag{25}$$

where large-scale overlapping of scatterers within subsets and between subsets can occur.

What distribution in actual compressibility and density variations does the above set of overlapping scatterers represent? Consider Equation 8.1.13 of Morse and Ingard[36] as a starting point, viz.,

$$p_\omega(\vec{r}) =$$

$$p_{inc}(\vec{r}) + \iiint_\Omega [k^2\gamma_{\kappa'}(\vec{r}')p_\omega(\vec{r}') + \gamma_{\rho'}(\vec{r}')\nabla'p_\omega(\vec{r}') \tag{26}$$

$$\cdot \nabla'g_\omega(\vec{r}, \vec{r}')] \, d\vec{r}'$$

where k is the wave number, ω is the frequency, $\vec{r}$ is the position vector of the point in space at which the total acoustic pressure, $p_\omega(\vec{r})$, and incident acoustic pressure, $p_{inc}(\vec{R})$, are defined, and $\gamma_\kappa \equiv (\kappa' - \kappa)/\kappa \equiv \Delta\kappa'/\kappa$ and $\gamma_{\rho'} \equiv (\rho' - \rho)/\rho \equiv \Delta\rho'/\rho'$. k and p are the uniform compressibility and density outside of the volume Ω_o and κ' and ρ' are the (variable) values inside Ω_o. $\Delta\kappa'$ and $\Delta\rho'$ are the variations of the compressibility and density from κ and ρ, respectively, inside Ω_o.

$$g_\omega(r, r') \equiv \frac{e^{ik|\vec{r} - \vec{r}'|}}{4\pi|\vec{r} - \vec{r}'|}$$

is the Green's function. The scattered pressure waves are given by the integral over Ω_o in Equation 26. The scattered wave amplitude is linear in $\gamma_{\kappa'}$ and $\gamma_{\rho'}$. Thus, the total amplitude due to superimposed scatterers is also given by Equation 26 in which

$$\gamma_{\kappa'}(\vec{r}') = \sum_{i=1}^{N_o} \gamma_{\kappa',i}(\vec{r}') = \sum_{i=1}^{N_o} \Delta\kappa_i'(\vec{r}')/\kappa \quad \text{and}$$

$$\gamma_{\rho'}(\vec{r}') = \sum_{i=1}^{N_o} \gamma_{\rho',i}(\vec{r}') = \sum_{i=1}^{N_o} \Delta\rho_i'(\vec{r}')/\rho'$$

where i is the label of the position coordinate of the ith scatterer and N_o is the total number of scatterers in Ω_o.

The backscatter coefficient for these scatterers is given by Equation 25. The question remains: What are the actual (total) distributions in compressibility, K_{TOT}, density, ρ_{TOT}, corresponding to these superior posed scatterers? $\kappa'_{TOT}(\vec{r}')$ is defined simply by

$$\sum_{i=1}^{N_o} \Delta\kappa_i' = \kappa_{TOT}' - \kappa; \text{ i.e., } \kappa_{TOT}' = \sum_{i=1}^{N_o} \Delta\kappa_i' + \kappa$$

$\rho'_{TOT}(r')$ can be found from

$$\sum_{i=1}^{N_o} \Delta\rho_i'/\rho = (\rho_{TOT}' - \rho)/\rho_{TOT}'$$

The result is

$$\rho_{TOT}' = N_o \prod_{i=1}^{N_o} \rho_i' \left[\sum_{i=1}^{N_o} \prod_{\substack{j=1 \\ j \neq i}}^{N_o} \rho_j' \right]^{-1}$$

In summary, the backscatter coefficient determined with the method described in Section III.B. applies to a continuum of scatterers as well as to a set of discrete scatterers. The meaning of the coherent term in the denominator of Equation 17 is not so clear, however, when the scatterers involved are comparable in size to the wavelength. It is probably negligible, but sufficiently large gate durations can be employed to assure that relation 18 applies.

VI. SIMULTANEOUS DETERMINATION OF BACKSCATTER AND ATTENUATION COEFFICIENTS

If it is reasonable to assume that both the backscatter coefficient and attenuation coefficient in an interrogated volume are uniform for any given frequency and that the attenuation coefficient everywhere between the transducer and uniform scattering medium is known, then the two coefficients can be simultaneously determined with the method described in Section IV.B. The idea for accomplishing this is rather simple: If the attenuation coefficient, α, used in the complex wavenumber, $k = \omega/c + i\,\alpha$ is smaller (larger) than that actually existing in the medium, the backscatter coefficient will appear to decrease (increase) with depth because the backscatter pressure waves would decrease in amplitude more rapidly with depth than assumed. Thus, to determine the backscatter and attenuation coefficients simultaneously, one can introduce trial values for α until the apparent backscatter coefficient becomes independent of depth.

ACKNOWLEDGMENTS

This work was supported in part by NIC grants R01CA39224, R01CA25634, and P30CA14520.

REFERENCES

1. **Sigelmann, R. A. and Reid, J. M.,** Analysis and measurement of ultrasound backscattering from an ensemble of scatterers excited by sine-wave bursts, *J. Acoust. Soc. Am.,* 53, 1351–1355, 1973.
2. **Morse, P. M. and Feshbach, H.,** *Methods of Theoretical Physics,* McGraw-Hill, New York, 1953, 1066.
3. **Burke, T. M.,** Quantitative characterization of the intrinsic scatter nature of tissue and tissue-mimicking materials, Ph.D. thesis, University of Wisconsin, Madison, 1982.
4. **Burke, T. M., Madsen, E. L., and Zagzebski, J. A.,** Differential scattering cross sections per unit volume to tissue-mimicking materials, *Ultras. Imag.,* 4, 174, 1982.
5. **O'Donnell, M. and Miller, J. G.,** Quantitative broadband ultrasonic backscatter: an approach to nondestructive evaluation in acoustically inhomogeneous materials, *Appl. Phys.,* 52, 1056–1065, 1981.
6. **Nicholas, D.,** Orientation and frequency dependence of backscattered energy and its clinical application, in *Recent Advances in Ultrasound in Medicine,* White, D. N., Ed., Research Studies Press, Forest Grove, OR, 1977, 29–54.
7. **Bamber, J. C. and Hill, C. R.,** Acoustic properties of normal and cancerous human liver. I. Dependence on pathological conditions, *Ultras. Med. Biol.,* 7, 121–133, 1981.
8. **Nicholas, D., Hill, C. R., and Nassiri, D. K.,** Evaluation of backscattering coefficients for excised human tissues: principles and techniques, *Ultras. Med. Biol.,* 8, 7–15, 1982.
9. **Lizzi, F. L., Greenebaum, M., Feleppa, E. J., and Elbaum, M.,** Theoretical framework for spectrum analysis in ultrasonic tissue characterization, *J. Acoust. Soc. Am.,* 73, 1366–1373, 1983.
10. **Campbell, J. A. and Waag, R. C.,** Normalization of ultrasonic scattering measurements to obtain average differential scattering cross sections for tissues, *J. Acoust. Soc. Am.,* 74, 393–399, 1983.
11. **Chivers, R. C. and Hill, C. R.,** A spectral approach to ultrasonic scattering from human tissues: methods, objectives and backscattering measurements, *Phys. Med. Biol.,* 20, 799–815, 1975 (see particularly pp. 808–810).
12. **Madsen, E. L., Insana, M. F., and Zagzebski, J. A.,** Method of data reduction for accurate determination of acoustic backscatter coefficients, *J. Acoust. Soc. Am.,* 76, 913–923, 1984.
13. **Madsen, E. L., Goodsitt, M. M., and Zagzebski, J. A.,** Continuous waves generated by focused radiators, *J. Acoust. Soc. Am.,* 70, 1508–1517, 1981.
14. **Beyer, R. T. and Letcher, S. V.,** *Physical Ultrasonics,* Academic Press, New York, 1969, 14.
15. **Morse, P. M. and Ingard, K. U.,** Theoretical Acoustics, McGraw-Hill, New York, 1968, 426. (Note: The authors used $\phi(\theta)$ instead of $\phi(K,\cos\theta)$ as we have.)
16. **Morse, P. M. and Ingard, K. U.,** Theoretical Acoustics, McGraw-Hill, New York, 1968, 336, 425. (Note particularly that $j_m (ka) = (1/2)[h_m(ka) + h_m*(ka)]$ where a is the radius of the spherical scatterer.
17. **Feller, W.,** *An Introduction to Probability Theory and its Applications,* Vol. I, 3rd ed., John Wiley & Sons, New York, 1968, 222.
18. **Sokolnikoff, I. A. and Redheffer, R. M.,** *Mathematics of Physics and Modern Engineering,* 2nd ed., McGraw-Hill, New York, 1960, 639.
19. **Glotov, V. P.,** Coherent scattering of sound from clusters of discrete inhomogeneities in pulsed emission, *Sov. Phys. Acoust.,* 8, 220–222, 1963.
20. **Twersky, V.,** Acoustic bulk parameters in distributions of pair-correlated scatterers, *J. Acoust. Soc. Am.,* 64, 1710, 1978.
21. **Shung, K. K.,** An alternative approach for formulating the backscattering equation in Sigelmann and Reid's method, *J. Acoust. Soc. Am.,* 73, 1384–1386, 1983.
22. **Shung, K. K.,** On the ultrasound scattering from blood as a function of hematocrit, *IEEE Trans. Sonics Ultras.,* SU-29, 327, 1982.

23. **Faran, J. J., Jr.**, Sound scattering by solid cylinders and spheres, *J. Acoust. Soc. Am.*, 23, 405–418, 1951.
24. **Insana, M. F., Madsen, E. L., Hall, T. J., and Zagzebski, J. A.**, Tests of the accuracy of a method of data reduction for determination of acoustic backscatter coefficients, *J. Acoust. Soc. Am.*, 79, 1230–1236, 1986.
25. **Hall, T. J., Madsen, E. L., Zagzebski, J. A., and Boote, E. J.**, Accurate depth-independent determination of acoustic backscatter coefficients with focused transducers, *J. Acoust. Soc. Am.*, 85, 2410–2416, 1989.
26. **Hall, T. J.**, Experimental methods for accurate determination of acoustic backscatter coefficients, Ph.D. thesis, University of Wisconsin, Madison, 1988.
27. **Madsen, E. L., Zagzebski, J. A., Banjavic, R. A., and Jutila, R. E.**, Tissue-mimicking materials for ultrasound phantoms, *Med. Phys.*, 5, 391–394, 1978.
28. **Burke, T. M., Goodsitt, M. M., Madsen, E. L., and Zagzebski, J. A.**, Angular distribution of scattered ultrasound from a single steel sphere in agar gel: a comparison between theory and experiment, *Ultras. Imag.*, 6, 342–347, 1984.
29. **Insana, M. F.**, Methods for measuring ultrasonic backscatter and attenuation coefficients for tissues and tissue-like media, Ph.D. thesis, University of Wisconsin, Madison, 1983.
30. **Bracewell, R. N.**, *The Fourier Transform and its Applications*, McGraw-Hill, New York, 1978.
31. **O'Neil, H. T.**, Theory of focusing radiators, *J. Acoust. Soc. Am.*, 21, 516–526, 1949.
32. **Kinsler, L. E., Frey, A. R., Coppens, A. B., and Sanders, J. V.**, *Fundamentals of Acoustics*, John Wiley & Sons, New York, 1982, 453.
33. **Ginzberg, V. L.**, Concerning the general relationship between absorption and dispersion of sound waves, *Sov. Phys. Acoust.*, 1, 32–41, 1955.
34. **Bevington, P. R.**, *Data Reduction and Error Analysis for the Physical Sciences*, McGraw-Hill, New York, 1969, 72, 116.
35. **Boote, E. J., Zagzebski, J. A., Madsen, E. L., and Hall, T. J.**, Instrument independent acoustic backscatter imaging, *Ultras. Imag.*, 10, 121–138, 1988.
36. **Morse, P. M. and Ingard, K. U.**, *Theoretical Acoustics*, McGraw-Hill, New York, 1968, 419.

Chapter 8

MEASUREMENT SYSTEM EFFECTS IN ULTRASONIC SCATTERING EXPERIMENTS

Robert C. Waag and Jeffrey P. Astheimer

TABLE OF CONTENTS

0-8493-6568-6/93/$0.00 + $.50

I. INTRODUCTION

Experimental design for the measurement of intrinsic properties such as average differential scattering cross-section or the power spectrum of medium variations is very important. In the case of tissue characterization, as well as in the characterization of other scattering media, not much may be known about the properties to be measured. Then detailed knowledge of the measurement system characteristics, especially the resolution, allows an estimate of medium properties such as the maximum rate of fluctuation in a given spatial-frequency range that the measurement system can be used to infer. Alternatively, knowledge of medium characteristics such as the pertinent spatial-frequency fluctuations allows the selection of experimental conditions for their efficient determination. These considerations coupled with the absence of a general model to describe the performance of an ultrasonic scattering measurement system have motivated the analysis and computations presented here.

The effects of transducer beams and time waveforms on ultrasonic measurements of scattering medium properties have been accommodated in a number of ways. The 3-dB beamwidth has been employed in some early studies[1,2] to bound the scattering volume while the 6-dB beamwidth has been used in other studies,[3,4] but more complex models of beams have subsequently been employed.[5-11] The incident waveform and the detector time gate have been characterized in some early analyses[1-4] by assuming the scattering is bounded in range by two planar surfaces although other analyses[5-12] have employed emitter waveform and detector gate models which lead to a non-uniform weighting of the scattering volume. The overall effect of the measurement system has also been considered in the estimation of ultrasonic attenuation where the words diffraction correction have been used to describe the process by which the measurement system weight is removed to obtain the intrinsic attenuation, but work[13-15] in this area only models the backscatter configuration and generally results in a frequency-independent scaling of the data as a function of distance. The emphasis in all the noted studies has, however, in contrast to the treatment here, been on spatial localization of the scattering volume without determination of the spatial-frequency spread that results from this localization. Also in contrast to the treatment here, the most extensive of these studies either, as in Reference 7, is limited to a narrow-band analysis in the space domain and employs only a far-field approximation for the emitter and detector beams or, as in Reference 8, is limited to the backscatter configuration and does not account for a finite correlation length. Furthermore, the treatment presented here extends the foregoing studies to show general relations that must exist between scattering medium properties and measurement system parameters to determine intrinsic scattering characteristics under practical conditions.

This chapter describes a general model for arbitrary emitter time waveforms and beam patterns and arbitrary detector time gates and sensitivity functions. Arbitrary fields are represented as angular spectra of plane waves and arbitrary time waveforms as frequency spectra of temporal harmonics in an integral expression that includes instrument effects on the spectrum of the scattered pressure. A transformation of variables then puts this integral into a form convenient for interpretation. Simplified expressions are developed for the effect of the emitter beam and detector sensitivity patterns when the detector time gate has an infinite duration. Simplified expressions are also developed for the effect of the emitter waveform and detector time gate using a narrow-band approximation and for the effect of the emitter beam and detector sensitivity patterns using a quasi plane wave approximation. The simplifications yield insightful forms for measurement system characteristics and these forms are evaluated analytically for Gaussian spatial apertures and Gaussian temporal waveforms. Numerical evaluations of general expressions for the weight of Gaussian, exponential, and uniform spatial apertures are also given. The results indicate the range of instrumentation parameters that can be employed to measure intrinsic characteristics of a scattering medium and thus may be used to design experiments from which intrinsic parameters of scattering media are obtained.

II. THEORY

This section develops families of expressions that describe the weight practical instrumentation imposes on the measurement of intrinsic scattering properties under a broad range of conditions. The first family, subscripted W, consists of wide-band quantities applicable in scattering measurements when the emitter waveform is a short pulse and the detector time gate is narrow. The second family, subscripted A, is for the limiting case of an infinite length detector time gate and describes the influence of the beam patterns alone on the measurement of scattering characteristics. The third family, subscripted N, is for a narrow-band case in which the effect of the detector time gate is described by a multiplication in real space. Each family of expressions consists of a system function $\Lambda_{(\cdot)}$ (Equations 15, 28, and 41) that describes the distribution of the measurement system weight in wave space, a transfer function $H_{(\cdot)}^V$ (Equations 18, 31, and 44) that applies when the scattering region is contained within the real space volume defined by the system parameters, a transfer function $|H_{(\cdot)}|^2$ (Equations 23, 35, and 46) that applies when the scattering region is defined by the system parameters, and a normalized factor $N_{(\cdot)}$ (Equations 25, 37, and 48) that has the physically appealing units of (power)·(length)^3 when divided by acoustic impedance. Within each family, a quasi plane wave approximation that yields a simplified expression for the combined weight of the emitter beam pattern and detector sensitivity function is used to obtain less complicated expressions (Table 1) denoted by corresponding sans serif characters. The quasi plane wave approximation is also used to obtain uncomplicated second-moment expressions (Table 2) that describe the resolution of scattering media power spectra measurements provided by a given experimental configuration.

TABLE 1
System Transfer Functions and Beam Power Factor Obtained for the Infinite Length Detector Gate, Narrow-Band, and Wide-Band Cases Each Under the Quasi Plane Wave Approximation

	Infinite length detector gate	Narrow-band approximation	Wide-band case
System transfer functions	$H_A^V(\mathbf{K}_0, w_0) = (2\pi)^4 C_A W_E(w_0)$ $\times \int_u \Psi(\mathbf{u}) d^3 u$	$H_N^V(\mathbf{K}_c, w_c) = (2\pi)^4 C_A c \tan\theta$ $\times \int_u \Phi_N(\mathbf{u}, \theta, w_c) d^3 u$	$H_W^V(\mathbf{K}_0, w_0) = (2\pi)^4 C_A c \tan\theta$ $\times \int_u \Phi_W(\mathbf{u}, \theta, w_0) d^3 u$
	$|H_A(\mathbf{K}_0, w_0)|^2 = (2\pi)^8 |C_A|^2 |W_E(w_0)|^2$ $\times \frac{1}{\sin 2\theta} \int_u |\Psi(\mathbf{u})|^2 d^3 u$	$|H_N(\mathbf{K}_c, w_c)|^2 = (2\pi)^8 |C_A|^2 \frac{c^2}{2} \frac{\sin\theta}{\cos^3\theta}$ $\times \int_u |\Phi_N(\mathbf{u}, \theta, w_c)|^2 d^3 u$	$|H_W(\mathbf{K}_0, w_0)|^2 = (2\pi)^8 |C_A|^2 \frac{c^2}{2} \frac{\sin\theta}{\cos^3\theta}$ $\times \int_u |\Phi_W(\mathbf{u}, \theta, w_0)|^2 d^3 u$
Beam power factor	$N_A(\mathbf{K}_0, w_0) = \frac{|C_A|^2}{k_0^4} \frac{1}{\sin 2\theta}$ $\times \int_u |\Psi(\mathbf{u})|^2 d^3 u$	$N_N(\mathbf{K}_c, w_c) = \frac{|C_A|^2 c^2/2}{k_c^4 |W_D(w_c)|^2} \frac{\sin\theta}{\cos^3\theta}$ $\times \int_u |\Phi_N(\mathbf{u}, \theta, w_c)|^2 d^3 u$	$N_W(\mathbf{K}_0, w_0) = \frac{|C_A|^2 c^2/2}{k_0^4 W_E(w_0)|^2} \frac{\sin\theta}{\cos^3\theta}$ $\times \int_u |\Phi_W(\mathbf{u}, \theta, w_0)|^2 d^3 u$
	$\Psi(\mathbf{u}) = \int_\zeta \sqrt{\vee_E\left(\frac{u_1}{2} + \zeta, u_2\right)}$ $\times \sqrt{\vee_D\left(\frac{u_1}{2} - \zeta, u_3\right)} d\zeta$	$\Phi_N(\mathbf{u}, \theta, w_c) = \int_q \Psi(u_1, u_2 - q, u_3 + q)$ $\times W(c\tan\theta q - w_c) dq$	$\Phi_W(\mathbf{u}, \theta, w_0) = \int_q \Psi(u_1, u_2 - q, u_3 + q)$ $\times W_E(c\tan\theta q) W_D(w_0 - c\tan\theta q)$

TABLE 2
Second Central Moment Expressions for the Squared Magnitude of the System Function

	Infinite length detector gate	Narrow-band approximation	Wide-band case
K_x Direction	$\sigma_1^2(\wedge_A) = \dfrac{\int_u u_1^2 |\Psi(u)|^2 d^3u}{m(\Psi)}$	$\sigma_1^2(\wedge_N) = \dfrac{\int_u u_1^2 |\Phi_N(u,\theta,w_c)|^2 d^3u}{m(\Phi_N)}$	$\sigma_1^2(\wedge_w) = \dfrac{\int_u u_1^2 |\Phi_w(u,\theta,w_0)|^2 d^3u}{m(\Phi_w)}$
K_y Direction	$\sigma_2^2(\wedge_A) = \dfrac{\cos^2\theta \int_u (u_2^2+u_3^2)|\Psi(u)|^2 d^3u}{m(\Psi)}$	$\sigma_2^2(\wedge_N) = \dfrac{\cos^2\theta \int_u (u_2^2+u_3^2)|\Phi_N(u,\theta,w_c)|^2 d^3u}{m(\Phi_N)}$	$\sigma_2^2(\wedge_w) = \dfrac{\cos^2\theta \int_u (u_2^2+u_3^2)|\Phi_w(u,\theta,w_0)|^2 d^3u}{m(\Phi_w)}$
K_z Direction	$\sigma_3^2(\wedge_A) = \dfrac{\sin^2\theta \int_u (u_2^2+u_3^2)|\Psi(u)|^2 d^3u}{m(\Psi)}$	$\sigma_3^2(\wedge_N) = \dfrac{\sin^2\theta \int_u (u_2^2+u_3^2)|\Phi_N(u,\theta,w_c)|^2 d^3u}{m(\Phi_N)}$	$\sigma_3^2(\wedge_w) = \dfrac{\sin^2\theta \int_u (u_2^2+u_3^2)|\Phi_w(u,\theta,w_0)|^2 d^3u}{m(\Phi_w)}$
Radial direction	$\sigma_K^2(\wedge_A) = \dfrac{\int_u u^2 |\Psi(u)|^2 d^3u}{m(\Psi)}$	$\sigma_K^2(\wedge_N) = \dfrac{\int_u u^2 |\Phi_N(u,\theta,w_c)|^2 d^3u}{m(\Phi_N)}$	$\sigma_K^2(\wedge_w) = \dfrac{\int_u u^2 |\Phi_w(u,\theta,w_0)|^2 d^3u}{m(\Phi_w)}$
	$m(\Psi) = \int_u |\Psi(u)|^2 d^3u$	$m(\Phi_N) = \int_u |\Phi_N(u,\theta,w_c)|^2 d^3u$	$m(\Phi_w) = \int_u |\Phi_w(u,\theta,w_0)|^2 d^3u$

Note: Obtained for the infinite length detector gate, narrow-band, and wide-band cases each under the quasi plane wave approximation and also the assumption that the angular spectrum of the emitter or detector gate is symmetric in the second variable in the infinite length detector gate case or $\Phi_{()}$ is symmetric in the second or third variable in the narrow-band and wide-band cases.

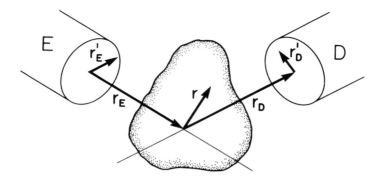

FIGURE 1. Scattering geometry. An emitter E and a detector D are directed at a point that is located at the intersection of the vector $\mathbf{r}_E$ along the axis of the emitter and the vector $\mathbf{r}_D$ along the axis of the detector and is the origin of the scattering volume (bounded by stippling). The vector $\mathbf{r}$ identifies a point within the scattering volume while the vectors $\mathbf{r}'_E$ and $\mathbf{r}'_D$ denote points on the emitter aperture and the detector aperture, respectively.

A. BASIC EXPRESSIONS

This analysis considers an emitter and detector directed at a scattering region as shown in Figure 1. The analysis, taken from References 16 and 17, begins with an expression for the temporal spectrum P_m of the signal that results from applying a time gate to the output of the detector. The expression is obtained by using the Born approximation, free-space Green's function, and angular spectrum representations in which evanescent waves and bandwidth limitations are neglected. The result may be written

$$P_m(\mathbf{r}_E, \mathbf{r}_D, \omega_0) = 2\pi \int_\omega \int_\nu \int_\eta \int_V A_E(\boldsymbol{\nu}, \omega) A_D(\boldsymbol{\eta}, \omega) W_E(\omega) W_D(\omega_0 - \omega)$$

$$\times \, k^2 \left[\gamma_\kappa(\mathbf{r}) + \frac{\mathbf{I} \cdot \mathbf{O}}{k^2} \gamma_\rho(\mathbf{r}) \right] e^{j(\mathbf{I} - \mathbf{O}) \cdot \mathbf{r}} \, d^3r \, d^2\eta \, d^2\nu \, d\omega \qquad (1)$$

$$A_E(\boldsymbol{\nu}, \omega) = C_E V_E(\boldsymbol{\nu}) \frac{1}{2\pi} \frac{j}{\sqrt{k^2 - \nu^2}} e^{j\mathbf{I}\cdot\mathbf{r}_E}, \quad C_E = -j\omega\rho 2\pi$$

$$A_D(\boldsymbol{\eta}, \omega) = C_D V_D(\eta_1, -\eta_2) \frac{1}{2\pi} \frac{j}{\sqrt{k^2 - \eta^2}} e^{j\mathbf{O}\cdot\mathbf{r}_D}, \quad C_D = \frac{1}{2}$$

$$\mathbf{I} = (\boldsymbol{\nu}, \nu_3), \quad \boldsymbol{\nu} = (\nu_1, \nu_2), \quad \nu_3 = \sqrt{k^2 - \nu^2}, \quad \nu^2 = \nu_1^2 + \nu_2^2$$

and

$$\mathbf{O} = (\boldsymbol{\eta}, \eta_3), \quad \boldsymbol{\eta} = (\eta_1, \eta_2), \quad \eta_3 = \sqrt{k^2 - \eta^2}, \quad \eta^2 = \eta_1^2 + \eta_2^2$$

In these expressions, A_E is the angular spectrum representation of the emitter beam perpendicular to the axis of the emitter at the center of the scattering volume, A_D is the angular spectrum of the detector sensitivity pattern perpendicular to the axis of the detector at the center of the scattering volume, W_E and W_D are the temporal spectra of the emitter waveform and the detector time gate, respectively, V_E is the angular spectrum of the normal component of the acoustic particle velocity at the emitting aperture, V_D is the angular spectrum of the detector sensitivity pattern, γ_κ and γ_ρ are the variations in compressibility and density, respectively, r_E and r_D are the distances to the center of the scattering volume from the emitter and detector respectively, and k is the wave number given by ω/c. The use of angular spectra for the emitter and detector spatial aperture functions permits focusing to be accommodated. The offset ω_o in W_D reflects the frequency domain convolution arising from multiplication of the temporal signal at the detector by a time gate which is assumed to be a lowpass function.

The quadruple integral given in Equation 1 for the time harmonic spectrum of the measured pressure is an extension of the textbook expression[18] that describes the scattered pressure at a given frequency as a factor times a volume integral over the medium variations under the assumptions of plane wave incidence and far-field point reception. The extension includes nonplane wave incidence by the ν integration, accommodates finite area detection at arbitrary distance by the η integration, and accounts for finite duration time waveforms by the ω convolution. The result is that, in addition to the usual integration over the scattering volume V, the measurement system output can be viewed as summations over the angular spectrum of the scattered pressure and of the incident pressure weighted by the emitter time harmonic spectrum, and then blurred by a convolution of the time harmonic spectrum of the detector gate function. When a single time harmonic frequency plane wave is incident and the receiver is a far-field point followed by an infinite duration time gate, the angular spectrum A_E and both of the frequency functions W_E and W_D become impulsive, and the integrations over the spatial-frequency and temporal-frequency variables ν, η, and ω can be evaluated, using stationary phase and far-field approximations for the η integration, to obtain the textbook result.

The system effects contained in the spatial-frequency functions A_E and A_D and in the temporal-frequency functions W_E and W_D can be represented compactly for further analysis by the parameterization

$$\boldsymbol{\nu} = \boldsymbol{\nu}(\mathbf{K}, \zeta)$$

$$\boldsymbol{\eta} = \boldsymbol{\eta}(\mathbf{K}, \zeta)$$

where the vector $\mathbf{K}$ is defined by the equation

$$\mathbf{K} = \mathbf{I} - \mathbf{O}$$

and ζ is a variable for the path on which the vector $\mathbf{K}$ is fixed. Using the result that

$$\mathbf{I} \cdot \mathbf{O} = k^2 - K^2/2$$

the time harmonic spectrum of P_m given in Equation 1 can be written

$$P_m(\mathbf{K}_0, \omega_0) = \int_{\mathbf{K}} [\Gamma_c(-\mathbf{K})\Lambda_1(\mathbf{K}, \mathbf{K}_0, \omega_0)$$
$$- (K^2/2)\Gamma_\rho(-\mathbf{K})\Lambda_2(\mathbf{K}, \mathbf{K}_0, \omega_0)] \, d^3K \quad (2)$$

In this expression, the relevant medium characteristics are given by

$$\Gamma_{(\cdot)}(\mathbf{K}) = \frac{1}{(2\pi)^3} \int_V \gamma_{(\cdot)}(\mathbf{r}) e^{-j\mathbf{K}\cdot\mathbf{r}} \, d^3r$$

in which $(\cdot)$ is c, κ, or ρ, and

$$\gamma_c(\mathbf{r}) = \gamma_\kappa(\mathbf{r}) + \gamma_\rho(\mathbf{r})$$

Also, the measurement system characteristics are given by

$$\Lambda_1(\mathbf{K}, \mathbf{K}_0, \omega_0) = (2\pi)^4 \int_\omega k^2 W_E(\omega)\Lambda(\mathbf{K}, \mathbf{K}_0, \omega)W_D(\omega_0 - \omega) \, d\omega \quad (3)$$

$$\Lambda_2(\mathbf{K}, \mathbf{K}_0, \omega_0) = (2\pi)^4 \int_\omega W_E(\omega)\Lambda(\mathbf{K}, \mathbf{K}_0, \omega)W_D(\omega_0 - \omega) \, d\omega \quad (4)$$

$$\Lambda(\mathbf{K}, \mathbf{K}_0, \omega) = \int_\zeta A_E[\mathbf{\nu}(\mathbf{K}, \zeta), \omega]A_D[\mathbf{\eta}(\mathbf{K}, \zeta), \omega]|J(\mathbf{\nu}, \mathbf{\eta}|\mathbf{K}, \zeta)| \, d\zeta \quad (5)$$

$\mathbf{K}_0$ is the difference between the vector $\mathbf{I}_0$ of magnitude k_0 directed along the axis of the emitter toward the scattering volume and the vector $\mathbf{O}_0$ of magnitude k_0 directed along the axis of the detector from the scattering volume, and J is the Jacobian of the transformation from $\mathbf{\nu}, \mathbf{\eta}$ to $\mathbf{K}, \zeta$. The temporal harmonic spectrum of the measured pressure is, thus, represented as the sum of two terms, each of which is a weighted average in three-dimensional wave space. The medium variation function being averaged in one term is the spatial-frequency spectrum of the sum of the compressibility and density variations, while only the power spectrum of the density variations is averaged in the other term.

The weighting functions Λ_1 given in Equation 3 and Λ_2 given in Equation 4 are expressed in the spatial-frequency domain in terms of an integral over

The form of Equation 14 for the wide-band system function Λ_W is a convolution different by a factor of k^2 from the convolution defining Λ_2 and different only by a scale factor k_0^2 from the convolution defining Λ_1. In each case, the convolution spreads the measurement system weight throughout wave space enlarging the extent of the beam intersection function Λ.

The integration over $\mathbf{K}$ in Equation 14 can also be expressed in a form similar to a convolution by introducing a shifted version of Λ_W given by

$$\Lambda_W^{(O)}(\mathbf{K}, \mathbf{K}_0, \omega_0) = \Lambda_W(\mathbf{K} - \mathbf{K}_0, \mathbf{K}_0, \omega_0)$$

Then Equation 14 can be written in terms of $\Lambda_W^{(O)}$ as

$$P_m(\mathbf{K}_0, \omega_0) = \int_K \Gamma(\mathbf{K}, \omega_0) \Lambda_W^{(O)}(\mathbf{K}_0 - \mathbf{K}, \mathbf{K}_0, \omega_0) \, d^3K \qquad (16)$$

The difference between Equation 16 and the usual form of a convolution is that the system function $\Lambda_W^{(O)}$ has an additional dependence on $\mathbf{K}_0$. This dependence arises from the varying intersection of the fields produced by the emitter and detector as the angle between the emitter and detector changes.

The spectrum of the measured pressure given in Equation 14 may be further simplified under the additional assumption that both the spectrum of the compressibility variations and the product of K^2 and the spectrum of the density variations vary slowly where the weight of the system is significant. Then, P_m can be expressed

$$P_m(\mathbf{K}_0, \omega_0) = \Gamma(-\mathbf{K}_0, \omega_0) H_W^V(\mathbf{K}_0, \omega_0) \qquad (17)$$

where

$$H_W^V(\mathbf{K}_0, \omega_0) = \int_{\mathbf{K}} \Lambda_W(\mathbf{K}, \mathbf{K}_0, \omega_0) \, d^3K \qquad (18)$$

The effect of the system is, thus, represented by the wide-band system transfer function H_W^V that depends on both the reference scattering vector and temporal frequency.

An expression for the mean-square magnitude of temporal harmonic spectrum of the measured pressure can be obtained from Equation 1 to show measurement system effects when second-order statistics of the scattering medium are sought. Assuming the scattering medium is statistically homogeneous or stationary, i.e., the second-order statistics of the medium variations are independent of origin in space, the relevant correlation functions of the medium parameters are

$$B_{\kappa\rho}(\mathbf{r}) = \langle \gamma_\kappa(\mathbf{r}_1) \gamma_\rho(\mathbf{r}_1 + \mathbf{r}) \rangle$$

$$B_{c\rho}(\mathbf{r}) = \langle \gamma_c(\mathbf{r}_1) \gamma_\rho(\mathbf{r}_1 + \mathbf{r}) \rangle$$

and

$$B_{(\cdot)}(\mathbf{r}) = \langle (\gamma_{(\cdot)}(\mathbf{r}_1) \gamma_{(\cdot)}(\mathbf{r}_1 + \mathbf{r}) \rangle$$

in which $(\cdot)$ is equal to c, κ, or ρ. The corresponding spectral power densities are

$$S_{(\cdot)}(\mathbf{K}) = \frac{1}{(2\pi)^3} \int_V B_{(\cdot)}(\mathbf{r}) e^{-j\mathbf{K}\cdot\mathbf{r}} \, d^3r$$

in which $(\cdot)$ is equal to $\kappa\rho$, $c\rho$, c, κ, or ρ. Then, the mean-square value of the measured temporal harmonic spectrum can be expressed

$$\langle |P_m(\mathbf{K}_0, \omega_0)|^2 \rangle = \int_\mathbf{K} [S_c(\mathbf{K})|\Lambda_1(\mathbf{K}, \mathbf{K}_0, \omega_0)|^2$$

$$- S_{c\rho}(\mathbf{K}) K^2 \Lambda_1(\mathbf{K}, \mathbf{K}_0, \omega_0) \Lambda_2^*(\mathbf{K}, \mathbf{K}_0, \omega_0) \quad (19)$$

$$+ S_\rho(\mathbf{K})(K^4/4)|\Lambda_2(\mathbf{K}, \mathbf{K}_0, \omega_0)|^2] \, d^3K$$

Thus, the mean-square magnitude of the temporal harmonic spectrum of the measured pressure, or more briefly, the power spectrum of the measured pressure, is equal to a sum of terms each of which is a weighted average in three-dimensional wave space. In this general form, three different weighting functions, each expressed in terms of the integrals defined in Equations 4 through 6, are needed to account for the effect of the measurement system on the observation of scattering.

The assumption that k^4 varies slowly where the system has significant weight also allows Equation 19 to be written in terms of the wide-band system function Λ_W as

$$\langle |P_m(\mathbf{K}_0, \omega_0)|^2 \rangle = \int_\mathbf{K} S_\gamma(\mathbf{K}, \omega_0)|\Lambda_W(\mathbf{K}, \mathbf{K}_0, \omega_0)|^2 \, d^3K \quad (20)$$

where

$$S_\gamma(\mathbf{K}, \omega) = S_c(\mathbf{K}) - (K^2/k^2)S_{c\rho}(\mathbf{K}) + (K^4/4k^4)S_\rho(\mathbf{K})$$

or, equivalently,

$$S_\gamma(\mathbf{K}, \omega) = S_\kappa(\mathbf{K}) + 2(1 - K^2/2k^2)S_{\kappa\rho}(\mathbf{K}) + (1 - K^2/2k^2)^2 S_\rho(\mathbf{K})$$

Reasoning similar to that which yields Equation 16 allows Equation 20 for the mean-square value of the measured pressure spectrum to be written in the convolution-like form

$$\langle |P_m(\mathbf{K}_0, \omega_0)|^2 \rangle = \int_{\mathbf{K}} S_\gamma(\mathbf{K}, \omega_0) |\Lambda_W^{(0)}(\mathbf{K}_0 - \mathbf{K}, \mathbf{K}_0, \omega_0)|^2 \, d^3K \qquad (21)$$

Expressions in Equations 20 and 21 show that the power spectrum of the measured pressure may be interpreted as an average of the power spectrum of medium variations S_γ weighted by the squared magnitude of the system function and that the spatial-frequency spread of the system function magnitude determines the resolution achieved in measurements of S_γ.

The additional assumption that the medium statistics as well as the factor K^4 are slowly varying relative to the system functions allows Equation 20 to be written in a simplified, compact form analogous to Equation 17. The form is

$$\langle |P_m(\mathbf{K}_0, \omega_0)|^2 \rangle = S_\gamma(\mathbf{K}_0, \omega_0) |H_W(\mathbf{K}_0, \omega_0)|^2 \qquad (22)$$

where

$$|H_W(\mathbf{K}_0, \omega_0)|^2 = \int_{\mathbf{K}} |\Lambda_W(\mathbf{K}, \mathbf{K}_0, \omega_0)|^2 \, d^3K \qquad (23)$$

In these expressions, the system effect is represented by the wide-band transfer function $|H_W|^2$ that, like H_W^V, depends on the reference scattering vector and temporal frequency. The transfer function $|H_W|^2$ can be interpreted as the normalization constant for the spatial-frequency distribution defined by $|\Lambda_W|^2$.

The wide-band transfer function expression in Equation 23 can be re-written in a form that contains a normalized factor N_W. This is accomplished by introducing the quantity

$$(2\pi)^8 k_0^4 \left| \int_\omega W_E(\omega) W_D(\omega_0 - \omega) \, d\omega \right|^2$$

so that Equation 23 becomes

$$|H_W(\mathbf{K}_0, \omega_0)|^2 = (2\pi)^8 k_0^4 \left| \int_\omega W_E(\omega) W_D(\omega_0 - \omega) \, d\omega \right|^2 N_W(\mathbf{K}_0, \omega_0) \qquad (24)$$

in which

$$N_W(\mathbf{K}_0, \omega_0) = \frac{1}{(2\pi)^8 k_0^4 \left| \int_\omega W_E(\omega) W_D(\omega_0 - \omega) \, d\omega \right|^2} \int_{\mathbf{K}} |\Lambda_W(\mathbf{K}, \mathbf{K}_0, \omega_0)|^2 \, d^3K$$

$$\qquad (25)$$

and is called the wide-band system power factor. The normalization yielding the power factor N_W is chosen to make the product $W_E W_D$ act like a δ function in Equation 1 when $W_E W_D$ is a narrow-band function.

The form of the system function Λ_W in wave space can be described by various moments. Among the moments are those defined around the reference scattering vector $\mathbf{K}_O$, which is specified in terms of the angle between the emitter and detector axes and is usually considered to be the scattering vector at which measurements are made for a given emitter and detector position. A general expression for these moments of Λ_W is

$$m_n(K_i, \omega_0) = \left\{ \int (\mathbf{K} - \mathbf{K}_0)^n_i \Psi[\Lambda_W(\mathbf{K}, \mathbf{K}_0, \omega_0)] \, d^3K \right\}$$

$$\times \left\{ \int \Psi[\Lambda_W(\mathbf{K}, \mathbf{K}_0, \omega_0)] \, d^3K \right\}^{-1},$$

$$i = 1, 2, 3, \quad n = 1, 2, 3, 4 \qquad (26)$$

and

$$\Psi[\Lambda_W] = |\Lambda_W|$$

or

$$\Psi[\Lambda_W] = |\Lambda_W|^2$$

In these equations, the subscript i refers to the component of the vector being considered. The moments for $n = 1$ describe the system function offset from the scattering vector defined by the emitter and detector axes, while the moments for $n = 2$ give a measure of the system function spread about the vector $\mathbf{K}_0$. These spreads may be used to justify the assumption that the power spectrum of medium variations may be brought out from under the integral in Equation 20. A normalization of each moment for $n = 3$ by the three halves power of the corresponding coordinate moment for $n = 2$ yields the skewness, which is a measure of asymmetry. A normalization of each moment for $n = 4$ by the square of the corresponding coordinate moment for $n = 2$ yields the coefficient of excess, which is a measure of non-normality.

B. INFINITE LENGTH DETECTOR TIME GATE

The temporal-frequency integration that defines the system functions Λ_1 and Λ_2 and likewise the system function Λ_W can be carried out when the receiver gate has an indefinitely long time duration because the receiver gate temporal harmonic spectrum W_D then becomes impulsive. This leads to a number of specialized system expressions as described in Reference 16. The first expression, from which other results follow, is for the temporal harmonic spectrum of the measured pressure. The expression for P_m resulting from an infinite length time gate can be written

$$P_m(\mathbf{K}_0, \omega_0) = \int_{\mathbf{K}} \Gamma(-\mathbf{K}, \omega_0) \Lambda_A(\mathbf{K}, \mathbf{K}_0, \omega_0) \, d^3K \tag{27}$$

in which

$$\Lambda_A(\mathbf{K}, \mathbf{K}_0, \omega_0) = (2\pi)^4 W_E(\omega_0) k_0^2 \Lambda(\mathbf{K}, \mathbf{K}_0, \omega_0) \tag{28}$$

and k_0 is equal to ω_0/c. Thus, the beam intersection function Λ alone determines the distribution of the measurement system weight on the scattering medium characteristics as expected from physical considerations.

If experimental conditions are chosen so that the effect of the system function Λ_A is concentrated in the neighborhood of $\mathbf{K}_0$ relative to the spatial-frequency fluctuations in the spectrum of scattering medium variations in that neighborhood, the integral given by Equation 27 can be rewritten as

$$P_m(\mathbf{K}_0, \omega_0) = \Gamma(-\mathbf{K}_0, \omega_0) H_A^V(\mathbf{K}, \mathbf{K}_0, \omega_0) \tag{29}$$

in which

$$H_A^V(\mathbf{K}_0, \omega_0) = \int_{\mathbf{K}} \Lambda_A(\mathbf{K}, \mathbf{K}_0, \omega_0) \, d^3K \tag{30}$$

and H_A^V is the infinite length detector gate system transfer function corresponding to H_W^V in the wide-band case. Using Equation 28 for Λ_A allows the transfer function H_A^V to be expressed

$$H_A^V(\mathbf{K}_0, \omega_0) = (2\pi)^4 W_E(\omega_0) k_0^2 \int_{\mathbf{K}} \Lambda(\mathbf{K}, \mathbf{K}_0, \omega_0) \, d^3K \tag{31}$$

Since the integration of Λ over $\mathbf{K}$ can be written as the Fourier transform of Λ evaluated at $r = 0$ (i.e., the center of the scattering volume), this Fourrier transform λ, which is defined in Equation 7, can be used to write the transfer function H_A^V in the form

$$H_A^V(\mathbf{K}_0, \omega_0) = (2\pi)^4 W_E(\omega_0) k_0^2 \lambda(0, \mathbf{K}_0, \omega_0) \tag{32}$$

This expression shows explicitly that the product of the emitter and detector beams in real space must be a constant over the range where γ is nonzero for Equation 29 to be valid.

The assumption of an indefinitely long detector time gate also allows simplification of Equation 19 for the power spectrum of the measured pressure. The result can be expressed

$$\langle |P_m(\mathbf{K}_0, \omega_0)|^2 \rangle = \int_{\mathbf{K}} S_\gamma(\mathbf{K}, \omega_0) |\Lambda_A(\mathbf{K}, \mathbf{K}_0, \omega_0)|^2 \, d^3K \tag{33}$$

The additional assumption that the experimental conditions are arranged so the system function weighting effect is concentrated in a spatial-frequency region around $\mathbf{K}_0$ where the power spectrum S_γ changes slowly permits the power spectrum of the measured pressure to be written

$$\langle |P_m(\mathbf{K}_0, \omega_0)|^2 \rangle = S_\gamma(\mathbf{K}_0, \omega_0)|H_A(\mathbf{K}_0, \omega_0)|^2 \tag{34}$$

where

$$|H_A(\mathbf{K}_0, \omega_0)|^2 = \int_\mathbf{K} |\Lambda_A(\mathbf{K}, \mathbf{K}_0, \omega_0)|^2 \, d^3K \tag{35}$$

and is the infinite length detector gate system transfer function that corresponds to $|H_W|^2$ in the wide-band case. Using Equation 28 for Λ_A allows the transfer function $|H_A|^2$ to be written

$$|H_A(\mathbf{K}_0, \omega_0)|^2 = (2\pi)^8|W_E(\omega_0)|^2 k_0^4 N_A(\mathbf{K}_0, \omega_0) \tag{36}$$

where

$$N_A(\mathbf{K}_0, \omega) = \int_\mathbf{K} |\Lambda(\mathbf{K}, \mathbf{K}_0, \omega_0)|^2 \, d^3K \tag{37}$$

and is the infinite length detector gate system power factor analogous to N_W in the wide-band case. Parseval's theorem permits the power factor N_A to be expressed also in terms of the real space beam intersection function λ as

$$N_A(\mathbf{K}_0, \omega) = \frac{1}{(2\pi)^3} \int_V |\lambda(\mathbf{r}, \mathbf{K}_0, \omega)|^2 \, d^3K \tag{38}$$

Thus, the temporal harmonic power spectrum of the measured pressure becomes proportional to the power spectrum of medium variations with the constant of proportionality including a power factor N_A, which is the integrated square magnitude of the intersection of the emitter and detector patterns in real space, or of the corresponding spectrum in wave space.

C. NARROW-BAND APPROXIMATION

The development of narrow-band relations is based on the special emitter waveform $w_E(t)e^{-j\omega_c t}$ in which w_E is a lowpass function with a bandwidth much less than the tone or carrier frequency ω_c. This motivates use of the term narrow-band. Details of the analysis are given in Reference 17.

The narrow-band assumption allows first-order differences between k and k_c in multiplicative factors and second-order differences between k and k_c in exponential terms to be neglected. The result is that expression for P_m in Equation 1 can be written

$$P_m(\mathbf{r}_E, \mathbf{r}_D, \omega_c) = 2\pi \int_\omega \int_\nu \int_\eta \int_V A_E(\nu, \omega_c) A_D(\eta, \omega_c) W_E(\omega - \omega_c)$$

$$W_D(\omega_c - \omega) \times e^{j(\omega - \omega_c)(r_E + r_D + 2\cos\theta \mathbf{i}_3 \cdot \mathbf{r})/c}$$

$$\times k_c^2 \left[\gamma_\kappa(\mathbf{r}) + \frac{\mathbf{I}_c \cdot \mathbf{O}_c}{k_c^2} \gamma_\rho(\mathbf{r}) \right] \tag{39}$$

$$e^{j(\mathbf{I}_c - \mathbf{O}_c) \cdot \mathbf{r}} \, d^3 r d^2 \eta d^2 \nu d\omega$$

where W_E is the Fourier transform of the lowpass function W_E and θ is the angle between the emitter and detector axes. The approximations yielding Equation 39 have limited the ω dependence to W_E, W_D, and the exponential term containing $\cos\theta$ which also depends on the i_3 component of $\mathbf{r}$.

The expression for the measured pressure spectrum P_m given in Equation 39, which is the narrow-band analog of Equation 1, can be written in a form analogous to Equation 13 by using the transformations

$$\nu = \nu(\mathbf{K}_c, \zeta)$$

$$\eta = \eta(\mathbf{K}_c, \zeta)$$

$$\mathbf{K}_c = \mathbf{I}_c - \mathbf{O}_c$$

and

$$\mathbf{K} = \mathbf{K}_c + \frac{\omega - \psi_c}{c} 2\cos\theta \, \mathbf{i}_3$$

where ζ is now a parameter for the path on which $\mathbf{K}_c$ is constant, $\mathbf{I}_c$ is $\mathbf{I}_0$ evaluated at $\omega_0 = \omega_c$, and $\mathbf{O}_c$ is $\mathbf{O}_0$ evaluated at $\omega_0 = \omega_c$. Then, neglecting again differences between k and k_c in nonexponential terms, Equation 39 for P_m can be written

$$P_m(\mathbf{K}_c, \omega_c) = \int_\mathbf{K} \Gamma(-\mathbf{K}, \omega_c) \Lambda_N(\mathbf{K}, \mathbf{K}_c, \omega_c) \, d^3 K \tag{40}$$

where

$$\Lambda_N(\mathbf{K}, \mathbf{K}_c, \omega_c) = (2\pi)^4 k_c^2 \int_\omega \Lambda\left(\mathbf{K} - \frac{\omega - \psi_c}{c} 2\cos\theta \, \mathbf{i}_3, \mathbf{K}_c, \omega_c \right)$$

$$\times W(\omega - \omega_c) \, d\omega \tag{41}$$

and

$$W(\omega) = W_E(\omega) W_D(-\omega) e^{j\omega(r_E + r_D)/c} \tag{42}$$

Since the argument of the function Λ changes linearly with ω in the i_3 direction, the narrow-band system function Λ_N can be interpreted as a convolution of the beam intersection function Λ and the temporal spectrum W applied along the i_3 direction. Also, since $\mathbf{i}_3$ is in the direction of the reference scattering vector now designated $\mathbf{K}_c$ and since the product form of the temporal spectrum W corresponds in time to a correlation of the lowpass emitter waveform w_E and the detector gate w_D, the narrow-band measurement system weight is limited by the emitter and detector apertures in the plane perpendicular to the reference scattering vector and by the emitter and detector temporal functions along the axis of the reference scattering vector.

The result in Equation 41, which is the narrow-band analog of the result given in Equation 15, can be written as the product of a measurement system transfer function and the spectrum of the medium variations when the spectra of the medium compressibility variations and the product of the factor K^2 and the spectrum of the medium density variations vary slowly relative to the system function. The spectrum of the measured pressure can then be expressed

$$P_m(\mathbf{K}_c, \omega_c) = \Gamma(-\mathbf{K}_c, \omega_c)H_N^V(\mathbf{K}_c, \omega_c) \tag{43}$$

where

$$H_N^V(\mathbf{K}_c, \omega_c) = \int_{\mathbf{K}} \Lambda_N(\mathbf{K}, \mathbf{K}_c, \omega_c) \, d^3K \tag{44}$$

and is the narrow-band analog of the wide-band system transfer function H_W^V defined by Equation 18. Also, when the second-order statistics of the medium variations are sought, specialized narrow-band expressions for the transfer function $|H_N|^2$ and the power factor N_N can be obtained by a development parallel to that yielding Equation 24 for the wide-band system transfer function and Equation 25 for the wide-band system power factor. The result is that the mean square value of the measured pressure spectrum under the narrow-band assumption can be written in a form analogous to Equation 22 as

$$\langle |P_m(\mathbf{K}_c, \omega_c)|^2 \rangle = S_\gamma(\mathbf{K}_c, \omega_c)|H_N(\mathbf{K}_c, \omega_c)|^2 \tag{45}$$

in which

$$|H_N(\mathbf{K}_c, \omega_c)|^2 = \int_{\mathbf{K}} |\Lambda_N(\mathbf{K}, \mathbf{K}_c, \omega_c)|^2 \, d^3K \tag{46}$$

This transfer function is given in terms of the power factor N_N by the relation

$$|H_N(\mathbf{K}_c, \omega_c)|^2 = (2\pi)^8 k_c^4 \left| \int_\omega W(\omega) \, d\omega \right|^2 N_N(\mathbf{K}_c, \omega_c) \qquad (47)$$

where

$$N_N(\mathbf{K}_c, \omega_c) = \frac{1}{(2\pi)^8 k_c^4 \left| \int_\omega W(\omega) \, d\omega \right|^2} \int_\mathbf{K} |\Lambda_N(\mathbf{K}, \mathbf{K}_c, \omega_c)|^2 \, d^3K \qquad (48)$$

and may be used for comparisons with N_W as well as N_A.

D. QUASI PLANE WAVE APPROXIMATION

The beam intersection function Λ defined in Equation 6 can be approximated when the angular spectra A_E and A_D given after Equation 1 are concentrated inside a small region at the apex of their respective hemispheres. The approximation represents the axial spatial-frequency $\nu_3 = \sqrt{k^2 - (\nu_1^2 + \nu_2^2)}$ of the . incident wave vector $\mathbf{I}$ and $\eta_3 = \sqrt{k^2 - (\eta_1^2 + \eta_2^2)}$ of the scattered wave vector $\mathbf{O}$ by the first term of a binomial expansion in the determinators of the respective angular spectra A_E and A_D and in the parameterization for constant values of the local scattering vector $\mathbf{K}$, and by the first two terms of a binomial expansion in the phases of the respective angular spectra A_E and A_D. These simplifications, which implement the Fresnel approximation in wave space, motivate the name quasi plane wave approximation. The following is a summary of material in Reference 17.

The use of the first term in a binomial expansion of ν_3 and η_3 permits the coordinates of the corresponding local scattering vector $\mathbf{K}$ to be written in the same reference coordinate system employed in the narrow-band approximation as

$$K_1 = \nu_1 + \eta_1 \qquad (49a)$$

$$K_2 = (\nu_2 - \eta_2)\cos\theta \qquad (49b)$$

and

$$K_3 = (\nu_2 + \eta_2)\sin\theta + 2k\cos\theta \qquad (49c)$$

These relations indicate that ν_2 and η_2 are uniquely defined by K_2 and K_3 but that ν_1 and η_1 are not completely determined by K_1.

The use of the first two terms in binomial expansions of ν_3 and η_3 results in quadratic expressions for the phases $\mathbf{I} \cdot \mathbf{r}_E$ and $\mathbf{O} \cdot \mathbf{r}_D$ that appear in the angular spectra A_E and A_D, respectively. These quadratic phases may be written

$$\mathbf{I} \cdot \mathbf{r}_E = k z_E - \frac{z_E}{2k} (\nu_1^2 + \nu_2^2) \tag{50}$$

and   (51)

$$\mathbf{O} \cdot \mathbf{r}_D = k z_D - \frac{z_D}{2k} (\eta_1^2 + \eta_2^2)$$

in which z_E and z_D are the distances from the emitter and detector, respectively, to the center of the scattering volume.

A parameterization like the one used previously can now be employed to obtain expressions for beam intersection function Λ in terms of the local scattering vector $\mathbf{K}$ in three-dimensional wave space and the parameter ζ for the path on which $\mathbf{K}$ is fixed. However, because the restriction of $\mathbf{I}$ and $\mathbf{O}$ to planes uniquely defines the values of ν_2 and η_2 and because the sum of ν_1 and η_1 is fixed to be K_1, the angular spectrum components ν_1 and η_1 must vary in a complementary way as indicated in Figure 3, which shows the basic geometry. Thus, the parameter ζ is constrained to be the difference between ν_1 and η_1 and can be written

$$\zeta = \frac{1}{2} (\nu_1 - \eta_1) \tag{52}$$

This equation along with the relations given in Equations 49a to c, can be solved for ν_1, ν_2 and η_1, η_2 in terms of $K_1, K_2,$ and K_3 to obtain

$$\nu_1 = \frac{K_1}{2} + \zeta \tag{53a}$$

$$\nu_2 = \frac{K_2}{2 \cos \theta} + \frac{K_3}{2 \sin \theta} - k \cot \theta \tag{53b}$$

$$\eta_1 = \frac{K_1}{2} - \zeta \tag{53c}$$

and

$$\eta_2 = - \frac{K_2}{2 \cos \theta} + \frac{K_3}{2 \sin \theta} - k \cot \theta \tag{53d}$$

From these coordinate relations, the Jacobian of the transformation from $(\boldsymbol{\nu}, \boldsymbol{\eta})$ to $(\mathbf{K}, \zeta)$ is found to be

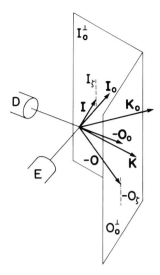

FIGURE 3. Geometric relations between the incident field and detector pattern that yield a constant scattering vector in the quasi plane wave approximation. The plane $I_O^{\perp}$, which is normal to the incident wave vector $\mathbf{I}_O$ along the axis of the emitter E, approximates the hemispherical region in which the angular spectrum of the emitter is assumed to be concentrated. The plane $O_O^{\perp}$, which is normal to the detector pattern wave vector $-\mathbf{O}_O$ along the axis of the detector D, approximates the hemispherical region in which the angular spectrum of the detector pattern is constant. The reference scattering vector $\mathbf{K}_O$ is the difference between $\mathbf{I}_O$ and $\mathbf{O}_O$. An arbitrary angular spectrum component $\mathbf{I}$ confined to the plane $I_O^{\perp}$ and an arbitrary angular spectrum component $-\mathbf{O}$ which is similarly confined to $O_O^{\perp}$ are shown. The trajectory of dashes I_{ζ} and the trajectory of dash-dot-dashes O_{ζ} define the paths over which $\mathbf{I}$ and $-\mathbf{O}$ may simultaneously vary while the local scattering vector $\mathbf{K}$ remains constant. The linear trajectories I_{ζ} and O_{ζ} are straight line approximations of circular segments in the exact analysis summarized in Figure 2.

$$J(\boldsymbol{v}, \boldsymbol{\eta}|\mathbf{K}, \zeta) = -\frac{1}{\sin 2\theta} \tag{54}$$

Substitution of Equations 50, 51, 53a to d, and 54 into Equation 6 for the beam intersection function yields

$$\wedge(\mathbf{K}, \mathbf{K}_0, \omega) =$$

$$C_A \frac{e^{jk(z_E + z_D)}}{k^2} \frac{1}{\sin 2\theta} \int_{\zeta} \vee_E \left(\frac{K_1}{2} + \zeta, \frac{K_2}{2\cos\theta} + \frac{K_3}{2\sin\theta} - k\cot\theta \right)$$

$$\times \vee_D \left(\frac{K_1}{2} - \zeta, \frac{K_2}{2\cos\theta} - \frac{K_3}{2\sin\theta} + k\cot\theta \right) d\zeta \tag{55}$$

where

$$\vee_E(\nu_1,\ \nu_2)\ =\ V_E(\nu_1,\ \nu_2)e^{jz_E(\nu_1^2+\nu_2^2)/2k} \qquad (56)$$

$$\vee_D(\eta_1,\ \eta_2)\ =\ V_D(\eta_1,\ \eta_2)e^{jz_D(\eta_1^2+\eta_2^2)/2k} \qquad (57)$$

and $\wedge$ has been used in place of Λ to designate the spatial approximation that has been made. The approximate result given in Equation 55 for the beam intersection function is the quasi plane wave analog of the exact result given in Equation 10.

E. COMBINED EXPRESSIONS

The quasi plane wave approximation for the beam intersection function can be employed to obtain expressions for the various system functions when the conditions leading to those expressions are satisfied. These expressions can then be used with variable changes in corresponding expressions for transfer functions, power factor, and second moments to show measurement system properties. The analysis, adapted from Reference 17, is summarized here for the narrow-band approximation since this is the most important case in practice and the steps for the infinite length detector gate and the wide-band cases are similar.

The narrow-band system function under the quasi plane wave approximation is obtained by substitution of Equation 55 for $\wedge$ into Equation 41 for Λ_N. The result is

$$\Lambda_N(\mathbf{K},\ \mathbf{K}_0,\ \omega_c)\ =$$

$$(2\pi)^4 C_A \frac{1}{\sin 2\theta} \int_\omega \int_\zeta \vee_E\left(\frac{K_1}{2}\ +\ \zeta,\ \frac{K_2}{2\cos\theta}\ +\ \frac{K_3}{2\sin\theta}\ -\ \frac{\omega}{c}\cot\theta\right)$$

$$\times\ \vee_D\left(\frac{K_1}{2}\ -\ \zeta,\ \frac{K_2}{2\cos\theta}\ -\ \frac{K_3}{2\sin\theta}\ +\ \frac{\omega}{c}\cot\theta\right)W(\omega\ -\ \omega_c)\ d\zeta\ d\omega \qquad (58)$$

Substitution of this expression into Equation 44 for H_N^V yields a corresponding result for the narrow-band system transfer function H_N^V that may be written compactly using the changes of variables

$$\begin{bmatrix} u_1 \\ u_2 \\ u_3 \end{bmatrix} = \begin{bmatrix} 1 & 0 & 0 \\ 0 & 1/(2\cos\theta) & 1/(2\sin\theta) \\ 0 & 1/(2\cos\theta) & -1/(2\sin\theta) \end{bmatrix} \begin{bmatrix} K_1 \\ K_2 \\ K_3 \end{bmatrix}$$

and

$$q = \frac{\omega}{c} \cot \theta$$

to extract much of the θ dependency. The result is that the transfer function H_N^V may be expressed

$$H_N^V(\mathbf{K}_0, \omega_c) = (2\pi)^4 C_A c \tan \theta \int_\mathbf{u} \Phi_N(\mathbf{u}, \theta, \omega_c) \, d^3u \qquad (59)$$

where

$$\Phi_N(\mathbf{u}, \theta, \omega_c) = \int_q \Psi(u_1, u_2 - q, u_3 + q) W(c \tan \theta q - \omega_c) \, dq \qquad (60)$$

and

$$\Psi(\mathbf{u}) = \int_\zeta V_E\left(\frac{u_1}{2} + \zeta, u_2\right) V_D\left(\frac{u_1}{2} - \zeta, u_3\right) d\zeta \qquad (61)$$

Analogous expressions for the narrow-band system transfer function $|H_N|^2$ and narrow-band system power factor N_N are found by the same substitutions in Equations 46 and 48, respectively.

The second central moments of the squared magnitude of the narrow-band system function for the quasi plane wave approximation are similarly obtained from Equation 26 which leads to the formula.

$$\sigma_n^2(\Lambda_N) = \int_\mathbf{K} (\mathbf{K} - \mathbf{K}_c)_i^2 |\Lambda_N(\mathbf{K}, \mathbf{K}_c, \omega_c)|^2 \, d^3K$$

$$\div \int_\mathbf{K} |\Lambda_N(\mathbf{K}, \mathbf{K}_c, \omega_c)|^2 \, d^3K \qquad (62)$$

where $i = 1$, 2, or 3. This formula yields

$$\sigma_1^2(\wedge_N) = \int_\mathbf{u} u_1^2 |\Phi_N(\mathbf{u}, \theta, \omega_c)|^2 \, d^3u/m(\Phi_N)$$

$$\sigma_2^2(\wedge_N) = \cos^2 \theta \int_\mathbf{u} (u_2 + u_3)^2 |\Phi_N(\mathbf{u}, \theta, \omega_c)|^2 \, d^3u/m(\Phi_N)$$

$$\sigma_3^2(\wedge_N) = \sin^2 \theta \int_\mathbf{u} (u_2 - u_3)^2 |\Phi_N(\mathbf{u}, \theta, \omega_c)|^2 \, d^3u/m(\Phi_N)$$

and

$$\sigma_K^2(\wedge_N) = \int_{\mathbf{u}} (u^2 + 2u_2 u_3 \cos 2\theta)^2 |\Phi_N(\mathbf{u}, \theta, \omega_c)|^2 \, d^3u/m(\Phi_N)$$

where

$$m(\Phi_N) = \int_{\mathbf{u}} \left| \int_q \Psi(u_1, u_2 - q, u_3 + q) W(c \tan \theta q - \omega_c) \, dq \right|^2 d^3u$$

$$\sigma_K^2(\Lambda_N) = \sum_{n=1}^{3} \sigma_n^2(\Lambda_N)$$

and

$$u^2 = u_1^2 + u_2^2 + u_3^2$$

The additional assumption that the integration over q produces a function symmetric in the second or third coordinate can be used to show the integration over the cross term $u_2 u_3$ is identically zero and further simplify the moment expressions for σ_2^2, σ_3^2, and σ_K^2. For statistically isotropic media, i.e., media with power spectra that are radial functions of the scattering vector $\mathbf{K}_O$, the only important second moment for resolution in measurements of S_γ is σ_3^2 in the direction of the scattering vector because no variation in the second-order statistics of these media exists in the orthogonal polar or azimuthal directions.

Quasi plane wave approximations for measurement system transfer functions and power factor obtained for the infinite length detector gate, narrowband, and wide-band cases are listed in Table 1 while second central moment expressions for the squared magnitude of the corresponding system functions obtained under the additional assumption that the emitter or detector angular spectrum is symmetric in the second coordinate for the infinite length detector gate case and the assumption that the function $\Phi_{(\cdot)}$ is symmetric for the narrowband and wide-band cases are listed in Table 2. The results for the infinite length time gate in Table 1 show that the transfer function H_A^V is independent of angle while the transfer function $|H_A|^2$ and the power factor N_A are inversely proportional to the sine of the angle between the emitter and detector and that the constant of proportionality is determined by the apertures. The results for the same case in Table 2 show that the second central moment in the direction orthogonal to the plane of the transducers is independent of angle and equal to another constant determined by the apertures while the second central moment in the direction of the scattering vector is proportional to $\cos^2\theta$ and the remaining moment is proportional to $\sin^2\theta$ with the constant of proportionality for these two moments being the same and determined only by the apertures. The results for the narrow-band case in Tables 1 and 2 show the angular dependence of the transfer functions, power factor, and second central moments also appears as a factor $\tan \theta$ which modifies the convolution in the direction of the scattering vector. The results in Tables 1 and 2 for the wide-band case show the same tan θ factor in the temporal spectra as in the narrow-band case but the form of the convolution is different and will generally have an effect in all coordinate directions. Although the convolutions in both the narrow-band and wide-band cases make the symmetry assumption needed for the elimination of the cross terms in the moment expressions difficult to translate into general condition on the apertures, the assumption of a narrow temporal bandwidth may be used to show that the contribution of the cross terms is small when either the emitter or detector angular spectrum is symmetric in the second coordinate.

III. SPECIAL ANALYTIC AND NUMERIC RESULTS

A. INFINITE LENGTH DETECTOR TIME GATE AND QUASI PLANE WAVE APPROXIMATION FOR GAUSSIAN SPATIAL APERTURES

Expressions resulting from the infinite length detector time gate assumption and the quasi plane wave approximation have been investigated in Reference 17 for Gaussian emitter and detector spatial aperture functions. The analysis, which is outlined here, assumes the normal component of the acoustic particle velocity v_E at the emitter has the circularly symmetric form

$$v_E(\mathbf{r}'_E) = A_0 e^{-\tau_E'^2/2\sigma^2}$$

and the detector spatial sensitivity has the similar form

$$v_D(\mathbf{r}'_D) = e^{-\tau_D'^2/2\sigma^2}$$

so that the corresponding angular spectra are given by

$$V_E(\boldsymbol{\nu}) = 2\pi\sigma^2 A_0 e^{-\sigma^2\nu^2/2} \tag{63}$$

and

$$V_D(\boldsymbol{\eta}) = 2\pi\sigma^2 e^{-\sigma^2\eta^2/2} \tag{64}$$

The analysis also assumes

$$z_E = z \tag{65a}$$

and

$$z_D = z \tag{65b}$$

Using these expressions for V_E and z_E in Equation 56 and for V_D and z_D in Equation 57 yields results that may be substituted in Equation 55 for the beam intersection function $\wedge$. The expression for $\wedge$ can then be evaluated in closed form to obtain

$$\wedge(\mathbf{K}, \mathbf{K}_0, \omega_0) = \frac{j\pi^{3/2}\rho c A_0 \sigma^3 e^{j2k_0 z}}{k_0(1 + jz/k_0\sigma^2)^{1/2}} \frac{1}{\sin 2\theta}$$

$$\times \ e^{-\sigma^2(1 + jz/k_0\sigma^2)[(K_1/2)^2 + (K_1/2\cos\theta)^2]}$$

$$\times \ e^{-\sigma^2(1 + jz/k_0\sigma^2)[(K_3/2\sin\theta) - k_0\cot\theta]^2} \qquad (66)$$

Thus, the beam intersection function $\wedge$ also has a Gaussian shape as a function of $\mathbf{K}$.

Expressions for the transfer functions H_A^V and $|H_A|^2$ and the power factor N_A given in Table 1 can be similarly evaluated using the form of the beam intersection function in Equation 66. The results are

$$H_A^V(\mathbf{K}_0, \omega_0) = \frac{j2^6\pi^7\rho A_0\omega_0 W_E(\omega_0)e^{j2k_0 z}}{(1 + jz/k_0\sigma^2)^2} \qquad (67)$$

$$|H_A(\mathbf{K}_0, \omega_0)|^2 = \frac{2^{17/2}\pi^{25/2}\rho^2 A_0^2\sigma^3\omega_0^2|W_E(\omega_0)|^2}{[1 + (z/k_0\sigma^2)^2]^{1/2}} \frac{1}{\sin 2\theta} \qquad (68)$$

and

$$N_A(\mathbf{K}_0, \omega_0) = \frac{2^{1/2}\pi^{9/2}\rho^2 c^2 A_0^2\sigma^3}{k_0^2[1 + (z/k_0\sigma^2)^2]^{1/2}} \frac{1}{\sin 2\theta} \qquad (69)$$

In these expressions, the normalized distance parameter $z/k_0\sigma^2$ describes the effect of distance from the scattering volume and the factor $\sigma^3/\sin2\theta$ is a measure of the real space volume defined by the product of the emitter and detector beam patterns.

Expressions for the second central moments given in Table 2 can also be evaluated similarly employing the form of the beam intersection function in Equation 66. The results may be written

$$\sigma_1(\wedge_A) = \frac{1}{\sigma} \qquad (70a)$$

$$\sigma_2(\wedge_A) = \frac{1}{\sigma}\cos\theta \qquad (70b)$$

$$\sigma_3(\wedge_A) = \frac{1}{\sigma}\sin\theta \qquad (70c)$$

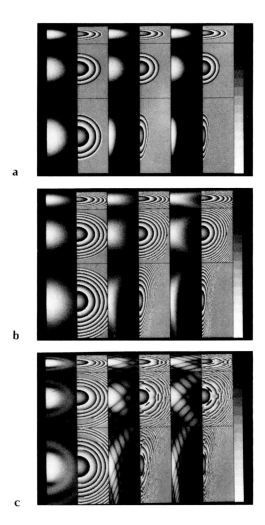

FIGURE 4. Beam function cross-sections in wave space for Gaussian (a), exponential (b), and uniform apertures (c). Relative magnitude (black background) and phase (gray background) for the beam function defined by Equation 7 as a function of the radial coordinate $K_T = K_1^2 + K_2^2)^{1/2}$ (horizontal axis) and the polar axis K_3 (vertical axis) for azimuthal angle α (right to left columns in pairs) equal to 0°, 45°, and 90°, and for θ (top to bottom row) equal to 15°, 45°, and 75°. The range of K_T is from 0.0 cy/mm to 0.65 cy/mm in all of the panels while the range of K_3 is from 6.20 cy/mm to 6.59 cy/mm for $\theta = 15°$, from 4.04 cy/mm to 5.24 cy/mm for $\theta = 45°$, and from 0.90 cy/mm to 2.55 cy/mm for $\theta = 75°$. The magnitudes are shown over a 60-dB range on a log scale in which the maximum value is white and the minimum value is black. The corresponding phases are displayed on a linear scale in which 180° is white and −180° is black.

to be determined by the product of the emitter and detector patterns. In the wide-band case, the ability to isolate a small region in real space results in a wave space blur that enlarges the limits set by the product of the beam patterns in all directions.

The quasi plane wave approximation is the spatial-frequency analog of the narrow-band approximation. The confinement of the incident wave vector to a plane normal to the axis of the emitter and the scattered wave vector to a plane normal to the axis of the detector simplifies the geometry of the wave space integration and, thus, enables additional analysis to characterize the effects of the measurement system. A general result of the quasi plane wave approximation particularly noteworthy for measurement of second-order scattering statistics of random media is that the power factor in the infinite length detector gate case becomes proportional to the cosecant of the angle between the emitter and detector. Also noteworthy for these measurements is that the second central moment of the squared magnitude of the system function in the direction orthogonal to the plane in which the emitter and detector rotate becomes a constant independent of θ. The second central moments in the other two directions have sine and cosine dependencies that result in their sum being a constant and so the second central moment in the radial direction is also a constant.

The assumption of Gaussian spatial apertures and the use of the quasi plane wave approximation along with an indefinitely long detector time gate yields simple expressions for the constant of proportionality in the expressions that give the angular dependence of the system transfer functions, power factor, and the second central Cartesian moments. The increase in the transfer function $|H_A|^2$ and power factor N_A as the angle between the emitter and detector becomes close to zero is physically caused by the increasing volume in real space as the emitter and detector patterns become coincident. As predicted by the general second central moment expression for the quasi plane wave approximation, this increase is accompanied by a commensurate decrease in the second central moment along the axis of the reference scattering vector.

The measurement system expressions which result from a finite length detector time gate and the narrow-band approximation along with the quasi plane wave approximation are not significantly more complicated when Gaussian temporal waveforms are employed with Gaussian spatial apertures. The finite temporal waveforms, consistent with the general narrowband theory, limit the increase of the transfer function $|H_N|^2$ and power factor N_N when the emitter and detector are coincident. Also consistent with the general narrowband theory, the second central moment in the direction of the reference scattering vector becomes nonzero in the backscatter configuration and is determined by the width of the Gaussian temporal waveforms.

An important assumption leading to the various measured pressure spectrum expressions in terms of transfer functions is that the effect of the system is concentrated relative to change in the spectrum of medium variations in the neighborhood of the reference scattering vector $\mathbf{K}_0$, which is defined by the operating frequency and the angle between the emitter and detector axis. This assumption allows the measurement system effect to be represented as a factor that scales the measurement of the medium variation function at the spatial-frequency $\mathbf{K}_0$ rather than as a distributed weighting of the medium variation function over a region of wave space.

B. MEASUREMENT SYSTEM TRANSFER FUNCTION RELATIONSHIPS

The family of transfer functions $H_{(\cdot)}^V$ given by Equations 18, 31, and 44 and the family of transfer functions $|H_{(\cdot)}|^2$ given by Equations 23, 35, and 46 have been obtained from different assumptions. The expressions for $H_{(\cdot)}^V$ assume that the product of the emitter and detector patterns is essentially constant, and, thus, require in practice that the scattering region be completely contained within the intersection of the emitter and detector patterns. The expressions for $|H_{(\cdot)}|^2$ assume that the scattering region extends beyond the intersection of the emitter and detector patterns, and, thus, the product of the emitter and detector patterns that, in this case, define the region of the scattering medium being studied, need not be constant throughout the scattering region. Only the unrealistic assumptions that the boundary of the scattering region and the boundary of the intersection of the emitter and detector patterns coincide and that the intersection of the patterns is constant throughout the scattering region can the expressions for $H_{(\cdot)}^V$ and $|H_{(\cdot)}|^2$ be related as noted in Reference 16.

When the boundary of the scattering region rather than the intersection of the emitter and detector patterns limits the measurement of scattering, the mean-square value of the scattered pressure spectrum may, however, be obtained in terms of $H_{(\cdot)}^V$. This is accomplished by recalling that the power spectrum S_γ can be defined

$$S_\gamma(\mathbf{K}) = \lim_{V \to \infty} \left[(2\pi)^3 \frac{|\Gamma(\mathbf{K})|^2}{V} \right] \tag{82}$$

Then, if V is large enough that the limit in Equation 82 is approximated, the relations for the mean square value of the measured pressure spectrum can be written in terms of V and $H_{(\cdot)}^V$ as

$$\langle |P_m(\mathbf{K}_0, \omega_0)|^2 \rangle = S_\gamma(\mathbf{K}_0, \omega_0) V |H_{(\cdot)}^V(\mathbf{K}_0, \omega_0)|^2 \tag{83}$$

where $(\cdot)$ is equal to W, N, or A.

The family of transfer functions $H_{(\cdot)}^V$ and the family of transfer functions $|H_{(\cdot)}|^2$ have different uses as a result of being based on different assumptions. The transfer functions $H_{(\cdot)}^V$ are important in coherent signal processing applications such as image reconstruction where the phase of scattered signals is employed. The transfer functions $|H_{(\cdot)}|^2$ are important in incoherent signal processing applications such as determination of second-order scattering statistics where the phase of scattered signals is not relevant when the medium is statistically stationary.

C. NUMERICAL CALCULATIONS

Numerical results included in this chapter for Gaussian, exponential, and uniform spatial apertures and an indefinitely long time gate based on parameters from published measurements have been noted[17] to agree with expressions obtained using the quasi plane wave approximation. This demonstrates the validity of the approximation for practical conditions. Thus, the quasi plane wave approximation alone can be a useful tool in the design of experiments that yield intrinsic parameters of scattering. Furthermore, the narrowband approximation can be employed to get useful expressions in the backscatter case where the quasi plane wave approximation leads to expressions that increase without bound.

The slower decay of the beam intersection function produced by the exponential and uniform apertures evident in Figure 4 is reflected in larger values of the second central moments. However, the constraints of equal radiated power and on-axis far-field amplitude produce equivalent values of the transfer function $|H_A|^2$ and power factor N_A for all these spatial apertures. In contrast to the monotonic decay in all directions resulting from Gaussian spatial apertures, the exponential spatial apertures produce curved smoothly decaying ridges in wave space and the uniform spatial apertures produce curved ridges of fluctuating amplitudes. Consequently, the weighting of intrinsic scattering characteristics by exponential and uniform apertures can extend unevenly over a considerable region of wave space.

The units of the power factor $N_{(\cdot)}$ are found from a dimensional analysis to be $(\text{force})^2 \cdot (\text{length})$ since the measured pressure spectrum P_m has units of $(\text{force}) \cdot (\text{time})$ as a result of integration over the detector aperture and subsequent temporal Fourier transformation. Division of the beam power factor by the acoustic impedance ρc yields units of $(\text{power}) \cdot (\text{volume})$. These units have been employed[16] because they have physical appeal in terms of the scaling required to obtain average differential scattering cross-section, which is the average power into an aperture normalized by the solid angle of the aperture, the incident intensity, and the volume of the scattering region.

D. EXPERIMENTAL DESIGN

The design of an experiment to measure intrinsic scattering characteristics requires choice of parameters that define an appropriate system function from the family $\Lambda_{(\cdot)}$, each of which depends on frequency and the two vector variables $\mathbf{K}_0$ and $\mathbf{K}$. While the variation of the local scattering vector $\mathbf{K}$ occurs during the irreversible summing process that yields P_m, the reference scattering vector $\mathbf{K}_0$ can be changed in a number of ways to place the system function in a given region of wave space and, thereby, obtain a measurement of the scattering characteristic in that neighborhood as described in Reference 16. Since the magnitude of $\mathbf{K}_0$ is proportional to $\cos \theta$ and also to temporal frequency ω, the magnitude of $\mathbf{K}_0$ and, thus, the distance of the system function from the origin in wave space is changed by varying either θ or ω. For a fixed temporal frequency ω_0, varying θ from $0°$ to $90°$ takes the magnitude of $\mathbf{K}_0$ through its entire range of 0 to $2k_0$. For a fixed angle θ, varying temporal frequency changes the magnitude of $\mathbf{K}_0$ from $2k_i \cos \theta$ to $2k_f \cos \theta$ where k_i is the initial wave number and k_f is the final wave number. Since the direction of $\mathbf{K}_0$ for a fixed angle between the emitter and detector depends on the spatial position of the emitter-detector combination, the direction of $\mathbf{K}_0$ may be changed by polar and azimuthal rotations of the rigid emitter-detector combination in space.

The second central moments can be employed, as shown in Reference 17, to guide the selection of system parameters for measurements of the power spectrum of scattering medium variations with an acceptable amount of blur. Suppose, for example, that the correlation length of the medium variations is L so that a measure of the spread in wave space is $1/L$. Then, if a wave space resolution of $0.1/L$ is desired, the second central moment of the system function squared magnitude should be less than $0.01/L^2$ in each of the co-ordinate directions. Thus, since the second central moment for identical Gaussian apertures as indicated by Equations 70a to d is inversely proportional to the square of the spatial width parameter σ and since the spatial width parameter is equal to 1/2 the radius r_A of an equivalent disk radiator, the desired resolution requires that $r_A \geq 20L$. Furthermore, if the scattering volume is beam limited (as required for the analysis to be valid), the volume must have a dimension on edge wider than the beam.

An estimate of the beam radius r at a distance z and frequency f in a medium with a sound speed c is obtained from the relation between spatial-frequency and distance[22]

$$\nu = 2\pi \frac{f}{c} \frac{r}{z}$$

combined with the exponential size requirement

$$\frac{1}{2} \sigma^2 \nu^2 \geq 4$$

Thus,

$$r \geq \frac{\sqrt{2}}{\pi} \frac{c}{f} \frac{z}{\sigma}$$

or, using $\sigma = 10L$,

$$r \geq \frac{\sqrt{2}}{10\pi} \frac{c}{f} \frac{z}{L}$$

For a correlation length of 0.1 mm estimated[23] in calf liver, this analysis implies that the radius of the transducer should be greater than 2 mm and that the radius of the sample volume should be greater than 18.2 mm for a sound speed of 1.5 mm/μsec, frequency of 5.0 MHz, and a distance z of 135 mm. These compare favorably with values of 3.125 mm for the transducer radius and 32 mm for the diameter of the cylindrical sample volume reported[20] for the collection of data from which the estimate was made.

V. CONCLUSION

A general model has been described to characterize the influence of beam patterns and time waveforms on the measurement of ultrasonic scattering in practical situations. Arbitrary emitter beam patterns and detector sensitivity functions are represented as angular spectra of plane waves, while arbitrary emitter waveforms and detector time gates are described by frequency spectra of temporal harmonics. The results show that measurements of scattering may be expressed as a spatial-frequency average of a quantity that is the product of a wave space description of the medium characteristics and a system weighting function. The results also show that the measurements become scaled values of intrinsic scattering characteristics when the system function weight is concentrated relative to the medium characteristics in wave space.

The analysis yields approximate expressions for measurement system effects when the temporal spectra are concentrated in a narrow-band of temporal frequency and when angular spectra are confined to a spatial-frequency region near the apex of their hemispheres. The development of approximate expressions for a narrow-band of temporal frequencies shows explicitly the conditions that must be satisfied to obtain intrinsic parameters of the scattering medium when tone bursts are used. Computations of spatial aperture effects show exponential and uniform spatial aperture functions designed to produce the same weight as Gaussian apertures have more slowly decaying angular spectra than the Gaussian aperture functions and result in beam intersection functions with broader spatial frequency widths. Computations also show that the beam function distribution and overall weight change with geometry, and that the blurring of scattering measurements by beam patterns can be minimized by the appropriate choice of experimental parameters.

The model described in this chapter may be used to design experiments to image and characterize scattering quantitatively for applications in medical diagnosis and for other applications as well as in other fields whenever intrinsic parameters of scattering media are desired.

ACKNOWLEDGMENTS

Gratefully acknowledged is assistance from Richard Phillips, who helped carry out the calculations on the Cyber 205 computer and IBM 4381 and 3090 computers, and William Haake, who was responsible for graphics software to display the results. Useful discussions with Adrian I. Nachman are acknowledged with pleasure. Some of the computations were performed with support from the University of Rochester Computing Center and the John von Neumann Computing Center at Princeton University. Other computations were performed at the Cornell National Supercomputing Facility, which is supported in part by the National Science Foundation, New York State, and the IBM Corporation. Funds from National Science Foundation Grant No. ECS 8414315; National Institutes of Health Grant No. CA 39516; Oak Ridge National Laboratory Contract No. 19X-55914C, Naval Research Laboratory Contract No. N00014-90-K-2011, NATO Grant No. 93 81, the Diagnostic Ultrasound Research Laboratory Industrial Associates, and the Alexander von Humboldt Foundation also supported the research.

REFERENCES

1. **Sigelmann, R. A. and Reid, J. M.,** Analysis and measurement of ultrasound backscattering from an ensemble of scatterers excited by sine-wave bursts, *J. Acoust. Soc. Am.,* 53(5), 1351–1355, 1973.
2. **O'Donnell, M. and Miller, J. G.,** Quantitative broadband ultrasonic backscatter: an approach to nondestructive evaluation in acoustically inhomogeneous materials, *J. Appl. Phys.,* 52(2), 1056–1065, 1981.
3. **Nichols, D.,** Orientation and frequency dependence of backscattered energy and its clinical application, in *Recent Advances in Ultrasound in Medicine,* White, D. N., Ed., Research Studies Press, New York, 1977, 29–54.
4. **Bamber, J. C. and Hill, C. R.,** Acoustic properties of normal and cancerous human liver. I. Dependence of pathological conditions, *Ultras. Med. Biol.,* 7, 121–133, 1981.
5. **Nichols, D. and Hill, C. R.,** Evaluation of backscattering coefficients for excised human tissues: principles and techniques, *Ultras. Med. Biol.,* 8, 7–15, 1982.
6. **Lizzi, F. L., Greenebaum, M., Feleppa, E. J., and Elbaum, M.,** Theoretical framework for spectrum analysis in ultrasonic tissue characterization, *J. Acoust. Soc. Am.,* 73(4), 1366–1373, 1983.
7. **Campbell, J. A. and Waag, R. C.,** Normalization of ultrasonic scattering measurements to obtain average differential scattering cross sections for tissues, *J. Acoust. Soc. Am.,* 74(2), 393–399, 1983.
8. **Madsen, E. L., Insana, M. F., and Zagzebski, J. A.,** Method of data reduction for accurate determination of acoustic backscatter coefficients, *J. Acoust. Soc. Am.,* 75(3), 913–923, 1984.

9. **Nassiri, P. K. and Hill, C. R.**, The differential and total bulk acoustic scattering cross sections of some human and animal tissues, *J. Acoust. Soc. Am.*, 79(6), 2034–2047, 1986.

10. **Peters, K. J. and Waag, R. C.**, Compensation for receiver bandpass effects on backscatter power spectra using a random medium model, *J. Acoust. Soc. Am.*, 84(1), 392–399, 1988.

11. **Insana, M. F, Wagner, R. F., Brown, D. G., and Hall, T. J.**, Describing small-scale structure in random media using pulse-echo ultrasound, *J. Acoust. Soc. Am.*, 87(1), 179–192, 1990.

12. **Chivers, R. C. and Hill, C. R.**, A spectral approach to ultrasonic scattering from human tissue: methods, objectives, and backscattering measurements, *Phys. Med. Biol.*, 20(5), 799–815, 1975.

13. **Fink, M. A. and Cardoso, J. F.**, Diffraction effects in pulse-echo measurements, *IEEE Trans. Sonics Ultras.*, SU-31(4), 313–329, 1984.

14. **Leeman, S., Ferrari, L., Jones, J. P., and Fink, M.**, Perspective on attenuation estimation from pulse-echo signals, *IEEE Trans. Sonics Ultras.*, SU-31(4), 352–361, 1984.

15. **Parker, K. J., Lerner, R. M., and Waag, R. C.**, Attenuation of ultrasound: magnitude and frequency dependence for tissue characterization, *Radiology,* 153(3), 785–788, 1984.

16. **Waag, R. C. and Astheimer, J. P.**, Characterization of measurement system effects in ultrasonic scattering experiments, *J. Acoust. Soc. Am.*, 88(5), 2418–2436, 1990.

17. **Waag, R. C., Astheimer, J. P., and Smith, J. F., III**, Analysis and computations of measurement system effects in ultrasonic scattering experiments, *J. Acoust. Soc. Am.*, 91(3), 1284–1297, 1992.

18. **Morse, P. M. and Ingard, K. U.**, *Theoretical Acoustics,* McGraw-Hill, New York, 1968, chap. 8. (See also Waag, R. C., A review of tissue characterization from ultrasonic scattering, *IEEE Trans. Biomed. Eng.*, BME-31(12), 884–893, 1984.

19. **Azároff, L. V., Kaplow, R., Kato, N., Weiss, R. J., Wilson, A. J. C., and Young, R. A.**, *X-Ray Diffraction,* McGraw-Hill, New York, 1974, chap. 1.

20. **Campbell, J. A. and Waag, R. C.**, Measurements of calf liver ultrasonic differential and total scattering cross sections, *J. Acoust. Soc. Am.*, 75(2), 603–611, 1984.

21. **Campbell, J. A. and Waag, R. C.**, Ultrasonic scattering properties of these random media with implications for tissue characterization, *J. Acoust. Soc. Am.*, 75(6), 1879–1886, 1984.

22. **Goodman, J. W.**, *Introduction to Fourier Optics,* McGraw-Hill, New York, 1968, chap. 5.

23. **Waag, R. C., Dalecki, D., and Smith, W. A.**, Estimates of wave front distortion from measurements of scattering by model random media and calf liver, *J. Acoust. Soc. Am.*, 85(1), 406–415, 1989.

Chapter 9

In Vitro Experimental Results on Ultrasonic Scattering in Biological Tissues

K. Kirk Shung

TABLE OF CONTENTS

0-8493-6568-6/93/$0.00 + $.50

I. INTRODUCTION

There exists a large body of *in vitro* experimental data on ultrasonic scattering from various biological tissues in the literature.[1-9] It would be extremely difficult, if not impossible, to do an exhaustive survey of all the experimental results. Therefore, this chapter will discuss only selective experimental results which the author feels are significant to the development of the field. Although different experimental methods have been used to obtain these results as addressed in Chapters 6 to 8 and as a result discrepancies sometimes are found to exist among results obtained by different approaches, consistent results have been obtained on a number of tissues under well-defined experimental conditions. Still there appears a compelling need to establish the validity and accuracy of these experimental methods. An interlaboratory comparison of the data obtained on a standardized medium seems to be a viable approach. In the following sections experimental results on ultrasonic scattering from blood, myocardium, liver, and other soft tissues will be discussed.

II. BLOOD

Ultrasonic scattering in blood is mainly determined by the red blood cells as addressed in Chapter 3 and has been found to be affected by a number of parameters which can be divided into two categories: those related to physical properties of red blood cells themselves and those related to the distribution of the red blood cells. The diameter relative to wavelength, size distribution, shape, and intrinsic acoustic properties, namely, density and compressibility, of the red cells belong to the former category whereas hematocrit, plasma proteins that promote red cell aggregation, and flow disturbance pertain to the latter category.

A. EFFECTS OF PHYSICAL PROPERTIES OF RED BLOOD CELLS ON SCATTERING

1. Size and Size Distribution

The fact that ultrasonic scattering in blood is affected by the size of the red cells was first demonstrated by Borders et al.[10] who showed that the Doppler power returned from blood is dependent upon animal species. These results are depicted in Figure 1 where the relative Doppler power reflected from goat and dog red cell suspensions in saline is plotted as a function of the number of red cells per mm^3. They found that the experimentally determined ratio of the Doppler powers throughout the cell concentration range was between 17 and 19 whereas the square of the ratio of dog to goat red cell volume, $(74/18)^2$, is 16.8. These findings are in agreement with theoretical prediction which indicates that when the diameter of red cells is much smaller than the wavelength (diameter/wavelength < 0.1), ultrasonic scattering in

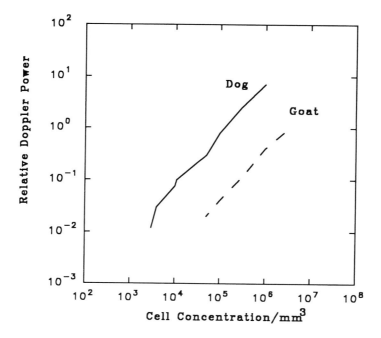

FIGURE 1. Relative Doppler power of ultrasound signal backscattered from goat (dashed line) and dog (solid line) red blood cell suspensions as a function of number of red cells per mm³.

blood assuming identical cells and that all other conditions remain unchanged should be proportional to the square of the volume of the red cells. These results were confirmed in a more recent investigation.[11] The theory on scattering of waves by small or tenuous scatterers, termed Rayleigh scattering, has been described in Chapters 4 and 5. Currently there exists little or no experimental data on the effect of size distribution of red cells on scattering although recent theoretical development has shown that cell size distribution can affect it to a great extent.[12,13]

Rayleigh scattering also predicts that scattering from scatterers much smaller than the wavelength is proportional to the fourth power of frequency. This was an experimental observation for human red blood cells suspended in saline.[15,18] These results are shown in Figure 2 where the dashed and solid lines denote, respectively, the backscattering coefficient in $cm^{-1} sr^{-1}$, defined as the power scattered in the backward direction per solid angle per unit incident intensity per unit volume of scatterers, for hematocrit = 26% and 8%. Also depicted in this figure is a line proportional to (frequency)⁴. However, it will be seen later that this relation may not hold in whole blood when there is red cell aggregation which causes the size of the scatterers to change. The phenomenon of red cell aggregation has been discussed in Chapter 3.

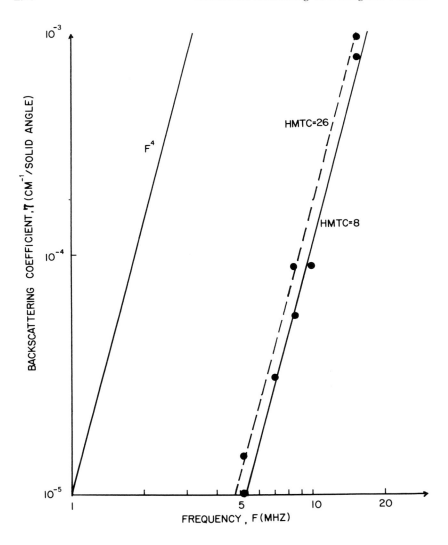

FIGURE 2. Frequency dependence of backscattering coefficient of outdated human red cell suspensions. Dashed and solid lines denote, respectively, results for 26 and 8% hematocrit. A f^4 line is also shown.

2. Shape

It is generally assumed that the shape of the scatterers does not play any significant role for small scatterers[14] in influencing scattering and thus red cells have been treated as spheres of equivalent volume.[15-18] This assumption seems to hold at low cell concentrations where there are minimal cell-to-cell interactions. However, as the hematocrit increases, interactions among the cells which modify the cell fluctuation pattern or cell packing in a unit volume

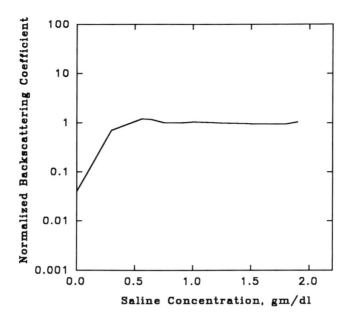

FIGURE 3. Backscattering coefficient of outdated human red cells suspended in saline solution of varying NaCl concentration normalized to that at 0.9 g/dl measured at 7.5 MHz.

must be considered. This point was obscured until the recent development of more advanced theoretical models[12,13,19,20] (see also Chapter 5).

Cell shape becomes an important factor as ultrasound frequency increases. Unfortunately, experimental data of this type are presently lacking. High frequency ultrasonic scattering in blood has recently gained considerable interest because of the development of intravascular imaging[21] which typically uses frequencies from 20 MHz up to 50 MHz. In order to be able to visualize the blood vessel wall, ultrasound pulses emitted by ultrasonic elements mounted on the tip of a catheter have to traverse through blood to reach the vessel lumen. As a result, an understanding of high frequency blood scattering properties is of great significance in design considerations. Roos et al.[22] measured ultrasonic scattering by individual red cells at 30 MHz in attempting to develop an acoustic method for characterizing and sizing microparticles.

3. Intrinsic Acoustic Properties

The red cell structurally can be considered to consist of a fluid, aqueous hemoglobin solution, enclosed by a thin membrane. The effect of red cell membrane on scattering seems to be negligible suggested by results obtained by Shung and Reid[23] on ultrasonic scattering from red cells suspended in saline of varying NaCl concentration. These results are shown in Figure 3

TABLE 1
Measured Ultrasonic Backscattering Cross-Sections of Human Erythrocytes at 5, 7.5, and 10 MHz

Frequency (MHz)	Backscattering cross-section (cm^2 sr^{-1})
5.0	0.4×10^{-14}
7.5	2.7×10^{-14}
10.0	10.8×10^{-14}

where it can be seen that scattering from red cell ghost at 7.5 MHz is only about 4% that of red cells suspended in normal saline of 5.7% hematocrit. Hence it appears to be reasonable to treat red cells as fluid spheres. As such, their scattering properties are dominated by the density and compressibility of the cell content which is mostly hemoglobin and water and should exhibit an angular distribution (see Equation 19 in Chapter 1). Experimental data on backscattering cross-section of a single cell agree well with theoretically computed values based upon published values for normal density and compressibility of red cells.[11,15] Table 1 lists the measured backscattering cross-sections for human red blood cells at 5, 7.5, and 10.0 MHz.[11] Although Roos et al.[22] were able to show that scattering at 30 MHz from red cells at 90° can be linearly correlated to the mean cellular hemoglobin concentration, the roles of cell size and size distribution were not separately examined. The effect of the increase in mean cellular hemoglobin concentration as saline concentration increases on scattering may be offset by the decrease in cell volume. This may explain why scattering does not change as saline concentration is increased beyond 0.9 g/dl in Figure 3.

Experimentally measured angular scattering distribution from red blood cells by Shung et al.[25] agrees reasonably well with the Rayleigh scattering theory[14] but it fits better to a more sophisticated model[25] that takes the viscosity of the suspending medium into consideration. This is shown in Figure 4 where θ is the angle between the directions of incident beam and receiving beam. Similar results were recently obtained by Nassiri and Hill.[9]

B. EFFECTS OF RED CELL DISTRIBUTION ON SCATTERING
1. Hematocrit

The assumption that ultrasonic scattering in blood was proportional to the number of red blood cells per unit volume or hematocrit defined as the volume concentration of red cells has been used by a number of investigators in designing and modeling ultrasonic Doppler flowmeter,[26,27] not realizing that in normal blood the volume concentration of red cells is so high that they can no longer be treated as independent scatterers or point scatterers. When cell concentration is less than a few percent, interactions among cells are

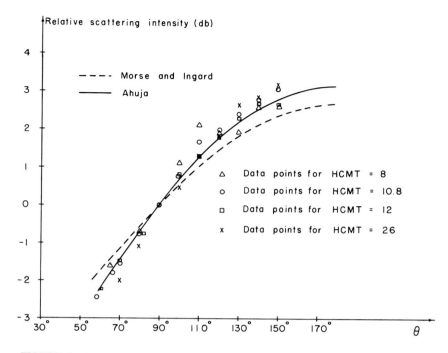

FIGURE 4. Measured angular dependence of scattering from outdated human red cell suspensions at different hematocrits over the scattering angular range from 60 to 150°. The solid and dashed lines represent, respectively, theoretical curves taken from References 14 and 25.

minimal and the cells can be considered independent scatterers. As concentration increases, the cells interact and the scattered waves from them may interfere constructively or destructively at the detector at a fixed instant of time depending upon the spatial distribution of these cells. The spatial distribution of the cells is usually characterized by some form of correlation functions.[12-14] The non-linear relationship between ultrasonic backscatter from red cells suspended in saline and hematocrit was first reported by Shung et al. in 1976.[15] For red cell suspensions under laminar flow using a mock flow loop it was found that the measured backscattering coefficient peaks at 13% as predicted by more recent theoretical models.[12,13] This is illustrated in Figure 5 for porcine red cells where the solid line indicates the theoretical curve for small spheres computed using the following data: red cell compressibility = 34.96×10^{-12} cm²/dyne, saline compressibility = 44.3×10^{-12} cm²/dyne, porcine red cell density = 1.078 g/cm³, saline density = 1.005 g/cm³, porcine red cell volume = 68 µm³, and the triangles denote the experimental data.[17] For whole blood where there is red cell aggregation, ultrasonic scattering is also a function of flow rate which affects the shearing forces acted on the red cells. Using a mock flow loop under laminar flow so that the mean shear rate or mean velocity gradient in the flow channel is known and in fact is pro-

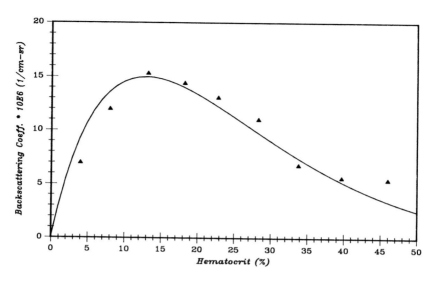

FIGURE 5.   Measured backscattering coefficient of porcine red cell suspension under steady laminar flow as a function of hematocrit denoted by solid triangles. The solid line is the theoretical curve (Equation 22b in Chapter 5).

portional to flow velocity, ultrasonic backscattering coefficient decreases as the mean shear rate increases. Data on porcine whole blood which has a red cell aggregation tendency similar to human for hematocrit of 4.5, 14.5, 25, 34, and 47% are shown in Figure 6 where the ordinate represents the mean of backscattering coefficients measured in the central zone and in the peripheral zone of the flow conduit since the shear rates in a vessel which vary as a function of the radial distance from the center of the vessel are different in these regions and the abscissa denotes the mean shear rate in these two regions. Clearly the relationship between scattering and hematocrit in whole blood no longer follows that described by Equation 22b in Chapter 5 or the solid line in Figure 5 in the shear rate range investigated. It is known that shear rate below 20 s^{-1} is not sufficient to completely disrupt red cell aggregation in normal whole blood[28] but whether the behavior dictated by Equation 22b in Chapter 5 can be restored as the shear rate is further increased remains to be seen. In bovine whole blood where there is little red cell aggregation the scattering is almost independent of shear rate.[17] These observations have recently been verified with a pulsed Doppler flowmeter. It was found that Doppler power from center stream of porcine red cell suspension under laminar flow measured as a function of hematocrit peaks also near 13% and the Doppler power of porcine whole blood under laminar flow is dependent upon shear rate.[29]

Boynard et al.[30] also measured scattering from red cell suspension, whole blood, and whole blood with higher sedimentation rate presumably due to

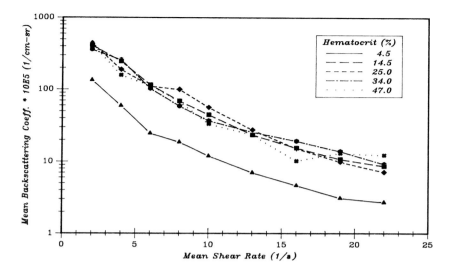

FIGURE 6. Measured mean backscattering coefficient of porcine whole blood under steady laminar flow as a function of mean shear rate at five different hematocrits.

increased red cell aggregation as a function of hematocrit and found that scattering peaks at between 20 to 30% in suspension and at between 30 and 40% in whole blood. It is unclear, however, from the paper whether attenuation was compensated and whether there is flow disturbance in the measurement chamber. An interesting conclusion from this series of experiments was that higher scattering can be correlated to a higher sedimentation rate. A-mode echo amplitude from human whole blood under laminar flow was found to decrease as shear rate was increased by Sigel et al.[31] It was later shown to decrease by more than 70% as the shear rate was increased from 5 to 13 s^{-1} by Kalllio and Alanen[32,33] using an approach in which background echoes due to vessel walls were eliminated or suppressed by temporal subtraction.

Results on echogenicity observed with real-time B-mode imaging of flowing whole blood as a function of shear rate corroborate these findings.[31,34-36] Blood echogenicity decreases as shear rate increases.

2. Plasma Proteins

Two plasma proteins, fibrinogen and serum globulins, have been known to cause red cell aggregation.[37] Shung and Reid[38] first demonstrated that an increase in fibrinogen concentration in blood causes scattering to increase. This observation was later verified by results from a more systematic study which are shown in Figure 7 where the mean backscattering coefficient of porcine whole blood of same hematocrit but different fibrinogen concentration is plotted as a function of mean shear rate.[18] Ultrasonic scattering has been

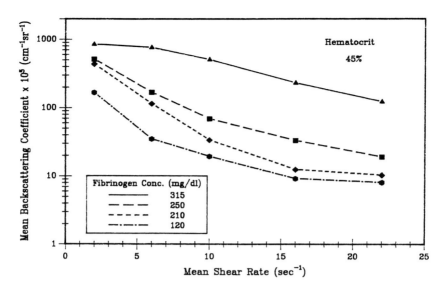

FIGURE 7. Measured mean backscattering coefficient of porcine whole blood at 45% hematocrit of varying fibrinogen concentration as a function of mean shear rate.

suggested and studied by several investigators as a means to quantitate the phenomenon of red cell aggregation.[30,31,38] Furthermore, since fibrinogen is the most dominant plasma protein in affecting red cell aggregation, ultrasonic scattering may potentially be useful as an assay for fibrinogen concentration in blood.[38,39]

Because the size of the scatterers in blood changes when there is aggregation and the dimension of these red cell aggregates may be so large that diameter/wavelength ratio is no longer much smaller than 1, this means that Rayleigh scattering may not hold and, thus, the frequency dependence of scattering from whole blood could well deviate from 4th power. This was confirmed by recent experimental results shown in Figure 8 where the mean ultrasonic backscattering coefficient is plotted as a function of frequency for porcine whole blood of 4.5% hematocrit and 210 mg/dl fibrinogen concentration at 3 shear rates.[18] Also shown in this graph is a (frequency)^4 line. It is apparent that at low shear rate (2 s^{-1}) the relationship between scattering and frequency deviates considerably from f^4 whereas at higher shear rates the f^4 dependence is maintained at this hematocrit.

3. Flow Disturbance

Although it was suggested by Shung et al.[15] in 1976 that flow disturbance may affect ultrasonic scattering in blood, there had been no experimental data to substantiate this hypothesis until 1984 when Shung et al.[40] showed in a mock flow loop where flow turbulent intensity could be well-controlled that

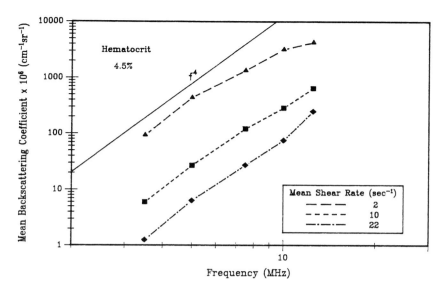

FIGURE 8. Measured mean backscattering coefficient of porcine whole blood under steady laminar flow as a function of frequency at three different mean shear rates. The solid line depicts (frequency)[4] dependence.

ultrasonic scattering from red cell suspensions is related to turbulent intensity. This relationship is depicted in Figure 9 where circles and squares represent, respectively, data for bovine red cell suspension at 40% hematocrit under laminar and turbulent flow. As turbulent intensity which is proportional to flow rate increases, scattering increases. These results can be readily explained by recent theoretical models discussed in Chapter 5 which assert that scattering from a distribution of small scatterers is related to fluctuation of the number of scatterers in a volume rather than the number itself. Turbulence which increases fluctuation causes scattering to increase. These results have also been recently validated by a study using a Doppler flowmeter.[29]

In spite of the fact that a substantial amount of data has been accumulated over the years on ultrasonic scattering in blood, there are still a number of areas that either are untouched or remain to be explored. First, the experimental results discussed above were all obtained under steady laminar flow conditions. It occurred only recently that a few investigators began to address the effects of pulsatile flow on ultrasonic scattering in blood.[41] Moreover, data on scattering in whole blood at shear rates beyond 50 s^{-1} which are more pertinent to *in vivo* conditions and on the relationship between flow disturbance and scattering in whole blood are totally lacking.

In addition to the aforementioned factors, diseases and biological processes in blood like clotting can also affect scattering. Ultrasonic scattering from sickled red cells in sickle cell disease and clotted blood was found to

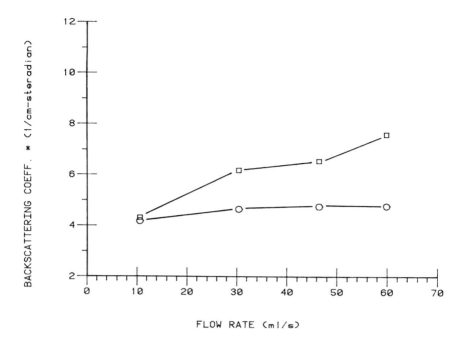

FIGURE 9. Measured backscattering coefficient of bovine red cell suspension at 40% hematocrit under steady laminar (open circles) and turbulent flow (open squares) as a function of flow rate.

increase probably due to increased cell rigidity and change in scatterer size, respectively.[42-44] Blood clots have also been observed by several investigators to be significantly more echogenic than unclotted blood.[45,46]

III. MYOCARDIUM

Earlier measurements on bovine myocardium using a standard substitution method described in Chapter 6 with the ultrasonic beam perpendicularly incident upon the myocardial fibers by Shung and Reid indicate that backscattering coefficient from bovine myocardium over the frequency range of 3 to 10 MHz can be fitted with a line of $7.6 \times 10^{-6} f^{3.3}$ where f is frequency in MHz.[6] Later measurements[8] by the same group in the frequency range of 2 to 7 MHz show a fitted line of $1.5 \times 10^{-5} f^{2.7}$. The measured frequency dependencies are reasonably close to the theoretically predicted value for small cylinders.[14] At 5 MHz the measured backscattering coefficients are 2.97 $\times 10^{-3}$ and 2.15×10^{-3} cm^{-1} sr^{-1} respectively. Similar frequency dependence was also observed by O'Donnell et al.[47] on canine myocardium. Their experimental results are slightly higher than those obtained by Shung et al.[6,8] probably because focused transducers were used in their measurements.[48]

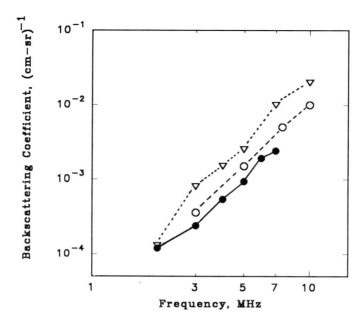

FIGURE 10. Measured backscattering coefficient for myocardium as a function of frequency. Open triangles, open circles, and solid circles represent, respectively, data on canine myocardium obtained by O'Donnell and Miller, and data on bovine myocardium from References 6 and 8.

These results are illustrated in Figure 10 where the triangles, open circles and filled circles are extracted from References 47, 6, and 8, respectively.

Since myocardial cells are preferentially aligned more or less parallel to the heart surface (see Chapter 2 for more detail), it can be easily understood why myocardium possesses the characteristics of an anisotropic medium.[49,50] The backscattered echoes are maximal when the ultrasonic beam is perpendicular to the myocardial cells and minimal when parallel.[50] The magnitude of this angular variation is frequency dependent.

It has been repeatedly demonstrated that heart diseases that affect anatomy and biological composition alter myocardial backscattering.[51,52] A comprehensive review of this topic can be found in Chapter 10. The increase in backscatter from infarcted canine myocardium correlates well the increase in collagen content in the myocardium.[47] The dashed line and the solid line in Figure 11 show, respectively, the backscattering coefficient for a 5- to 6-week-old infarct and normal myocardium as a function of frequency.[47] Both the magnitude and slope of the backscattering coefficient for infarcted myocardium can be seen to differ from those of normal myocardium.

IV. LIVER

A substantial amount of data on scattering has been collected for liver.[6-9,53,54] These results are summarized in Figure 12 where open circles,

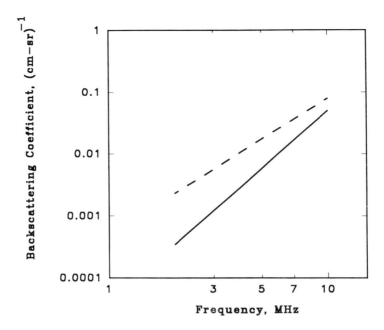

FIGURE 11. Measured backscattering coefficient of canine myocardium as a function of frequency from Reference 47. Solid line indicates data for normal myocardium. Dashed line indicates data for infarcted myocardium, 5 to 6 weeks following occlusion of coronary artery.

open triangles, solid circles, solid triangles, and open squares are obtained, respectively, from References 53, 7, 8, 54, and 9. Data from References 7, 8, and 53 were measured using a standard substitution method with nonfocused transducers but it is unclear what type of transducers were used by Nassiri and Hill[9] to obtain their results. Nonetheless, it seems all the data are in reasonable agreement except for those obtained by Campbell and Waag who used a different approach (see Chapter 8).[54] Moreover Campbell and Waag's data were obtained on calf liver for differential scattering cross-section at 165° angle. These data and power fitted curves wherever available are further tabulated in Table 2.

Angular scattering by liver has also been measured.[9,54-56] The results show that scattering in the forward direction is stronger than in the backward direction and the scattering pattern is almost frequency independent. At present the question as to what are the basic structures in liver responsible for ultrasonic scattering remains a controversial matter. The difficulty primarily lies in the fact that liver anatomy is much more complex than blood or myocardium. There are perhaps multiple scattering structures in liver. In this case, a simple power law would not be adequate in describing ultrasonic scattering behavior in liver. Several authors have used multiple term polynomial fit.[7,57,58] The problem on identifying multiple scattering sources in biological tissues is also addressed in Chapter 4.

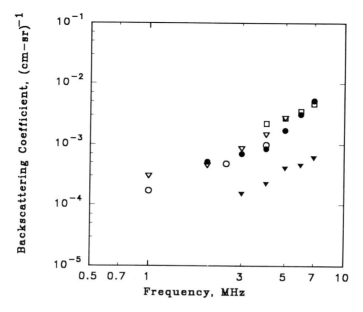

FIGURE 12. Backscattering coefficient for human liver from Reference 53 (open circles), Reference 7 (open triangles) and Reference 9 (open squares); and bovine liver from Reference 8 (solid circles) and Reference 54 (solid triangles) as a function of frequency.

The effects of diseases on liver scattering may vary depending upon the nature of the disease.[52,53] For instance, Bamber and Hill[53] demonstrated that solid tumors in human liver have a smaller scattering than normal liver.

V. KIDNEY

Structurally the peripheral layer of the kidney, the cortex, is very different from the inner portion, the medulla. The cortex is consisted of renal corpuscles surrounded by collecting tubes called medullary rays. Therefore it can be readily understood that kidney cortex is ultrasonically anisotropic. This has been experimentally demonstrated.[59,60] For ultrasonic beam perpendicular to the surface but approximately parallel to the medullary rays, the results on the backscattering coefficient for kidney cortex are summarized in Figure 13 where the solid circles represent the experimental results on bovine tissue obtained by Fei and Shung;[8] and the dashed and dotted lines denote respectively the power fit curves to data on human tissue obtained by Turnbull et al.[61] and Insana et al.[60] The equations for the power fit curves are tabulated in Table 3. Turnbull et al.[61] also measured backscattering from the medullary and found that it is similar to that from cortex. It can be seen from Figure 13 that the data obtained by Fei and Shung are one order of magnitude higher than those obtained by Turnbull and Insana. The reason for this discrepancy

TABLE 2
Measured Ultrasonic Backscattering Coefficient for Liver

Species	Frequency (MHz)	BSC[a] (cm^{-1} sr^{-1})	Power fit	Source
Human	1	1.7×10^{-4}	—	Ref. 53
	2.5	4.8×10^{-4}		
	4	9.8×10^{-4}		
Human	1	3.0×10^{-4}	$2.7 \times 10^{-4} f^{1.2}$	Ref. 7
	2	4.5×10^{-4}		
	3	8.5×10^{-4}		
	4	1.5×10^{-3}		
	5	2.7×10^{-3}		
Bovine	2	4.2×10^{-4}	$8.4 \times 10^{-5} f^{2.0}$	Ref. 8
	3	7.0×10^{-4}		
	4	9.2×10^{-4}		
	5	1.7×10^{-3}		
	6	3.1×10^{-3}		
	7	5.2×10^{-3}		
Human	4	2.2×10^{-3}	$3.2 \times 10^{-4} f^{1.4}$	Ref. 9
	5	2.7×10^{-3}		
	6	3.9×10^{-3}		
	7	4.6×10^{-3}		
Calf[b]	3	1.5×10^{-4}	—	Ref. 54
	4	2.2×10^{-4}		
	5	4.0×10^{-4}		
	6	4.5×10^{-4}		
	7	6.0×10^{-4}		

[a]   BSC = Backscattering coefficient
[b]   Measured at a scattering angle which deviates from the backward direction by 15°.

is probably due to the following two reasons: (1) In Turnbull and Insana's measurements focused transducers are used. It is known that results collected with focused transducers could deviate substantially from those measured with nonfocused transducers when the standard substitution method is employed for measuring scattering.[48] (2) There may be anatomical differences between human and bovine tissues.

Backscattering in various forms of cancer has been studied by Turnbull et al.[61] Renal cell carcinoma was found to have higher backscatter at 5 MHz than normal kidney whereas oncocytoma had a lower backscatter at the same frequency.

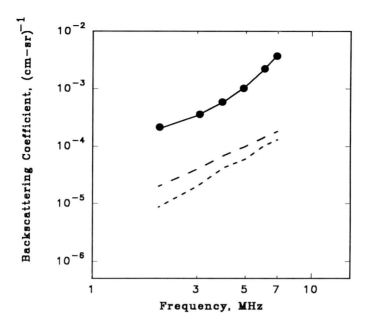

FIGURE 13. Measured backscattering coefficient for bovine kidney cortex as a function of frequency from Reference 8 is shown as solid circles. The least square fitted power curves for data from References 61 and 62 on human kidney cortex are shown as dashed and solid lines.

TABLE 3
Power Fitted Equations Obtained by Various Investigators for Ultrasonic Backscattering Coefficient for Kidney Cortex

Authors	Power fitted equation[a]
Fei and Shung	$3.1 \times 10^{-5} f^{2.3}$
Turnbull and Foster	$6.6 \times 10^{-6} f^{1.7}$
Insana and Hall	$1.9 \times 10^{-6} f^{2.2}$

[a] f denotes frequency in MHz.

VI. OTHER SOFT TISSUES

Experimental results on backscattering from spleen are shown in Figure 14 where solid and open circles represent respectively data obtained by Nicholas[7] on human tissue, Fei and Shung[8] on bovine tissue. The power fit equations to these data are $1.2 \times 10^{-4} f^{1.7}$ and $1.1 \times 10^{-5} f^{2.6}$. Experimental results for bovine skeletal muscle,[9] white matter of human brain[7] and bovine pancreas[8] are depicted in Figure 15 as solid circles, open triangles, and open circles.

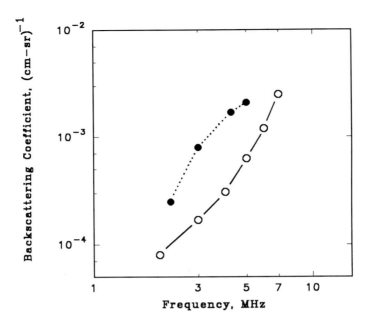

FIGURE 14.   Measured backscattering coefficient for spleen from Reference 7 on human tissue (solid circles) and from Reference 8 on bovine tissue (open circles) as a function of frequency.

Finally, the data reported in this chapter are all for fresh tissues. Ultrasonic scattering measurements have also been made on fixed tissues.[6,9,56] Fixation of tissues cause alteration of tissue structure and composition. The effect of fixation on scattering in tissues varies depending upon tissue type. Therefore, great caution should be taken when extrapolating tissue scattering properties from data on fixed tissues or comparing data on fixed tissues to those on fresh tissues.

VII. CONCLUSIONS

Data on ultrasonic scattering from a variety of tissues of different animal species can be found in the literature. Considering the complexity of the tissues and its dependencies on animal species, it is not surprising to find that the magnitude of scattering and its frequency dependence vary not only from one type of tissue to another from the same species of animal but also from animal to animal on the same type of tissues. The results are also technique dependent. However, when the same technique is used, the results obtained by different groups of investigators are remarkably consistent on the same type of tissues. It is fair to say at this point that a promising start in understanding the ultrasonic scattering process in tissues has been made but it also has to be recognized that there is still a long way to go before this phenomenon

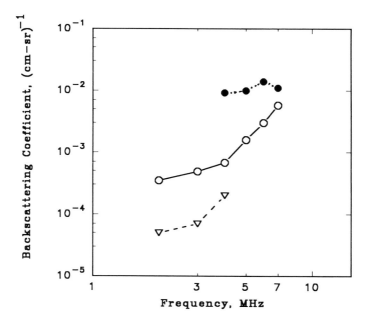

FIGURE 15. Measured backscattering coefficient for bovine skeletal muscle from Reference 9 (solid circles), white matter of human brain from Reference 7 (open triangles) and bovine pancreas from Reference 8 (open circles) as a function of frequency.

can be fully understood. A better understanding of the scattering phenomenon in tissues is critical for the realization of the ultimate goal of ultrasonic imaging, quantitative tissue characterization.

REFERENCES

1. **Linzer, M.,** Ultrasonic Tissue Characterization I, Spec. Publ. 453, National Bureau of Standards, Gaithersburg, MD, 1976.
2. **Linzer, M.,** Ultrasonic Tissue Characterization II, Spec. Publ. 525, National Bureau of Standards, Gaithersburg, MD, 1979.
3. **Greenleaf, J. M.,** *Tissue Characterization with Ultrasound,* CRC Press, Boca Raton, FL, 1986.
4. **Goss, S. A., Johnston, R. L., and Dunn, F.,** Comprehensive compilation of empirical ultrasonic properties of mammalian tissues, *J. Acoust. Soc. Am.,* 64, 423, 1978.
5. **Goss, S. A., Johnston, R. L., and Dunn, F.,** Comprehensive compilation of empirical ultrasonic properties of mammalian tissues. II, *J. Acoust. Soc. Am.,* 68, 93, 1980.
6. **Shung, K. K. and Reid, J. M.,** Ultrasonic scattering from tissues, *1977 IEEE Ultras. Symp. Proc.,* (IEEE Cat. #77CH1264-ISU) 203, 1977.
7. **Nicholas, D.,** Evaluation of backscattering coefficients for excised human tissues: results, interpretations, and associated measurements, *Ultras. Med. Biol.,* 8, 17, 1982.

8. **Fei, D. Y. and Shung, K. K.**, Ultrasonic backscatter from mammalian tissues, *J. Acoust. Soc. Am.*, 78, 871, 1985.

9. **Nassiri, D. K. and Hill, C. R.**, The differential and total bulk scattering cross sections of some human and animal tissues, *J. Acoust. Soc. Am.*, 79, 2034, 1986.

10. **Borders, S. E., Fronek, A., Kemper, W. S., and Franklin, D.**, Ultrasonic energy backscattered from blood — an experimental determination of the variation of sound energy with hematocrit, *Ann. Biomed. Eng.*, 6, 83, 1978.

11. **Yuan, Y. W.**, Ultrasonic Scattering from Blood, Ph.D. thesis in Bioengineering, Pennsylvania State University, University Park, PA, 1988.

12. **Twersky, V.**, Low-frequency scattering by mixtures of correlated nonspherical particles, *J. Acoust. Soc. Am.*, 84, 409, 1988.

13. **Berger, N. E., Lucas, R. J., and Twersky, V.**, Polydisperse scattering theory and comparisons with data for red blood cells, *J. Acoust. Soc. Am.*, 89, 1394, 1991.

14. **Morse, P. M. and Ingard, K. U.**, *Theoretical Acoustics*, McGraw-Hill, New York, 1968.

15. **Shung, K. K., Sigelmann, R. A., and Reid, J. M.**, Ultrasonic scattering by blood, *IEEE Trans. Biomed. Eng.*, BME-23, 460, 1976.

16. **Atkinson, P. and Woodcock, J. P.**, *Doppler Ultrasound and Its Use in Clinical Measurement*, Academic Press, London, 1982.

17. **Yuan, Y. W. and Shung, K. K.**, Ultrasonic backscatter from flowing whole blood. I. Dependence on shear rate and hematocrit, *J. Acoust. Soc. Am.*, 84, 52, 1988.

18. **Yuan, Y. W. and Shung, K. K.**, Ultrasonic backscatter from flowing whole blood. II. Dependence on frequency and fibrinogen concentration, *J. Acoust. Soc. Am.*, 84, 1195, 1988.

19. **Mo, L. Y. L. and Cobbold, R. S. C.**, A stochastic model of the backscattered Doppler ultrasound from blood, *IEEE Trans. Biomed. Eng.*, BME-33, 20, 1986.

20. **Mo, L. Y. L., Cobbold, R. S. C., and Shung, K. K.**, Common misconceptions about the scattering of ultrasound by blood, *Proc. 12th Annu. Int. Conf. IEEE Eng. Med. Biol. Soc.*, (IEEE Cat. #90CH2936-3) 291, 1990.

21. **Pandian, N. G.**, Intravascular and intracardiac ultrasound imaging: an old concept, now on the road to reality, *Circulation*, 4, 1091, 1989.

22. **Roos, M. S., Apfel, R. E., and Wardlaw, S. C.**, Application of 30 MHz acoustic scattering to the study of human red blood cells, *J. Acoust. Soc. Am.*, 83, 1639, 1988.

23. **Shung, K. K. and Reid, J. M.**, The effect of hypotonicity upon the ultrasonic scattering properties of erythrocytes, in *Ultrasound in Medicine*, Vol. 4, White, D. N. and Lyons, E. A., Eds., Plenum Press, New York, 1978.

24. **Ahuja, A. S.**, Effect of particle viscosity on propagation of sound in suspensions and emulsions, *J. Acoust. Soc. Am.*, 51, 182, 1972.

25. **Shung, K. K., Sigelmann, R. A., and Reid, J. M.**, Angular dependence of scattering of ultrasound from blood, *IEEE Trans. Biomed. Eng.*, BME-24, 325, 1977.

26. **Brody, W. R. and Meindl, J. D.**, Theoretical analysis of the CW Doppler ultrasonic flowmeter, *IEEE Trans. Biomed. Eng.*, BME-21, 183, 1974.

27. **Hottingger, C. F. and Meindl, J. D.**, Blood flow measurement using the attenuation-compensated volume flowmeter, *Ultras. Imag.*, 1, 1, 1979.

28. **Schmid-Schonbein, H., Gaeghtgens, P., and Hirsh, H.**, On the shear rate dependence of red cell aggregation in vitro, *J. Clin. Invest.*, 47, 1447, 1968.

29. **Shung, K. K. and Lim, C.**, The effect of hematocrit and shear rate on the Doppler spectrum under steady and pulsatile flow, *Proc. 12th Annu. Int. Conf. IEEE Eng. Med. Biol. Soc.*, (IEEE Cat. #90CH2936-3), 306, 1990.

30. **Boynard, M., Leilierve, J. C. and Guillet, R.**, Aggregation of red cells studied by ultrasound backscattering, *Biorheology*, 24, 451, 1987.

31. **Sigel, B., Machi, J., Beitler, J. C., and Justin, J. R.**, Red cell aggregation as a cause of blood-flow echogenicity, *Radiology*, 148, 799, 1983.

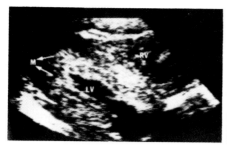 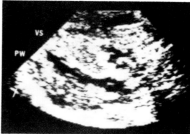

FIGURE 1.   Type IIA myocardial texture. Subcostal views. Multiple, discrete, and small highly reflective echoes (HREs) involve all walls of both ventricles in a patient with cardiac amyloidosis (left) and in another patient with Pompe's disease (right). The ventricular walls are markedly thickened in both. AO = aorta; LV = left ventricle; M = myocardium; PW = posterior wall; RV = right ventricle; VS = ventricular septum. (From Bhandari, A. K. and Nanda, N. C., *Am. J. Cardiol.*, 51, 817, 1983. With permission.)

A. ANALYSIS OF CONVENTIONAL TWO-DIMENSIONAL ECHOCARDIOGRAPHIC IMAGES

Conventional two-dimensional echocardiographic images contain significant qualitative information that may facilitate clinical diagnosis of cardiovascular disease. The granular sparkling patterns of amyloidosis and Pompe's disease (Figure 1) represent examples of qualitative observations derived from conventional images that reflect the physical composition of tissues.[4] Other common examples familiar to practiced echocardiographers are the highly echogenic appearance of mitral annular calcification and the hypoechoic appearance of an aortic ring abscess.

Although such qualitative observations may provide valuable diagnostic information, characterization of the fundamental physical properties of tissue based on conventional echocardiographic images is influenced by the data processing algorithms implemented in commercially available imagers. Electronic pre- and post-processing functions usually are applied to enhance the data display, and include variable transmission gain and attenuation, data compression for convenient mapping of the full dynamic range of the instrument to the gray-scale monitor, and post-process filtering. This ''processed'' image data then may be analyzed further off-line to generate numerical indexes of tissue acoustic properties. Two such off-line analytic procedures are quantitative videodensitometry and color coding of images. Both processes permit rough estimation of the amplitude of backscatter from myocardial tissues.

1. Videodensitometry

Videodensitometry provides a method for converting analog videotaped two-dimensional images into an array of digitized picture elements (pixels), typically with 6 to 8 bit gray-scale resolution. The process converts conventional two-dimensional videotape images to binary data with the use of a video frame digitizer (or "framegrabber") that is under computer control. After each frame is digitized, the binary data can be stored on magnetic media for later analysis off-line. Each digitized pixel within a stored image frame can be accessed independently to assess the gray level value of the image at a single point in time and space. The intensity of scattering then can be judged roughly according to the pixel gray level. Multiple frames of image data can be digitized over several heart cycles and stored. In addition, selected regions of interest can be analyzed and statistical parameters of scattering computed.

Several groups have used digitized two-dimensional video image data to compute "moment statistics" that depict the physiologic state of myocardial tissue.[5,6] The hypothesis underlying the use of moment statistics is that pathologic alterations of tissue composition will change the distributional characteristics of pixel intensities within a region of interest. In particular, computation of the third and fourth central moments of the distribution of intensities allows assessment of skewness and kurtosis of pixel intensities: skewness characterizes the asymmetry of the distribution and indicates the extent to which one or the other tail of a curve is splayed; kurtosis identifies a departure from the normal distribution in terms of the peakedness of a curve. Although the statistical distribution of pixel intensities has been shown to change after myocardial infarction in experimental animals, the pathologic correlates of altered skewness and kurtosis have not yet been defined.

Quantification of regional backscatter amplitude by videodensitometry is technically simple to implement because standard half-inch videotaped two-dimensional images serve as data input and no special modifications of the echocardiographic imaging system are required for data acquisition. The data acquired, however, are subject to the electronic characteristics of the particular imager used. The echocardiographic image processing parameters that are under operator control may not remain constant throughout a series of studies in an individual patient or from study to study in different patients, precluding accurate comparisons among different data sets. Echocardiographic data compression schemes applied at the level of the imager or the video digitizer also may influence the accuracy of comparisons of regional backscatter among patients. Degradation of image quality may occur because of the additional processing steps required for conversion of the image to analog videotape before digitization and quantitative estimation of backscatter intensity.

2. Color Coding

Color coding of two-dimensional images produces visual enhancement of gray-scale image contrast.[7,8] The standard presentation of pixel intensity as gray levels is replaced by a range of colors that are mapped to discrete pixel intensities (Plates 1 and 2*). Color coding of pixel intensity serves to enhance the contrast between gray levels of similar intensities that may not

---

* Plates 1 and 2 follow page 374.

be well differentiated by eye. Furthermore, the programmable components of color (hue, saturation, and luminescence) theoretically permit encoding of additional information in ultrasound images beyond the one-dimensional representation of intensity by gray-scale encoding. Color coding may be performed in real time, as in color Doppler imaging, or applied off-line to digitized image data.

The potential disadvantages of color coding are similar to those of video-densitometry. Because color coding is applied to the data after pre- and post-processing by the imager, color mappings of pixel intensities are not easily standardized. Color provides visual cues to tissue characteristics and permits qualitative estimation of physical properties of tissue during the performance of the study but may be less useful for longitudinal evaluation of the acoustic properties of tissue, an evaluation that may require more objective quantitative indexes.

B. ANALYSIS OF IMAGE TEXTURE

Complex wave interference patterns that arise as a consequence of scattering from myocardial tissue are responsible to a large extent for the appearance of acoustic "speckle" in conventional two-dimensional echocardiograms. Measurement of image texture provides a method for quantification of speckle by the description of the regional spatial distribution of echo amplitudes in two-dimensional echocardiograms.

Bhandari and Nanda[4] reported a set of qualitative textural patterns derived from analysis of videotaped two-dimensional images from patients with a wide range of cardiac structural abnormalities. They proposed a classification of textural patterns that comprised type I (uniform, low intensity, fine speckle: normal), type IIA (multiple, discrete, 3 to 5 mm, bright echoes in all walls), type IIB (one or more, but not all, walls with type IIA pattern), and type IIC (uniform very bright echoes in one wall or region). Despite the appeal of this simple classification, differentiation of cardiac disease with the use of such subjective descriptors suffers a lack of specificity that may vitiate this approach.

More quantitative methods for analysis of texture have used digitized image data that are processed according to the methods described above (see Section II.A.2). Selected regions of interest within single frames of digitized two-dimensional image data are analyzed to determine the intensity and the distribution of gray levels (Figure 2A). Gray-level histogram statistics, edge count and run-length, and difference statistics represent a few of the measurements of texture that have been used.[9] Measures of gray-level heterogeneity (gray-level difference and run-length statistics) have been shown to differentiate normal from contused myocardium in dogs (Figure 2B).

The principal caveat applicable to these measurements is that textural parameters may register characteristics of the ultrasound imager itself rather than the physical composition of the tissue imaged. Skorton and colleagues[10]

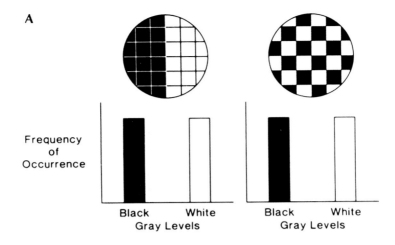

FIGURE 2. (A) The two-dimensional spatial pattern of gray levels in a region of interest may contain information not apparent on analysis of the overall distribution of gray levels in that region. Two schematic regions of interest are shown in the top panels. Since each region is half black and half white, the gray-level histograms (amplitude distributions) are identical (lower panels). The obvious difference in the pattern of blacks and whites (i.e., the texture) is not appreciated by examination of the overall gray-level histograms. (B) The results of three representative texture calculations are shown for normal and contused regions in the post-trauma images; values are means ($\pm$ SD). The mean gray level and selected run-length and gray-level difference statistics were significantly altered in regions of myocardial contusion. *, $p = 0.0001$; †, $p = 0.0018$; ●, p $= 0.0015$; LRE-$0°$ = long-run emphasis calculated along $0°$ (in azimuth); ($\Delta = 2$) = gray-level difference contrast calculated for a pixel separation of two. (From Skorton, D. J. et al., *Circulation*, 68, 217, 1983. With permission.)

have reported that image texture manifests both range- and azimuth-dependent variability as well as variability encountered when different echocardiographic systems are used to image a stable tissue-mimicking phantom. Speckle and textural parameters also depend on system wavelength so that the use of ultrasonic transducers with different center frequencies could yield disparate measurements from the same tissue.

C. ANALYSIS OF INTEGRATED BACKSCATTER

The measurement of integrated backscatter is based on analysis of unprocessed radio frequency signals to derive quantitative ultrasonic indexes capable of differentiating normal from pathologic tissue structure and function.[1-3] The concept of integrated backscatter was first introduced by O'Donnell, Miller, and colleagues at Washington University to provide a single estimate of the energy backscattered from myocardial tissue after insonification with a broadband piezoelectric transducer.[11-17] The technique comprises

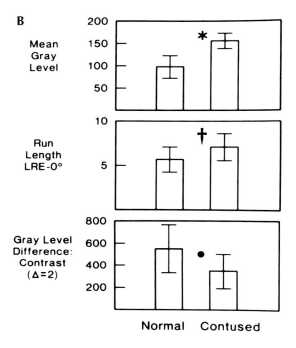

FIGURE 2B.

recording of raw radio frequency data backscattered from a myocardial region of interest, performing spectral analysis by fast-Fourier transformation (FFT), and integrating the received power spectrum over the usable bandwidth of frequencies produced by the broadband transducer to derive a frequency-averaged index of backscatter amplitude.[18] The rationale for computing frequency-averaged backscatter stems from the presence of significant random fluctuations in the spectral content of ultrasound scattered from normal tissue that occur partly as a consequence of wave interference and phase-cancellation effects at the piezoelectric receiver due to the acoustical inhomogeneity of tissue (Figure 3A). By averaging backscattered spectra over the useful range of frequencies, statistical fluctuations are minimized. Thus, integrated back-scatter "smooths" the speckle pattern apparent in conventional two-dimensional images and emphasizes the average scattering properties of a region, whereas textural analysis (see above) emphasizes the distribution of gray levels. Replicate sampling at adjacent myocardial sites further reduces the potential impact of texture on the measurement of average backscatter (Figure 3B). Extensive laboratory evaluation by researchers at Washington University has demonstrated that the magnitude of integrated backscatter correlates with the physical state of myocardium and can differentiate normal myocardial tissue structure from pathologic structure associated with cardiomyopathy, ischemia, stunned myocardium, and myocardial infarction.[1-3]

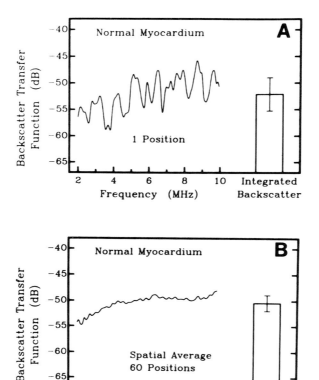

FIGURE 3. (A) The backscatter transfer function measured at a single position within normal dog myocardium. (B) The average of 60 measurements obtained by physically translating the transducer by 1 mm. (From Miller, J. G., Pérez, J. E., and Sobel, B. E., *Prog. Cardiovasc. Dis.*, 28, 85, 1985. With permission.)

The paradigm for analysis of backscatter was developed initially for application in experimental studies and is based on a convolution model that equates convolutions of several ultrasonic properties in the time domain with products in the frequency domain.[17] The power spectrum of a signal scattered from a material that produces both scattering and attenuation can be written as $|E(f)|^2 = |P(f)|^2|I(f)|^2|G(f)|^2|S(f)|^2$, where $|P(f)|^2$ represents the power spectrum of an interrogating ultrasonic signal, $|I(f)|^2$ is the attenuation of ultrasound through tissue intervening between the surface of the specimen and the front of the gated volume, $|G(f)|^2$ is the attenuation within the gated volume itself, and $S(f)$ is the intrinsic backscatter transfer function of the material. A generalized substitution technique is used to produce a calibrated measurement of the scattering properties of the cardiac tissue.[1,11,19] If the scattering material is replaced with a reflector of known properties positioned at the focal point of the transducer, a reference power spectrum can be calculated as $|E_{ref}(f)|^2 = |P(f)|^2|R_{ref}(f)|^2$, where the reference spectrum is determined only by the transducer spectrum, $|P(f)|^2$, and the reflection coefficient of the planar reflector, $|R_{ref}(f)|^2$. In practice, the stainless steel plate used is a nearly perfect reflector. Thus $|R_{ref}(f)|^2$ is taken as unity in the following formulation. By normalizing the spectrum measured from a scattering material to the spectrum from the "perfect" reflector and expressing the result on a logarithmic scale, one can calculate the frequency-dependent "apparent" backscatter transfer function $|B(f)|^2$ expressed in units of decibels below the backscatter from a reflector as $10\cdot\log|B(f)|^2 = 10\cdot\log(|E_{tissue}(f)|^2/|E_{ref}(f)|^2)$. The intrinsic backscatter transfer function, $|S(f)|^2$, may be computed by correcting the apparent backscatter transfer function by the total attenuation, or $10\cdot\log(|I(f)|^2|G(f)|^2)$. This formulation assumes that transmission coefficients at boundaries are unity. To generate a single relative index of backscatter efficiency, the frequency-averaged or integrated backscatter is computed by averaging $|S(f)|^2$ over the useful bandwidth of the transducer, which is typically 3 to 7 MHz for a 5 MHz transducer (a more detailed discussion on this subject is given in Chapter 6).

Because the apparent backscatter transfer function, $|B(f)|^2$, incorporates the effects of attenuation, calculation of the true backscatter transfer function, $|S(f)|^2$, requires correction for the frequency-dependent attenuation. Correction of $|B(f)|^2$ for attenuation requires either measurement or estimation of the frequency-dependent amplitude attenuation coefficient, $\alpha(f)$. The attenuation coefficient of myocardium is measured by the transmission of ultrasound through tissue and then detection of the attenuated pulses on the opposite side to derive the signal loss per centimeter of tissue for each frequency. Although quantitative characterization of ultrasonic attenuation can be used to delineate the physical state of myocardial tissue independent of its scattering properties,[12,14,15] the transmission method for measurement of attenuation has no practical application *in vivo* at present. Methods for assessment of attenuation from backscattered ultrasound have been used in liver,[20] but heart tissue appears less amenable to these approaches.

The standard clinical approach for correction of attenuation is time-gain compensation (TGC), a method that allows the echocardiographic operator to adjust the signal gain for different depths of tissue to produce a subjectively pleasing image. Unfortunately, the apparent scattering properties of tissue may be influenced greatly by the particular attenuation profile chosen. Furthermore, the attenuation profile differs along every line of ultrasound used to produce a two-dimensional echocardiographic image. Melton and Skorton[21] have proposed an intermediate solution to this problem called "rational gain compensation" that applies realistic attenuation coefficients previously measured for myocardium and blood to appropriate regions along individual lines of sight. A real-time algorithm for detection of the interface between myocardium and blood is required. Although this approach permits estimation of intrinsic scattering properties, it depends on an accurate estimate of attenuation

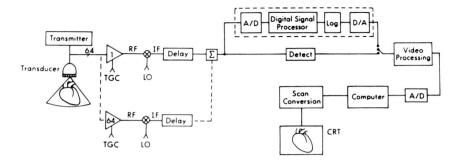

FIGURE 4.   Image processing path for integrated backscatter imaging. For most applications, time-gain compensation (TGC) is used. Once the summed intermediate frequency signal (IF) has been formed, it can be steered to the conventional imaging path onto the backscatter processor. (From Vered, Z. et al., *Circulation,* 76, 1067, 1987. With permission.)

under diverse clinical conditions. Nevertheless, rational gain compensation has been used in pilot clinical studies for estimation of integrated backscatter.[22]

The backscatter coefficient, $\eta(f)$, provides the most objective measure of the intrinsic scattering properties of a material because it is independent of the characteristics of the imaging system used.[1,17] It is computed by correcting the backscatter transfer function, $|S(f)|^2$, for the frequency-dependent volume of tissue interrogated by the ultrasonic beam. Higher frequencies tend to be more tightly focused at the focal point but diverge more sharply outside of the focal zone than do lower frequencies. Thus, the higher frequencies in a broadband ultrasonic beam may interrogate relatively fewer scatterers at the focal point than do lower frequencies. Measurement of the backscatter coefficient is not possible with currently available two-dimensional backscatter imaging modalities, but its estimation would offer the most objective comparison of tissue characteristics among patients.

D. REAL-TIME INTEGRATED BACKSCATTER IMAGING

Thomas and co-workers[23,24] at Washington University recently have implemented the measurement of apparent integrated backscatter in a commercially available two-dimensional phased-array imager that provides real-time quantification of myocardial backscatter at the patient's bedside without the necessity for extensive computer analysis off-line. The technique comprises modification of the data processing pathway of a conventional two-dimensional imager to permit input of intermediate frequency data to a custom-designed integrated backscatter processor (Figure 4). The integrated backscatter processor samples the incoming data in real time, and converts it to a log power representation of integrated backscatter. These data are then sent to the scan converter and a two-dimensional image of quantitative backscatter is displayed on the monitor. Either rational-gain or time-gain compensation protocols may be selected. The operator may switch easily between conven-

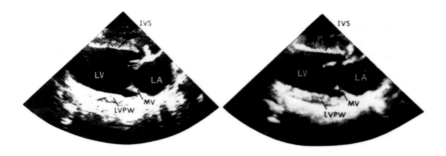

FIGURE 5. Conventional echocardiographic image (left panel) and integrated backscatter image (right panel) obtained nearly sequentially at end-diastole in one dog through a closed chest. Anatomic detail is comparable in both imaging modalities. IVS = interventricular septum; LV = left ventricle; LA = left atrium; LVPW = left ventricular posterior wall; MV = mitral valve. (From Thomas, L. J., III et al., *IEEE Trans. Ultrason. Ferroelectr. Frequency Control,* 36, 466, 1989. With permission.)

tional and integrated backscatter images. Automated algorithms have been incorporated to estimate the magnitude of backscatter from user-selected regions of interest in single-image frames.

The integrated backscatter processor produces an output in real time that approximates the frequency-averaged apparent backscatter transfer function, $|B(f)|^2$.[24] Approximation of the frequency-domain integration by a time-domain convolution is based on a practical application of the Parseval (Rayleigh) theorem. The resultant real-time two-dimensional backscatter image enables satisfactory identification of cardiac anatomic landmarks without resort to the conventional image (Figure 5). Thus, selected regions of interest can be tracked on-line throughout the cardiac cycle to measure the amplitude of backscatter.

Despite incorporation of rational-gain compensation into the data processing pathway of the backscatter imager, quantification of absolute scattering from tissue is compromised by the lack of a clinically applicable scheme for calibration of the imager. The standard method for calibration by substitution of a standard planar reflector for tissue (see above) results in saturation of the front-end electronics of commercially available imagers. Thus, at present integrated backscatter imaging is most useful clinically for the measurement of scattering among different regions of myocardial tissue or among different patients.

One such relative measure of scattering that reflects functional properties of myocardial tissue is the magnitude of cardiac cycle-dependent variation of backscatter (Figure 6), which has been shown by Madaras and co-workers[25] to be related to contractile function in experimental animals and humans. The observation of a variation in backscatter with cardiac cycle indicates that myocardial tissue scatters ultrasound more strongly at end-diastole than at end-systole. Fortunately, measurement of the amplitude of cyclic variation

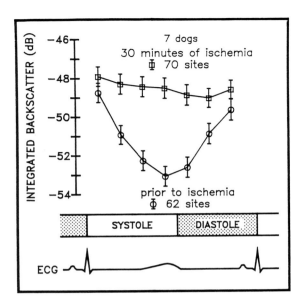

FIGURE 6. Systematic variation of integrated backscatter during the cardiac cycle (lower curve) and blunting of the cyclic pattern of backscatter and elevation of the time-averaged integrated backscatter (upper curve) after 30 minutes of ischemia. (From Miller, J. G., Pérez, J. E., and Sobel, B. E., *Prog. Cardiovasc. Dis.*, 28, 85, 1985. With permission.)

of backscatter does not depend on the availability of an absolute calibration standard because only relative changes of myocardial scattering properties from diastole to systole are considered.

Initial real-time measurements of the magnitude of cyclic variation of apparent integrated backscatter have been reported by Thomas et al.[23] in canine and human myocardium. This first real-time M-mode backscatter acquisition and analysis system was based on the use of an acoustoelectric power detector to convert radio frequency data received from selected myocardial regions to integrated backscatter. The time-varying, integrated backscatter curve demonstrated the typical decrease at end-systole when displayed in juxtaposition to a physiologic waveform such as the electrocardiogram. The subsequent real-time two-dimensional backscatter imaging system developed by Thomas et al. uses the expanded imaging capabilities of two-dimensional sector scanning to select regions of interest for analysis of the cyclic variation of backscatter (Figure 7).[24]

Rhyne et al.[26] have developed an approach that is similar to the tissue characterization method based on analysis of backscatter described by Thomas et al.[24] in which a clinically applicable M-mode method is used for quantification of backscatter. They have proposed two ultrasonic indexes similar to those developed previously.[1-3] The "integrated backscatter Rayleigh 5 MHz" (IBR5) is similar to the measurement of average integrated backscatter with

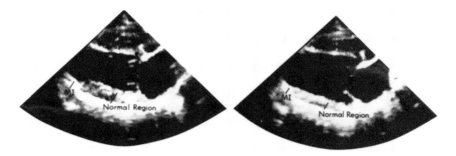

FIGURE 7. End-diastolic (right panel) and end-systolic (left panel) still-frame images demonstrating normal cardiac cycle-dependent cyclic variation of integrated backscatter in normal myocardium and lack of appreciable cyclic variation of backscatter in scar tissue. MI = region of remote myocardial infarction. (From Thomas, L. J., III et al., *IEEE Trans. Ultrason. Ferroelectr. Frequency Control*, 36, 466, 1989. With permission.)

additional corrections for beam diffraction, attenuation, and spectral whitening. The "Fourier amplitude modulation" of backscatter (FAM) is similar to the measurement of the amplitude of cyclic variation of integrated backscatter. Sagar et al.[27,28] have demonstrated that IBR5 and FAM are useful for delineation of regional cardiac contractile performance.

Landini et al.[29] have developed an M-mode approach useful for tissue characterization *in vivo* based on time-domain analysis of integrated backscatter. Conventional M-mode imaging techniques are used for initial localization of regions of myocardial tissue deemed appropriate for tissue characterization. The separate backscatter system comprises a 2.25 MHz transducer, amplifier, analog-to-digital converter (8 bit), and microprocessor. The absolute values of radio frequency data gated from tissue are integrated over the gate duration (3 μs) and referenced to the absolute value of radio frequency from pericardial echoes integrated over the same gate duration to provide a relative measure of scattering amplitude. The relative measure of backscatter can be expressed on a decibel scale by multiplying the log of the ratio of tissue and pericardial scattering amplitudes by 20. Because this method depends on a stable specular pericardial reflection for calibration, the angle of insonification of pericardium can influence the reference amplitude and may render comparisons among patients problematic.

E. ANALYSIS OF BACKSCATTERED RADIO FREQUENCY DATA

Analysis of unprocessed backscattered radio frequency data offers another promising approach to tissue characterization that can be applied in a clinical environment. The magnitude of backscatter manifests a dependence on frequency that can be characterized by a power law relationship between the backscatter coefficient and frequency. The backscatter coefficient, $\eta(f)$, is computed by correcting the backscatter transfer function, $S(f)$, for the beam-dependent volume of tissue interrogated. The frequency-dependent backscatter

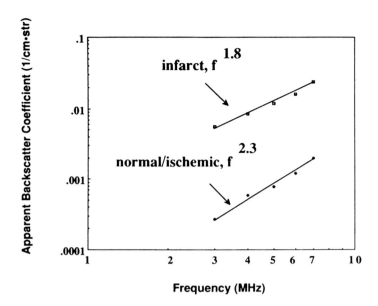

FIGURE 8.   Apparent backscatter coefficients (not corrected for attenuation within the tissue) for infarct and normal/ischemic myocardium. The backscatter coefficient for infarcts is larger in magnitude and rises less rapidly with frequency. (From Wear, K. A. et al., *J. Acoust. Soc. Am.*, 85, 2634, 1989. With permission.)

coefficient is assumed to obey a simple power law: $\eta(f) = Af^n$, where A is a constant and n describes the frequency dependence. A least-squares linear fit of the log of the backscatterer coefficient vs. the log of the frequency permits computation of the slope of the line, n, or the frequency dependence (Figure 8). The slope n is expected to range from 0 (for ka $\gg$ 1) to 4 (for ka $\ll$ 1), where k is the wave vector ($2\pi f/c$) and a is the scatterer radius. The case in which n $=$ 4 represents Rayleigh scattering and is associated with scatterers that have diameters much smaller than a wavelength of sound. A thorough discussion of this subject can be found in Chapters 3 to 5. The case in which n $=$ 0 represents specular scattering and is associated with scatterers that are much larger than a wavelength of sound. For normal myocardial tissue, n is approximately 3, which indicates that the scatterers are smaller than a wave length of sound. Pathologic alterations in scatterer geometry (size, shape, or concentration) should elicit alterations in this relationship between backscatter coefficient and frequency. Thus, as scatterer size decreases, n will approach the limit of 4, and as scatterer size increases, n will approach 0.

Wear et al.[30] have shown that measurement of the power law dependence of scattering on frequency permits differentiation of pathologic changes in scatterer geometry that are associated with myocardial infarction. Mature canine myocardial infarction *in vivo* is characterized by a flatter frequency

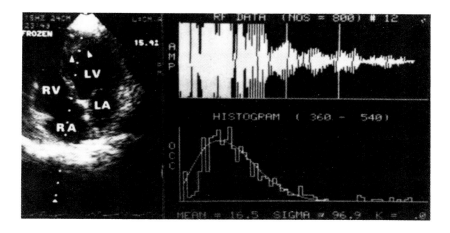

FIGURE 9. Composite of data obtained from a patient with an apical left ventricular thrombus confirmed at autopsy. The left panel shows a two-dimensional echocardiographic image in the apical four-chamber projection. The M-mode cursor passes through the thrombus (arrows) identifying the scan line from which the ultrasonic signal is sampled. The upper right panel is the 12th of 40 unprocessed signal samples and consists of 800 data points. Signal amplitude is represented on the ordinate and tissue depth on the abscissa. The operator-determined region of interest is enclosed by two vertical lines. The panel on the lower right depicts a histogram representing the number of occurrences (ordinate) of the digitized component amplitudes (abscissa) of the signal from the region of interest. A theoretical probability density function, derived from the mean amplitude and variance of the signal, is superimposed on the histogram. The histogram is described by k = O, suggesting a random, independent collection of small scatterers. AMP = amplitude; LA = left atrium; LV = left ventricle; OCC = number of occurrences of digitized component amplitude; RA = right atrium; rf = radio frequency; RV = right ventricle. (From Green, S. E. et al., *Am. J. Cardiol.*, 51, 231, 1983. With permission.)

dependence of the backscatter coefficient than is either acutely ischemic or normal myocardium (Figure 8). These authors also have reported that the frequency dependence of backscatter manifests a cardiac cycle-related alteration in slope of frequency dependence that reflects cyclic alterations in the geometry of scatterers determined by physiologic cardiac contractile activity.[31] Clinical application of this method would require access to the raw radio frequency data before any internal processing is begun within the echocardiographic imager. Because the measurement of frequency dependence incorporates corrections for several factors that include the beam-dependent volume of interrogation and attenuation, the specific value for n will reflect the value chosen for the slope of attenuation $[\beta(f)]$ and the particular method used to compensate for the beam-dependent volume of interrogation.

Another approach to tissue characterization based on analysis of raw radio-frequency data was used by Green et al.[32] to examine the distributional characteristics of radio frequency amplitudes in patients with suspected intracardiac masses. A region of interest along a single digitized radio frequency line was selected and a histogram was plotted to depict the frequency of occurrence of amplitudes of the digitized radio frequency within the region (Figure 9).

The probability density function of the data was characterized by the ratio of the mean amplitude and the variance of the data (mean-to-standard ratio, or MSR) and then compared with the MSR of a theoretical probability density function. The reference theoretical probability density functions reflected either Rayleigh distribution (independent scatterers randomly distributed) or non-Rayleigh distribution (nonindependent, nonrandom scattering with a correlated component). Rayleigh distribution indicates the presence of diffuse small scatterers typical of normal myocardium, whereas non-Rayleigh distributions indicate the presence of larger, more specular scatterers. The MSR appears to depend less on the depth or azimuthal parameters of the scan than do textural parameters or integrated backscatter. However, when MSR is used, signal-to-noise ratio is poor for regions of interest less than 5 mm. Nevertheless, mural thrombi can be distinguished from artifact and from atrial myxomas with the use of this method, although artifacts may not be differentiable from myxomas.[32]

III. TISSUE CHARACTERIZATION OF SPECIFIC DISEASE PROCESSES

A number of important clinical problems have been studied with the use of tissue characterization methods outlined above. As may be appreciated from the preceding discussion, the appropriate application of these methods depends on the specific clinical objective, which can range from the qualitative diagnosis of a pathologic condition for the purposes of initial screening to the quantitative evaluation of the composition of tissues. The following compendium of clinical applications is not intended to be exhaustive but to define clinical problems that appear most amenable to tissue characterization. Extensive experimental evaluation of the fundamental determinants of backscatter and attenuation have facilitated the application of tissue characterization methods for clinical investigation of myocardial disease. Comprehensive reviews of these experimental data are available.[1,2]

A. ISCHEMIC HEART DISEASE
1. Structure
Ultrasonic tissue characterization of ischemic heart disease has been extensively investigated. One of the initial uses of ultrasound for characterization of human infarction was reported in 1957 by Wild et al.[33] who measured backscatter from a zone of myocardial infarction in a fresh human heart obtained at autopsy with the use of a prototypical two-dimensional imaging apparatus applied directly to the myocardium. Rasmussen et al.[34] subsequently used a commercially available M-mode system to image patients with infarction and observed a qualitative increase in echogenicity from the interventricular septum or posterior wall of patients with mature infarct scars in these distributions.

Tanaka and collaborators[35] reported a more objective echotomographic approach for tissue characterization that used backscatter from pericardium as a reference for comparison with backscatter from infarcted myocardium. They found that the intensity of backscatter from infarcted myocardium more closely approximated the bright echogenicity of pericardium (to within 10 dB), whereas the intensity of backscatter from normal myocardium was at least 20 to 25 dB below that from pericardium.

In patients with coronary artery disease, Logan-Sinclair et al.[7] used the magnitude of the posterior pericardial reflection in the parasternal long-axis view as a reference amplitude for comparison with myocardial backscatter by adjusting the instrument gain so that the pericardial echo just achieved the maximum gray level. Color coding on-line of seven available gray levels in two-dimensional images facilitated selection of myocardial regions of interest for computation of average pixel intensity. Septal backscatter was significantly greater in patients with coronary disease than in normal subjects (71 vs. 33% for septal/pericardial intensity ratio). Subsequent studies of patients with myocardial fibrosis revealed that the observed increase in echo intensity reflected the regional tissue content of collagen measured biochemically.[36]

Hoyt et al.[37] measured integrated backscatter in excised human hearts with mature infarctions with 2.25 MHz ultrasound. Tissue backscatter was computed from digitized radio frequency data and referenced to the backscatter from a steel plate. Backscatter magnitude was compared to the content of tissue collagen measured by hydroxyproline assay. The average magnitude of integrated backscatter correlated well with collagen content at single transmural sites (r = 0.78).

Chandraratna et al.[8] determined the average intensity of scattering from digitized videotaped two-dimensional echocardiographic images obtained in patients with acute and mature infarctions. Images were color coded according to 64 discrete gray-level pixel intensities. Comparison of myocardial segments that manifested abnormal wall motion with adjacent segments that manifested normal wall motion revealed no significant difference between normal tissue and that in areas of acute infarct within 48 h of infarction in terms of pixel intensity (Plate 1). However, a significant increase in pixel intensity was observed in the regions of mature infarct after 4 weeks as compared with scattering from adjacent tissue with normal motion (Plate 2).

2. Function

Each of the studies reviewed above represents efforts to characterize the structural composition of myocardial tissue with ultrasound. Cardiac mechanical function also may be characterized by measuring the magnitude of cardiac cycle-dependent variation of integrated backscatter.[38,39] Madaras et al.[25] were the first to observe that myocardial contraction in open-chest dogs caused a progressive decline of integrated backscatter throughout mechanical systole with minimal values at end-systole approximately 5 dB lower than values at end-diastole (see Figure 6). Subsequent studies of the cyclic variation of integrated backscatter under experimental conditions of acute myocardial ischemic injury demonstrated that the magnitude of cyclic variation was severely blunted after the onset of ischemia and returned toward

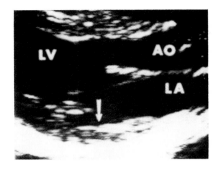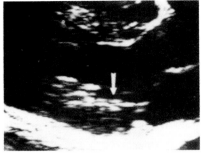

FIGURE 10. Parasternal long-axis two-dimensional echocardiographic images of the normal human heart at end-diastole (left) and end-systole (right). These images demonstrate a decrease in echo intensity (gray level) in the left ventricular posterior wall (arrows) from end-diastole to end-systole. AO = aorta; LV = left ventricle; LA = left atrium. (From Olshansky, B. et al., *Circulation,* 70, 972, 1984. With permission.)

normal values after coronary artery reperfusion.[40-42] The extent of recovery of cyclic variation depended on the antecedent duration of ischemia.[42]

Wickline et al.[39] showed that the magnitude and rate of change of cyclic variation are determined in part by intrinsic regional myocardial contractile performance in canine myocardium. Endocardial regions manifest greater magnitudes and rates of change of backscatter than do epicardial regions, which reflects a relatively greater contractile performance of endocardium at baseline. Paired pacing and beta-blockade increase and decrease the maximal rate of change of left ventricle pressure, $(dP/dt)_{max}$, respectively, and concomitantly increase and decrease the maximal rate of change of backscatter. These authors have proposed a physiologic model for the mechanism of cyclic variation based on cyclic alterations of local cardiac elastic properties, such as series and parallel elastic elements, which are determined by regional contractile function.[38] Other possible mechanisms of cyclic variation of backscatter include cardiac cycle-dependent alterations of myocardial geometry, attenuation, or anisotropy.[43]

Olshansky et al.[44] observed a cardiac cycle-dependent variation of regional gray-level intensity in digitized two-dimensional echocardiograms from 16 patients with normal hearts. Average gray level was measured in the left ventricular posterior wall in the parasternal long-axis view at end-diastole and at end-systole (Figure 10). The mean gray-level in the posterior wall decreased significantly from end-diastole to end-systole, which confirmed the previous observation by Madaras et al.[25] of cyclic variation of backscatter in open-chest dogs and indicated its potential clinical utility.

Vered et al.[45] used the real-time two-dimensional backscatter imaging system developed at Washington University to study 15 patients with remote myocardial infarction. The cyclic variation from myocardial sites that demonstrated normal wall motion manifested 3.2 ± 0.2 dB of cyclic variation

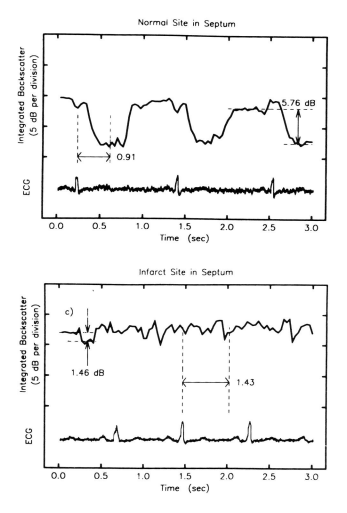

FIGURE 11. Top panel: Integrated backscatter data from a normal septal size exhibiting a normal pattern of diastolic to systolic cyclic variation. **Dotted horizontal lines** depict the method of measuring the magnitude of the variation. **Dotted vertical lines** demonstrate the R wave to nadir delay. Bottom panel: Infarct segment from a septal wall, manifesting no detectable cyclic variation. ECG = electrocardiogram. (From Vered, Z. et al., *J. Am. Coll. Cardiol.*, 13, 84, 1989. With permission.)

(Figure 11, left panel). The magnitude of cyclic variation in segments involved in the infarct was significantly lower (1.1 ± 0.2 dB). No cyclic variation was detectable in 20 of 42 total infarct sites. Ultrasonic characterization of the timing of regional contractile function was described in terms of the "delay" of the nadir of the backscatter waveform from the peak of the R-wave, corrected for the QT interval (Figure 11, right panel). This dimensionless measure represents an extension of the concept of "phase" of cyclic variation originally reported by Wickline et al.[38,39] and later by Fitzgerald et al.[46] The delay of cyclic variation was markedly increased in areas of infarct as compared with normal tissue (1.47 ± 0.12 vs. 0.87 ± 0.03, infarct vs. normal, respectively; p <0.0001). Thus, ultrasonic tissue characterization with a real-time two-dimensional echocardiographic imaging system differentiates normal from infarcted tissue based on analysis of the magnitude and delay of cyclic variation.

Milunski et al.[47] studied 21 patients admitted to the cardiac intensive care unit with acute myocardial infarction with the use of the real-time backscatter imaging system. The mean time after onset of symptoms to the initial study was 11.3 h. The mean magnitude of cyclic variation in normal myocardial regions not involved with acute infarction (as judged by the presence of normal wall motion) was 4.8 ± 0.5 dB in the 21 patients (Figure 12A). The magnitude of cyclic variation in regions of infarction was 0.8 ± 0.3 dB (p <0.05 for infarct vs. normal). The majority of patients with acute infarction had been treated upon admission with a thrombolytic agent or with emergency coronary artery angioplasty in an attempt to induce arterial recanalization and salvage jeopardized tissue. Fifteen of these patients underwent coronary arteriography later during their hospitalization to define patency of the infarct-related vessel. In patients with patent infarct-related vessels, the magnitude of cyclic variation increased over time from 1.3 ± 0.6 to 2.5 ± 0.5 dB from the initial to the final study (p <0.05) (Figure 12B). Five patients with occluded infarct-related arteries exhibited no significant recovery of cyclic variation (0.3 ± 0.3 to 0.6 ± 0.3 dB, p = NS). The mean time from the initial to the final ultrasonic study was 7.1 days. Analysis of standard two-dimensional echocardiographic images revealed no significant recovery of regional wall thickening in the segments for patients with either patent or occluded infarct-related arteries over this time interval. Thus, ultrasonic tissue characterization based on analysis of the magnitude of cyclic variation detects acute myocardial infarction. Furthermore, the recovery of integrated backscatter appears to precede the restoration of regional wall thickening and may be useful for the detection of the presence of "stunned" (i.e., viable but transiently dysfunctional) myocardium in patients who undergo prompt arterial recanalization. These data also confirm previous observations in open-chest canine experiments that indicate that the magnitude of cyclic variation of backscatter recovers more promptly than does regional wall motion under conditions of stunned myocardium.[48]

B. IDIOPATHIC DILATED CARDIOMYOPATHY

Idiopathic cardiomyopathy is a form of congestive heart failure that commonly involves both ventricles, and culminates in severely depressed ventricular function. The syndrome may be preceded by a viral infection that is followed by an acute inflammatory response in myocardial tissue, with infiltration with lymphocytes, myocyte degeneration, and varying degrees of fibrosis. A more chronic, less inflammatory appearance characterized by

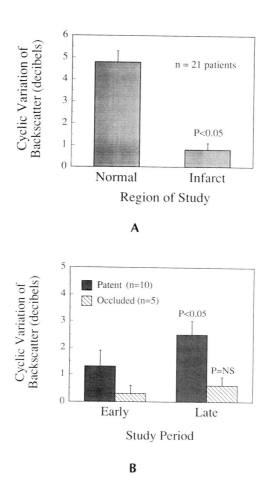

FIGURE 12. (A) Plot of the magnitude of cyclic variation of integrated backscatter in all patients during the initial study period. Comparison of nonischemic and infarct zones within 24 h of onset of symptoms of acute myocardial infarction. n = 21 for both normal and infarct regions. $p < 0.05$ for normal compared with infarct region cyclic variation. (B) Plot of the evolution of cyclic variation of integrated backscatter in patients with patent and occluded infarct-related vessels. There is a significant recovery of cyclic variation during the study period in 10 patients with patent vessels ($p < 0.05$, early compared with late mean cyclic variation in injured regions) but not in five patients with occluded vessels (p = ns). (From Milunski, M. R. et al., *Circulation*, 80, 491, 1989. With permission.)

myofibril degeneration, patchy fibrosis, and hypertrophy of remaining viable myocytes also may be seen on endomyocardial biopsy.

Early application of tissue characterization techniques to the diagnosis of dilated cardiomyopathy were reported by Davies et al.[49] in studies of hyper-eosinophilic cardiomyopathy. Conventional two-dimensional echocardiographic images were color coded to reveal increased pixel intensity in nine patients with biopsy-proven endomyocardial fibrosis secondary to eosinophilic infiltration.

Picano et al.[50] studied 16 patients with idiopathic cardiomyopathy with a 2.25 MHz M-mode echocardiographic system. They computed integrated backscatter from the absolute value of digitized gated radio frequency signals and referenced all values to the amplitude of backscatter from a pericardial interface. The percentage of tissue fibrosis was calculated by computer-assisted planimetric analysis of left ventricular endomyocardial biopsy specimens from each patient. A significant although weak correlation (r = 0.55) was observed between the intensity of backscatter and percent connective tissue area. A similar correlation was observed between fibrosis and backscatter after radio frequency signals were compensated for the effects of attenuation along the path to the myocardial region of interest.

Vered et al.[22] measured the cyclic variation of backscatter in septal and posterior wall segments in patients with idiopathic cardiomyopathy and compared it with that in normal subjects. Patients with cardiomyopathy manifested significantly blunted cyclic variation in the septum (0.9 ± 0.8 dB) and in the posterior wall (1.8 ± 1.2 dB). Normal magnitudes of cyclic variation were observed in control patients in both septum (4.6 ± 1.4 dB) and posterior wall (5.3 ± 1.5 dB). Thus, the measurement of the magnitude of cyclic variation may provide a sensitive index of deranged contractile function in patients with congestive heart failure due to idiopathic cardiomyopathy.

C. AMYLOIDOSIS

Cardiac amyloidosis is a unique form of nonischemic cardiomyopathy caused by the deposition of amyloid protein in the interstitial regions of cardiac tissue. Clinical manifestations include valvular dysfunction, conduction system disease, and diastolic dysfunction that reflects the excess stiffness conferred by the amyloid deposits. Chiaramida et al.[51] first reported that cardiac texture changes as a consequence of deposition of amyloid. They described a "diffuse patchy distribution of highly refractile echoes" in one patient with biopsy-documented amyloidosis. Siqueira-Filho et al.[52] also reported similar textural changes with amyloidosis. Bhandari and Nanda[4] subsequently proposed qualitative criteria for diagnosis of amyloidosis based on the appearance of cardiac texture in two-dimensional echocardiograms. They found that patients with cardiac amyloidosis manifested predominantly either type IIA or type IIB texture, but that these features were not specific for amyloidosis and were observed in other conditions such as hypertrophic cardiomyopathy, hemochromatosis, chronic renal failure, and left ventricular hypertrophy (see Figure 1).

A prospective study of the diagnostic power of two-dimensional echocardiography for the identification of patients with amyloid heart disease was conducted by Nicolosi et al.[53] Two thousand and seventy-eight consecutive echocardiograms were analyzed, and 30 of these contained discrete, small, highly reflective echoes. Results of gingival biopsy procedures performed in

15 of the 30 patients in whom amyloid heart disease was indicated by the appearance of the two-dimensional echocardiogram were positive for amyloid in 11 cases, supporting the clinical utility of qualitative textural parameters provided by conventional two-dimensional echocardiography for prospective identification of patients with cardiac amyloidosis.

Chandrasekaran et al.[54] recently reported that quantitative analysis of textural parameters permitted fairly specific identification of patients with amyloid infiltration. Standard two-dimensional videotaped images were digitized into 8 byte digital format from parasternal long- and short-axis views. Regions of interest were selected from left ventricular septal and posterior wall zones at end-diastole and end-systole. Textural parameters included gray-level run-length and gray-level difference measures (see above). Selected gray-level run-length measures differentiated amyloid from normal myocardium (Figure 13). Pinamonti et al.[55] also reported successful differentiation of amyloid tissue from normal myocardium with the use of textural parameters. These observations provide a basis for quantitative estimation of textural indexes of cardiac amyloidosis.

D. HYPERTROPHIC CARDIOMYOPATHY

Ventricular hypertrophy comprises a number of clinical entities, some that represent primary pathologic alterations in cardiac structure, and others that represent physiologic adaptations to external stresses. Ultrasonic tissue characterization has been used to study hypertensive ventricular hypertrophy and idiopathic hypertrophic subaortic stenosis (IHSS). Pathologic alterations of cardiac structure associated with IHSS include intracellular disarray of myofibrils and of myocytes, and interstitial fibrosis. Symptoms of congestive heart failure may predominate due to the excess tissue stiffness conferred by the abnormal architecture. Asymmetric septal hypertrophy is characteristic of IHSS and may be associated with dynamic left ventricular outflow tract obstruction, which may compromise ventricular systolic function. In contrast to IHSS, left ventricular hypertrophy due to systemic hypertension represents a physiologic adaptation to increased wall stress. The left ventricular wall thickness is increased due to an increase in volume of individual cells and this increase serves to reduce the average level of wall stress toward more normal levels. Patients with left ventricular hypertrophy also manifest increased tissue stiffness, which may be partially attributable to a mild increase in interstitial fibrous tissue, particularly in subendocardial regions. Systolic function usually appears grossly normal in hypertensive hypertrophy, however.

Bhandri and Nanda[4] observed that the textural appearance of myocardial tissue in the presence of uncomplicated ventricular hypertrophy typically is fine, speckled, and uniform, i.e., much like that of normal myocardium (type I texture). Some hearts demonstrated type IIb texture (multiple highly reflective echoes in one or more walls of the ventricle) resembling that observed with hypertrophic cardiomyopathy.

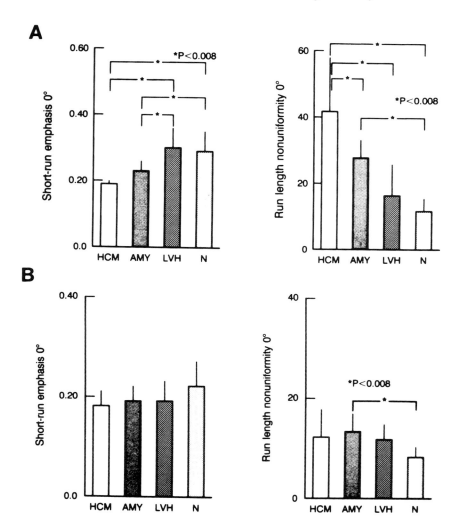

FIGURE 13. Differentiation of cardiomyopathy from normal myocardium on clinical echo-cardiograms by quantitative texture analysis. The graphs show selected gray-level run-length data obtained from end-diastolic long-axis clinical echocardiographic views of the ventricular septum (A) and posterior wall (B) in patients with hypertrophic cardiomyopathy (HCM), amyloidosis (AMY), left ventricular hypertrophy (LVH), and in normal subjects (N). Quantitative parameters differentiated both cardiomyopathies from normal and from each other. (From Chandrasekaran, K. et al., *J. Am. Coll. Cardiol.*, 13, 832, 1989. With permission.)

Chandrasekaran et al.[54] delineated quantitative tissue textural features in eight patients with hypertrophic cardiomyopathy and in six patients with left ventricular hypertrophy due to hypertension. Hypertensive hypertrophy was indistinguishable from normal myocardium on the basis of textural parameters, but gray-level run-length texture statistics from end-diastolic long-axis views of the ventricular septum differentiated hypertrophic cardiomyopathy from both left ventricular hypertrophy and normal tissue (Figure 13). Furthermore, selected gray-level run-length data differentiated hypertrophic cardiomyopathy from amyloid infiltration.

Masuyama et al.[56] measured the cyclic variation of integrated backscatter in 12 patients with uncomplicated pressure-overload hypertrophy and in 13 patients with hypertrophic cardiomyopathy. The cyclic variation of backscatter was used to delineate regional cardiac function (see above). The magnitude of cyclic variation measured in the posterior wall did not differ significantly among any of these three groups (range = 5.7 to 6.7 dB). However, the magnitude of cyclic variation in the septum was significantly lower both in patients with pressure-overload hypertrophy (2.8 ± 1.3 dB) and those with hypertrophic cardiomyopathy (3.1 ± 2.3 dB) compared with cyclic variation in normal subjects (4.9 ± 1.0 dB). Thus, cyclic variation of backscatter in hypertrophied hearts differed from that in hearts of normal subjects only in the septum in the majority of patients. The authors concluded that the measurement of cyclic variation was not useful in distinguishing hypertrophic cardiomyopathy from pressure overload hypertrophy.

An additional point of interest raised by this report is the nonlinear relationship between the magnitude of cyclic variation and percent systolic thickening (Figure 14). Modest reductions in percent systolic thickening are associated with preserved magnitude of cyclic variation, with an inflection point for reduced cyclic variation at around 30% systolic thickening. The observation of this non-linear relationship in patients is in accord with that previously observed by Wickline et al.[42] in canine hearts with ischemic injury. Collectively these data indicate that cyclic variation of backscatter provides more than a simple measure of wall motion, and may reflect preserved regional contractile function despite abnormal wall motion.

E. VALVULAR VEGETATIONS

Infective endocarditis often produces valvular vegetations that can be detected with two-dimensional echocardiography. Its clinical diagnosis usually is established when multiple blood cultures indicate the presence of bacterial growth in association with a compatible clinical history and physical examination. The ability of a noninvasive technique to detect bacterial invasion of a valve with either modest thickening or frank vegetation might facilitate the decision for early aggressive therapy.

Tak et al.[57] studied 22 patients with newly diagnosed infective endocarditis prospectively with two-dimensional echocardiography. Videotaped images of vegetations were digitized and displayed on a color monitor. Color coding of 64 available gray levels was performed to delineate the intensity of scattering from vegetations (Plate 3*). The mean intensity of scattering from valvular vegetations in the active stage was 20.6 ± 3.6 and increased to 34.4 ± 4.3 in the chronic stage, which correlated with a diminution in the mean size of

---

* Plate 3 follows page 374.

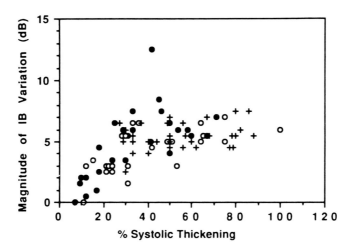

FIGURE 14.   Plot of the relation between percent (%) systolic thickening of myocardium and magnitude of cardiac cycle-dependent variation of integrated backscatter. As percent-systolic thickening of wall decreases, magnitude of variation of integrated backscatter tends to decrease, and there is a weak but significant correlation between these two parameters ($r = 0.55$, $p < 0.01$, $n = 82$ for all data) by linear regression analysis. Polynominal regression analysis showed a curvilinear relation of these data ($r = 0.67$, $p < 0.01$). (+) Normal subjects; (O) patients with pressure-overload hearts; (●) patients with hypertrophic cardiomyopathy. (From Maysuyama, T. et al., *Circulation,* 80, 925, 1989. With permission.)

vegetation from 0.73 cm² in the active stage to 0.56 cm² in the chronic stage. The increased pixel intensity reflected healing associated with fibrosis and endothelialization of the vegetation. One patient with recurrent infection manifested a concomitant reduction in mean pixel intensity in the acute phase of recurrent endocarditis, with an increase after cure to levels compatible with chronic vegetation.

F. CARDIAC TRANSPLANT REJECTION

Cardiac transplantation for intractable congestive heart failure is an effective long-term therapy for selected patients. A major problem limiting long-term success is acute and/or chronic transplant rejection. The 5-year survival rates for cardiac allograft recipients currently are in excess of 60% due to improved detection of rejection phenomena by endomyocardial biopsy and improved treatment with agents such as cyclosporine. Noninvasive diagnosis of incipient transplant rejection by ultrasonic tissue characterization would facilitate early aggressive therapeutic intervention.

Masuyama et al.[58] compared results of ultrasonic tissue characterization in 23 cardiac allograft recipients and 18 normal subjects. Tissue characterization was performed before and during moderate acute rejection episodes in 11 recipients with 14 episodes of rejection. The magnitude of cyclic variation decreased from 6.7 ± 1.3 to 5.1 ± 1.4 dB in the posterior wall, and from

4.2 ± 2.1 to 2.9 ± 1.8 dB in the septum as a consequence of rejection (p <0.05 for both). With resolution of acute rejection after therapy, the magnitude of cyclic variation tended to recover. Thus, measurement of cyclic variation of integrated backscatter may be useful for the diagnosis of acute rejection and for serial evaluation after therapeutic intervention.

G. MURAL THROMBUS

Left ventricular mural thrombus formation usually occurs after anterior myocardial infarction in a dyskinetic portion of the ventricular apex due to stasis of blood, which predisposes to local clot formation. The clinical significance of mural thrombus resides in an increased incidence of stroke associated with myocardial infarction, particularly in the presence of mural thrombi that are highly mobile. The diagnosis of mural thrombosis often is obscured by inadequate visualization of the left ventricular apex by two-dimensional echocardiography, and by the inability to distinguish mural thrombus from underlying myocardial tissue.

Green et al.[32] quantified ultrasonic tissue characteristics of mural thrombi based on measurement of the probability density function of digitized radio frequency data. They observed that mural thrombi were associated with a Rayleigh-type probability density function, that is, scattering from an array of independent scatterers much like those present in normal tissue (see Figure 9). Mural thrombi could be differentiated from atrial myxoma and artifact, both of which manifested a non-Rayleigh probability density function.

Lloret et al.[59] examined 38 patients with left ventricular mural thrombi diagnosed by conventional two-dimensional echocardiography. Eight patients had clinically apparent systemic embolic events. Textural features of these thrombi were determined from digitized images and included gray-level distribution measures such as run-length and gray-level difference statistics. Statistical analysis indicated that selected textural features could predict which thrombi were associated with a high risk of systemic embolization.

Pavan et al.[60] reported qualitative ultrasonic characteristics of fresh cardiac thrombi determined by two-dimensional echocardiography. Two-dimensional echocardiographic images of 30 thrombi obtained from autopsy specimens were recorded *in vitro* after subjective adjustment of imaging parameters to provide a pleasing image. Thrombi were examined in a water bath, and were classified by histopathologic analysis as "fresh" or "not fresh", based on the presence of cellular elements and collagen. Thrombi were described as "dense and nonhomogeneous structures with coarse texture and granular characteristics." The authors concluded that qualitative tissue characterization of mural thrombi could be used to differentiate fresh from fibrotic thrombi.

H. CONGENITAL HEART DISEASE

The evaluation of patients with congenital heart disease represents an increasing challenge for contemporary cardiovascular diagnosis in adults. Serial assessment of ventricular function after operative repair is an area in which tissue characterization may play a role. After open heart surgery the ventricular septum typically moves paradoxically or anteriorly during left ventricular contraction, despite the presence of presumably normal regional contractile function. Milunski et al.[61] measured cardiac cycle-dependent variation of integrated backscatter in the septa of eight children with surgically corrected congenital cardiac lesions associated with paradoxical septal motion and in six children with no history of cardiac disease who served as a control group. Cyclic variation in the septum in the study and control groups was 8.3 ± 1.0 and 5.7 ± 0.4 dB, respectively. The authors concluded that measurement of cyclic variation was not distorted by altered regional wall motion and that intrinsic contractile function was normal. Thus, measurement of cyclic variation may be useful for determining left ventricular function in pediatric patients after open heart surgery despite potentially confounding wall motion abnormalities.

IV. PROSPECTS FOR CLINICAL CARDIOVASCULAR TISSUE CHARACTERIZATION

Ultrasonic characterization of myocardium in real time with integrated backscatter imaging is designed to provide robust measurements of structural and functional properties of heart muscle by quantifying the fundamental acoustic properties of myocardial tissue rather than wall motion. Both experimental and clinical studies indicate that ultrasonic backscatter from cardiac tissue depends on physical properties such as elasticity, density, and fiber geometry. Alterations of these properties associated with myocardial disease have been delineated with quantitative backscatter imaging in experimental animals and human beings. The recent development of a real-time backscatter imaging system by modification of a commercially available two-dimensional echocardiographic imager should facilitate widespread clinical application of ultrasonic tissue characterization. It is hoped that further experience with backscatter imaging will provide the clinical experience necessary to maximize its diagnostic and prognostic power.

Several potentially confounding issues require further experimental and clinical evaluation before backscatter imaging assumes a role as a standard diagnostic modality. The principal limitation of ultrasonic tissue characterization in a clinical environment is image quality, which can be affected both by the operator and by the subject. Objective criteria for establishing the quality of images necessary for ultrasonic tissue characterization are required so that inadequate primary data do not compromise measurements of tissue acoustic parameters.

Another potential difficulty entails the lack of a suitable standard for absolute calibration of integrated backscatter measurements. As mentioned above, current real-time backscatter imaging systems are most suitable for measurement of relative alterations in integrated backscatter throughout the cardiac cycle. The ability to measure absolute scattering from tissue would facilitate the quantitative characterization of structural changes in tissue, such as those accompanying edema, fibrosis, and cellular disarray, and permit reliable comparisons among different patients. Measurement of absolute scattering requires accurate compensation for attenuation of ultrasound encountered along the path to a myocardial region of interest and compensation for beam diffraction effects.

Ultrasonic anisotropy also requires consideration for proper analysis of backscatter from clinical images. Recent reports indicate that myocardial tissue manifests a directional dependence of scattering with maximal backscatter for insonification perpendicular to fibers and minimal backscatter for insonification parallel to fibers.[43,62] Attenuation, in contrast, is maximal parallel to fibers and minimal perpendicular to fibers.[63] Because clinical echocardiographic imaging is carried out from multiple angles of interrogation with respect to intramural fiber orientation, compensation for the angle dependence of scattering and attenuation is required.

The mechanism of cardiac cycle-dependent variation of backscatter has yet to be established. Potential sources of the cyclic variation of backscatter in myocardial tissue include alterations in local acoustic properties (elasticity and density), in scatterer geometry (size, shape, and concentration), and in ultrasonic attenuation over the heart cycle. In addition, the effect of ultrasonic anisotropy should be considered as a potential contributing mechanism. Several hypotheses have been advocated that include alterations in local elastic properties and scatterer geometry.[38,64] Development of a robust model for cyclic variation will be crucial for sensitive interpretation of clinical data that reflect both normal and pathologic structure and function.

Further improvements of data acquisition and analysis systems also will accelerate clinical dissemination of ultrasonic tissue characterization techniques. Backscatter imaging systems that include automated selection and tracking of regions of interest and automated analysis of cyclic variation and delay are in development. Future modifications of this emerging technology will benefit from its widespread clinical application and, it is hoped, will serve to establish ultrasonic tissue characterization as a technique that may be performed by clinicians, technical personnel, and researchers alike.

ACKNOWLEDGMENT

The authors appreciate the assistance of Burton E. Sobel, M.D., for constructive criticism of this work and for his efforts devoted to the development of ultrasonic tissue characterization.

REFERENCES

1. **Miller, J. G., Pérez, J. E., and Sobel, B. E.,** Ultrasonic characterization of myocardium, *Prog. Cardiovasc. Dis.,* 28, 85, 1985.
2. **Pérez, J. E., Miller, J. G., Wickline, S. A., Milunski, M. R., Barzilai, B., and Sobel, B. E.,** Myocardial tissue characterization, *Prog. Cardiol.,* 3, 83, 1990.
3. **Wickline, S. A. and Sobel, B. E.,** Ultrasonic tissue characterization: prospects for clinical cardiology, *J. Am. Coll. Cardiol.,* 14, 1709, 1989.
4. **Bhandari, A. K. and Nanda, N. C.,** Myocardial texture characterization by two-dimensional echocardiography, *Am. J. Cardiol.,* 51, 817, 1983.
5. **Skorton, D. A., Melton, H. E., III, Pandian, N. G., Nichols, J., Koyanagi, S., Marcus, M. L., Collins, S. M., and Kerber, R. E.,** Detection of acute myocardial infarction in closed-chest dogs by analysis of regional two-dimensional echocardiographic gray-level distributions, *Circ. Res.,* 52, 36, 1983.
6. **Tak, T., Gamage, N., Shao-Lin, L., Mahler, C., Steen, S. N., Coletti, P., Rahimtoola, S. H., and Chandraratna, P. A. N.,** Detection of early myocardial tissue changes in acute canine myocardial infarction by ultrasonic tissue characterization methods, *Can. J. Cardiol.,* 5, 408, 1989.
7. **Logan-Sinclair, R., Wong, C. M., and Gibson, D. G.,** Clinical application of amplitude processing of echocardiographic images, *Br. Heart J.,* 45, 621, 1981.
8. **Chandraratna, P. A. N., Ulene, R., Nimalasuriya, A., Reid, C. L., Kawanishi, D., and Rahimtoola, S. H.,** Differentiation between acute and healing myocardial infarction by signal averaging and color encoding two-dimensional echocardiography, *Am. J. Cardiol.,* 56, 381, 1985.
9. **Skorton, D. J., Collins, S. M., Nichols, J., Pandian, N. B., Bean, J. A., and Kerber, R. E.,** Quantitative texture analysis in two-dimensional echocardiography: Application to the diagnosis of experimental myocardial contusion, *Circulation,* 68, 217, 1983.
10. **Skorton, D. J., Collins, S. M., Woskoff, S. D., Bean, J. A., and Melton, H. E.,** Range- and azimuth-dependent variability of image texture in two-dimensional echocardiograms, *Circulation,* 68, 834, 1983.
11. **O'Donnell, M., Bauwens, D., Mimbs, J. W., and Miller, J. G.,** Broadband integrated backscatter: An approach to spatially localized tissue characterization *in vivo, Proc. IEEE Ultrason. Symp.,* 79 CH 1482-9, 175, 1979.
12. **O'Donnell, M., Mimbs, J. W., and Miller, J. G.,** The relationship between collagen and ultrasonic attenuation in myocardial tissue, *J. Acoust. Soc. Am.,* 65, 512, 1979.
13. **O'Donnell, M., Mimbs, J. W., and Miller, J. G.,** The relationship between collagen and ultrasonic backscatter in myocardial tissue, *J. Acoust. Soc. Am.,* 69, 580, 1981.
14. **Mimbs, J. W., O'Donnell, M., Miller, J. G., and Sobel, B. E.,** Changes in ultrasonic attenuation indicative of early myocardial ischemic injury, *Am. J. Physiol.,* 236, H340, 1979.
15. **Mimbs, J. W., O'Donnell, M., Bauwens, D., Miller, J. G., and Sobel, B. E.,** The dependence of ultrasonic attenuation and backscatter on collagen content in dog and rabbit hearts, *Circ. Res.,* 47, 49, 1980.
16. **Mimbs, J. W., Bauwens, D., Cohen, R. D., O'Donnell, M., Miller, J. G., and Sobel, B. E.,** Effects of myocardial ischemia on quantitative ultrasonic backscatter and identification of responsible determinants, *Circ. Res.,* 49, 89, 1981.
17. **Miller, J. G., Pérez, J. E., Mottley, J. G., Madaras, E. I., Johnston, P. H., Blodgett, E. D., Thomas, L. J., III, and Sobel, B. E.,** Myocardial tissue characterization: an approach based on quantitative backscatter and attenuation, *Proc. IEEE Ultrason. Symp.,* 83 CH 1947-1, 782, 1983.
18. **Miller, J. G., Pérez, J. E., and Sobel, B. E.,** Ultrasonic characterization of myocardium, *Prog. Cardiovasc. Dis.,* 28, 85, 1985.

19. **Sigelmann, R. A. and Reid, J. M.**, Analysis and measurement of ultrasonic backscattering from an ensemble of scatterers excited by sine-wave bursts, *J. Acoust. Soc. Am.*, 53, 1351, 1973.

20. **Ophir, J., Shawker, T. H., Maklad, N. F., Miller, J. G., Flax, S. W., Narayana, P. A., and Jones, J. P.**, Attenuation estimation in reflection: Progress and prospects, *Ultrason. Imag.*, 6, 349, 1984.

21. **Melton, H. E. and Skorton, D. J.**, Rational gain compensation for attenuation in cardiac ultrasonography, *Ultrason. Imag.*, 5, 214, 1983.

22. **Vered, Z., Barzilai, B., Mohr, G. A., Thomas, L. J., III, Genton, R., Sobel, B. E., Shoup, T. A., Melton, H. E., Miller, J. G., and Pérez, J. E.**, Quantitative ultrasonic tissue characterization with real-time integrated backscatter imaging in normal human subjects and in patients with dilated cardiomyopathy, *Circulation*, 76, 1067, 1987.

23. **Thomas, L. J., III, Wickline, S. A., Pérez, J. E., Sobel, B. E., and Miller, J. G.**, A real-time integrated backscatter measurement system for quantitative cardiac tissue characterization, *IEEE Trans. Ultrason. Ferroelectr. Frequency Control*, UFFC-33, 27, 1986.

24. **Thomas, L. J., III, Barzilai, B., Pérez, J. E., Sobel, B. E., Wickline, S. A., and Miller, J. G.**, Quantitative real-time imaging of myocardium based on ultrasonic integrated backscatter, *IEEE Trans. Ultrason. Ferroelectr. Frequency Control*, 36, 466, 1989.

25. **Madaras, E. I., Barzilai, B., Pérez, J. E., Sobel, B. E., and Miller, J. G.**, Changes in myocardial backscatter throughout the cardiac cycle, *Ultrason. Imag.*, 5, 229, 1983.

26. **Rhyne, T. L., Sagar, K. B., Wann, S. L., and Haasler, G.**, The myocardial signature: Absolute backscatter, cyclical variation, frequency variation, and statistics, *Ultrason. Imag.*, 8, 107, 1986.

27. **Sagar, K. B., Rhyne, T. L., Warltier, D. C., Pelc, L., and Wann, S.**, Intramyocardial variability in integrated backscatter: effects of coronary occlusion and reperfusion, *Circulation*, 75, 436, 1987.

28. **Sagar, K. B., Pelc, L. E., Rhyne, T. L., Wann, S., and Warltier, D. C.**, Influence of heart rate, preload, afterload, and inotropic state on myocardial ultrasonic backscatter, *Circulation*, 77, 478, 1988.

29. **Landini, L., Mazzarisi, A., Salvadori, M., and Benassi, A.**, On-line evaluation of ultrasonic integrated backscatter, *J. Biomed. Eng.*, 7, 301, 1985.

30. **Wear, K. A., Milunski, M. R., S. A., W., Pérez, J. E., Sobel, B. E., and Miller, J. G.**, Differentiation between acutely ischemic myocardium and zones of completed infarction in dogs on the basis of frequency-dependent backscatter, *J. Acoust. Soc. Am.*, 85, 2634, 1989.

31. **Wear, A., Milunski, M. R., Wickline, S. A., Pérez, J. E., Sobel, B. E., and Miller, J. G.**, Contraction-related variation in frequency dependence of acoustic properties of canine myocardium, *J. Acoust. Soc. Am.*, 86, 2067, 1989.

32. **Green, S. E., Joynt, L. F., Fitzgerald, P. J., Rubenson, D. S., and Popp, R. L.**, In vivo ultrasonic tissue characterization of human intracardiac masses, *Am. J. Cardiol.*, 51, 231, 1983.

33. **Wild, J. J., Crafford, H. D., and Reid, J. M.**, Visualization of the excised human heart by means of reflected ultrasound or echography, *Am. Heart J.*, 54, 903, 1957.

34. **Rasmussen, S., Corya, B. C., Feigenbaum, H., and Knoebel, S. B.**, Detection of myocardial scar tissue by M-mode echocardiography, *Circulation*, 57, 230, 1978.

35. **Tanaka, M., Teresawa, Y., and Hikichi, H.**, Qualitative evaluation of the heart tissue by ultrasound, *J. Cardiogr.*, 7, 515, 1977.

36. **Shaw, T. R. D., Logan-Sinclair, R. B., Surin, C., McAnulty, R. J., Heard, B., Laurent, G. J., and Gibson, D. J.**, Relation between regional echo intensity and myocardial connective tissue in chronic left ventricular disease, *Br. Heart J.*, 51, 46, 1984.

37. **Hoyt, R. H., Collins, S. M., Skorton, D. J., Ericksen, E. E., and Conyers, D.,** Assessment of fibrosis in infarcted human hearts by analysis of ultrasonic backscatter, *Circulation*, 71, 740, 1985.

38. **Wickline, S. A., Thomas, L. J., III, Miller, J. G., Sobel, B. E., and Pérez, J. E.,** A relationship between ultrasonic integrated backscatter and myocardial contractile function, *J. Clin. Invest.*, 76, 2151, 1985.

39. **Wickline, S. A., Thomas, L. J., III, Miller, J. G., Sobel, B. E., and Pérez, J. E.,** The dependence of myocardial ultrasonic backscatter on contractile performance, *Circulation*, 72, 183, 1985.

40. **Barzilai, B., Madaras, E. I., Sobel, B. E., Miller, J. G., and Pérez, J. E.,** Effects of myocardial contraction on ultrasonic backscatter before and after ischemia, *Am. J. Physiol.*, 247, H478, 1984.

41. **Glueck, R. M., Mottley, J. G., Sobel, B. E., Miller, J. G., and Pérez, J. E.,** Changes in ultrasonic attenuation and backscatter of muscle with state of contraction, *Ultrasound Med. Biol.*, 11, 605, 1985.

42. **Wickline, S. A., Thomas, L. J., III, Miller, J. G., Sobel, B. E., and Pérez, J. E.,** Sensitive detection of the effects of reperfusion on myocardium by ultrasonic tissue characterization with integrated backscatter, *Circulation*, 74, 389, 1986.

43. **Madaras, E. I., Pérez, J. E., Sobel, B. E., Mottley, J. G., and Miller, J. G.,** Anisotropy of the ultrasonic backscatter of myocardial tissue. II. Measurements *in vivo*, *J. Acoust. Soc. Am.*, 83, 762, 1988.

44. **Olshansky, B., Collins, S. M., Skorton, D. J., and Prasad, N. V.,** Variation of left ventricular myocardial gray level on two-dimensional echocardiograms as a result of cardiac contraction, *Circulation*, 70, 972, 1984.

45. **Vered, Z., Barzilai, B., Gessler, C. J., Wickline, S. A., Wear, K. A., Shoup, T. A., Weiss, A. N., Sobel, B. E., Miller, J. G., and Pérez, J. E.,** Ultrasonic integrated backscatter tissue characterization of remote myocardial infarction in human subjects, *J. Am. Coll. Cardiol.*, 13, 84, 1989.

46. **Fitzgerald, P. J., McDaniel, M. D., Rolett, E. L., Strohben, J. W., and James, D. H.,** Two-dimensional ultrasonic tissue characterization: backscatter power, endocardial motion, and their phase relationship for normal, ischemic, and infarcted myocardium, *Circulation*, 76, 850, 1987.

47. **Milunski, M. R., Mohr, G. A., Gessler, C. J., Vered, Z., Wear, K. A., Sobel, B. E., Miller, J. G., Pérez, J. E., and Wickline, S. A.,** Ultrasonic tissue characterization with integrated backscatter. Acute myocardial ischemia, reperfusion, and stunned myocardium in patients, *Circulation*, 80, 491, 1989.

48. **Milunski, M. R., Mohr, G. A., Wear, K. A., Sobel, B. E., Miller, J. G., and Wickline, S. A.,** Early identification with ultrasonic integrated backscatter of viable but stunned myocardium in dogs, *J. Am. Coll. Cardiol.*, 14, 462, 1989.

49. **Davies, J., Gibson, D. G., Foale, R., Heerk, K., Spry, C. J. F., Oakley, C. M., and Goodwin, J. F.,** Echocardiographic features of eosinophilic endomyocardial disease, *Br. Heart J.*, 48, 434, 1982.

50. **Picano, E., Gualtiero, P., Marzilli, M., Lattanzi, F., Benassi, A., Landini, L., and L'Abbate, A.,** In vivo quantitative ultrasonic evaluation of myocardial fibrosis in humans, *Circulation*, 81, 58, 1990.

51. **Chiaramida, S. A., Goldman, M. A., Zema, M. J., Pizzarello, R. A., and Goldberg, H. M.,** Real-time cross-sectional echocardiographic diagnosis of infiltrative cardiomyopathy due to amyloid, *J. Clin. Ultrason.*, 8, 58, 1980.

52. **Siqueira-Filho, A. G., Cunha, C. L. P., Tajik, A. J., Seqard, J. B., Schatternberg, T. T., and Giuliani, E. R.,** M-mode and two-dimensional echocardiographic features in cardiac amyloidosis, *Circulation*, 63, 188, 1981.

53. **Nicolosi, G. L., Pavan, D., Lestuzzi, C., Burelli, C., Zardo, F., and Zanuttini, D.,** Prospective identification of patients with amyloid heart disease by two-dimensional echocardiography, *Circulation*, 70, 432, 1984.

54. **Chandrasekaran, K., Aylward, P. E., Fleagle, S. R., Burns, T. L., Seward, J. B., Tajik, A. J., Collins, S. M., and Skorton, D. J.,** Feasibility of identifying amyloid and hypertrophic cardiomyopathy with the use of computerized quantitative texture analysis of clinical echocardiographic data, *J. Am. Coll. Cardiol.,* 13, 832, 1989.

55. **Pinamonti, B., Picano, E., Ferdeghina, E. M., Lattanzi, F., Slavich, G., Landini, L., Camerini, F., Benassi, A., Distante, A., and L'Abbate, A.,** Quantitative texture analysis in two-dimensional echocardiography: application to the diagnosis of myocardial amyloidosis, *J. Am. Coll. Cardiol.,* 14, 666, 1989.

56. **Masuyama, T., St. Goar, F. G., Tye, T. L., Oppenheim, G., Schnittger, I., and Popp, R. L.,** Ultrasonic tissue characterization of human hypertrophied hearts in vivo with cardiac cycle-dependent variation in integrated backscatter, *Circulation,* 80, 925, 1989.

57. **Tak, T., Rahimtoola, S. H., Kumar, A., Gamage, N., and Chandraratna, P. A. N.,** Value of digital image processing of two-dimensional echocardiograms in differentiating active from chronic vegetations of infective endocarditis, *Circulation,* 78, 116, 1988.

58. **Masuyama, T., Hannah, A., Valentine, M. D., Gibbons, R., Schnittger, I., and Popp, R. L.,** Serial measurement of integrated ultrasonic backscatter in human cardiac allografts for the recognition of acute rejection, *Circulation,* 81, 829, 1990.

59. **Lloret, R. L., Corda, X., Bradford, J., Metz, M. N., and Kinney, E. L.,** Classification of left ventricular thrombi by their history of systemic embolization using pattern recognition of two-dimensional echocardiograms, *Am. Heart. J.,* 110, 761, 1985.

60. **Pavan, D., Nicolosi, G. L., Lestuzzi, C., Burelli, C., Zardo, F., Collazo, R., Pizzolitto, S., and Zanuttini, D.,** Qualitative tissue characterization of fresh cardiac thrombi by two-dimensional echocardiography, *J. Cardiovasc. Ultrason.,* 5, 341, 1986.

61. **Milunski, M. R., Canter, C. E., Wickline, S. A., Sobel, B. E., Miller, J. G., and Pérez, J. E.,** Cardiac cycle-dependent variation of integrated backscatter is not distorted by abnormal myocardial wall motion in human subjects with paradoxical septal motion, *Ultrasound Med. Biol,* 15, 311, 1989.

62. **Mottley, J. G. and Miller, J. G.,** Anisotropy of the ultrasonic backscatter of myocardial tissue. I. Theory and measurements *in vitro, J. Acoust. Soc. Am.,* 83, 755, 1988.

63. **Mottley, J. G. and Miller, J. G.,** Anisotropy of the ultrasonic attenuation in soft tissue: measurements in vitro, *J. Acoust. Soc. Am.,* 88, 1203, 1990.

64. **Wear, K. A., Shoup, T. A., and Popp, R. L.,** Ultrasonic characterization of canine myocardial contraction, *IEEE Trans. Ultrason. Ferroelectr. Frequency Control,* UFFC-33, 347, 1986.

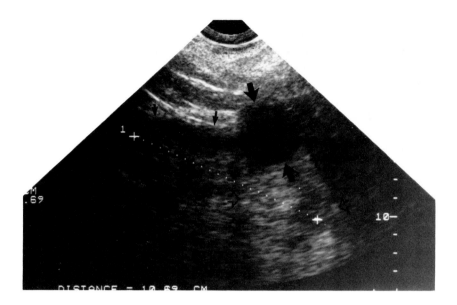

FIGURE 4. Cyst (large arrows) arising from the mid-portion of the left kidney (small arrows). The zone of acoustic enhancement posterior to the cyst is marked with hollow arrows. The size of the kidney is measured with electronic calipers.

The other reason for interest in ultrasonic tissue characterization is the fact that ultrasonic B-scans display only echo amplitude. No attempt is made to analyze the ultrasound signal in terms of frequency content or phase shifts prior to display. Since the ultrasound beam is generally coherent, scattering from tissues produces a granular appearance on the image known as speckle, which is caused by interference of the scattered acoustic waves. The speckle pattern contains information about the distribution of scatterers which cannot be deciphered by observers from the B-scan image alone. Quantitative analysis methods which can recover the information lost when a B-scan image is made could provide much additional information about tissue structure and pathology.

C. DOPPLER APPLICATIONS

Doppler ultrasound and color Doppler imaging were the first methods using the frequency content of the returning ultrasound signal to gain wide clinical application. Although Doppler ultrasound is based on backscattering (similar to the B-scan), the shift in frequency of the returning beam relative to the outbound beam is measured. The process involves a coherent detection method known as quadrature detection which removes the carrier frequency (2 to 10 MHz) leaving the Doppler shift frequencies which are usually in the audible kilohertz (KHz) range. The Doppler shift signal may then be decomposed into its constituent frequency components using Fourier transformation

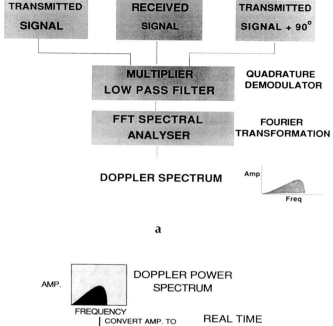

FIGURE 5.   Doppler signal processing. The real time spectral display (b) is composed of many individual doppler power spectra (a) generated by a fast Fourier transform spectral analyzer from the quadrature detected Doppler frequency shift signal.

(Figure 5a). The process of fast Fourier transformation is typically applied many times per second yielding a clinical Doppler spectral display in which time is along the horizontal axis, frequency shift is along the vertical axis, and the magnitude of each frequency component is indicated by the brightness of the pixel at each frequency shift (y-axis) value (Figure 5b). Doppler ultrasound is thus a good method of characterizing blood flow.

While standard pulsed Doppler ultrasound can usually analyze only one region of interest at a time, color Doppler imaging produces an image in which flow at many points of the image are analyzed in real time. The Doppler shift signals from many tiny subregions in the image are analyzed by autocorrelation methods resulting in estimates of average velocity for the subre-

TABLE 1
Reported Criteria for Doppler Diagnosis of Rejection

Author	N	Resistive index	Sensitivity	Specificity
Rifkin et al.[13]	81	>0.8	.69	.86
Allen et al.[17]	55	>0.75	.76	.83

gions which are then color coded and displayed on the B-scan image (Plate 1*). As Doppler signal processing is already well covered in many texts, the reader is referred to these[5-7] for further discussion of methods.

Because of its ability to detect and characterize blood flow, tissue and disease characterization by Doppler ultrasound has been attempted in many organs. For many years, work has been progressing on the detection of early breast carcinoma using continuous Doppler,[8] pulsed Doppler,[9,10] and, lately, color Doppler imaging.[11] Liver lesions have also been studied using pulsed Doppler ultrasound. Vascularity and blood flow velocity have been shown to be increased around primary breast malignancies, around hepatocellular carcinomas, and around some metastatic lesions.[12] Despite the promise that this work has shown, the method has yet to gain widespread clinical acceptance, possibly because of the operator dependence of the method and the need for sonographers trained in abdominal and small parts Doppler examinations.

Another application for Doppler untrasound has been in the detection of renal transplant rejection. Initial studies focused on the resistive index (RI) and pulsatility index (PI) as means of distinguishing transplant rejection from hydronephrosis and acute tubular necrosis in patients with declining transplant function.[13-15] The resistive index,

$$Resistive\ Index = \frac{V_{systolic} - V_{diastolic}}{V_{systolic}}$$

is a measure of diastolic flow and is elevated when diastolic flow is decreased as in transplant rejection (Table 1). Unfortunately, the resistive index may also be elevated in acute tubular necrosis, chronic rejection, renal vein thrombosis, pyelonephritis, and hydronephrosis[16-18] and is sensitive primarily to vascular (not cellular) rejection. Since many cases of transplant rejection consist mostly of cellular rejection, the RI has been far less useful than was originally hoped.

Other applications for Doppler ultrasound include evaluation for renal artery stenosis by measurement of the acceleration of blood velocity in early systole,[19] the evaluation of portal venous flow to detect passive congestive of the liver,[20,21] and the study of hepatic venous flow in the detection of Budd-Chiari syndrome.[22] These are all limited applications aimed at the detection of primarily vascular disorders.

---

* Plate 1 appears after page 374.

D. ULTRASONIC TISSUE CHARACTERIZATION

Although Doppler ultrasound is clearly a successful method for characterizing blood flow disturbances, it cannot directly analyze pathological changes in tissue structure. Tissue characterization is a term that usually refers to the quantitative estimation of tissue or image features leading to a more accurate distinction of normal from abnormal tissue. The results of tissue characterization may be quantitatively interpreted using numerical values or may be displayed as an image for qualitative interpretation by an observer. Tissue characterization aims to provide additional information about tissues not ordinarily available by simple viewing of the ultrasound B-scan. Doppler ultrasound may be thought of as a form of tissue characterization for vascular abnormalities. The information gained from tissue characterization is usually quantitative and is far less operator dependent than is the usual B-scan image.

Although work in ultrasonic tissue characterization has been underway since the late 1970s, tissue characterization as yet has had no widespread clinical impact. This delay has been caused by several factors. First, the interaction of ultrasound with tissue is relatively complex; both tissue attenuation and scattering are nonlinear functions of ultrasound frequency for many tissues. In addition to these effects, the polychromatic nature of the ultrasound beam and the effects of transducer focusing and diffraction make it difficult to separate one effect from the others. Also, the nature of the actual scatterers in tissue and the effects of scatterer arrangement on the ultrasound image are not well understood. A second problem is the fact that certain ultrasonic tissue features, such as attenuation, may not provide much discrimination between normal and abnormal tissues. A third problem is the complex nature of many of the computations that must be carried out to perform tissue characterization. These time-intensive computations delay the display of the features and make real-time implementation difficult. In spite of these problems, a number of tissue characterization methods have achieved a degree of success. Tissues from several organs (liver, eye, pancreas, kidney, spleen, heart, skeletal muscle, breast) have been analyzed using ultrasonic tissue characterization. The liver has been the most popular organ because of its large size, easy accessibility, and relatively homogeneous structure. It is therefore a good place to begin an overview of methods and results.

II. LIVER TISSUE CHARACTERIZATION FROM SCATTERING

A. METHODS OVERVIEW

If one includes analysis of both the ultrasound signal and the image produced by detection and display of that signal, a large number of potential tissue characterization features can be calculated. The methods may be divided into three categories: (1) methods which *analyze the image data*; (2) methods which attempt to *estimate ultrasonic properties* of tissue such as sound speed or acoustic attenuation; and (3) methods which use ultrasound to *detect a nonacoustic physical property* of tissue, such as hardness.

Image analysis methods typically use statistics calculated from pixel gray level values in the image or from intensity values calculated from the envelope detected radio frequency signal. These statistics typically describe the spatial variation in signal intensity or the average signal intensity from a region of interest. Fractal analysis of image data has also been studied.[23] The exact nature of the underlying tissue/ultrasound interaction is not studied in this group of methods.

On the other hand, methods which *estimate acoustic properties* are entirely concerned with estimating a single acoustic property from the backscattered ultrasound tissue by eliminating system effects and the effects of other interactions. Examples of specific acoustic properties include frequency-dependent backscatter coefficients, acoustic attenuation, and speed of sound. Other acoustic properties such as the frequency dependence of backscatter may be used to calculate tissue properties such as the effective size of the scatterers producing the ultrasound signal and their density in tissue.[24] The integral over all received frequencies of the backscatter coefficients (integrated backscatter) has been widely used to evaluate normal and ischemic myocardium.[25-27]

The use of ultrasound to determine *nonacoustic tissue properties* has focused on estimation of tissue elasticity from the motion of tissues in response to externally applied vibrations[28,29] or to internal motion such as cardiac pulsations.[30-32] Abnormal tissue is often stiff and nonelastic, causing it to move differently in response to external pressure when compared to normal elastic tissue. Detection of the tissue motion has been accomplished by Doppler analysis and by application of correlation techniques to M-mode data. The techniques have been applied to fetal lung, myocardium, skeletal muscle, liver, and to the prostate gland, with varying degrees of success.

B. LIVER TISSUE CHARACTERIZATION BY STATISTICAL IMAGE DATA ANALYSIS

The simplest form of image data analysis is that of the pixel data histogram which is a display of the occurrence frequency of gray levels in a region or along a line in the image. This information has been described by Julesz[33] as first-order texture statistics (i.e., giving information about gray level frequencies but not about spatial location). This form of analysis has been implemented on several commercial ultrasonic imagers and usually allows calculation of the mean intensity value and variance of the pixels along a given line of interest or from within a region of interest. Usually a histogram of the gray level distribution is displayed on the screen. Although this form of analysis has been available on many scanners, it has been little used since mean pixel gray level is strongly dependent on the gain settings used and no method for standardization or calibration of the intensity levels is usually available.

    Several groups have supplemented the gray level histogram with higher-order statistical features. Second-order statistical features give not only occurrence frequencies of gray levels but also spatial interdependencies between the image elements (pixels). Raeth et al.[34] analyzed ultrasound images from 71 patients with biopsy-proven diffuse and focal liver disease and compared them with 20 normals. The first-order image features which they examined included mean gray level, variance, skewness (deviation of the pixel intensity distribution from symmetry), kurtosis (steepness of the distribution relative to a normal distribution), and percentiles of the gray level distribution.

    Other features used by the Raeth group contain second-order statistical information and include features derived from the co-occurrence matrix. This matrix is a two dimensional histogram characterizing the occurrence of gray level combinations in pairs of spatially related pixels. An entry in the co-occurrence matrix, $Cd(i,j)$ specifies the frequency that pixels separated by distance d display gray levels i and j, respectively. From the normalized co-occurrence matrix, several features may be calculated. These are *contrast* — a measure of how many large gray level differences are present in the region of interest (frequent large gray level differences increase contrast); *angular 2nd moment* — a measure of the degree of clustering of co-occurrence matrix values around major gray level transitions (increases when only few gray level transitions exist); *entropy* — a measure of the uniformity of matrix values which increases with increasing coarsensss of the image texture; and *correlation* — a measure of the linearity of the gray level relationship in d related pixels.

    Additional features used include those from the gray level run-length histogram which is a count of the number of gray level runs by length and gray level range. A run is a set of vertically or horizontally contiguous pixels displaying nearly identical gray levels. Features derived from the run-length histogram include: *runpercentage* — a descriptor of the percentage of runs present (increases with increasing homogeneity); *long-run emphasis* — an indicator of the number of long runs present; *gray level distribution* — an estimator of the uniformity of runs with respect to gray level; and *run length distribution* — a measure of the uniformity of runs with respect to length.

    The Raeth group also used features derived from the two-dimensional (2-D) power spectrum obtained by Fourier transformation of the 2-D autocorrelation function.[35] The inner ring sum which measures the low spatial frequency content in the image signal and the outer ring sum which measures the high spatial frequency content may both be calculated from the 2-D power spectrum.

    Combinations of the features were tested for their ability to detect disease by using linear discriminant analysis to find an appropriate linear Bayesian classification rule. Using a combination of the above features, an accuracy of 98% was achieved for the detection of diffuse liver disease and an accuracy of 89% for malignancy. Table 2 summarizes the performance of the system. In later *in vitro* work, the same group[36] found that using ten of the above

TABLE 2
Diagnostic Accuracy for Various Diagnoses

Diagnostic class	N	2-D image analysis (%)	Subjective evaluation (%)
All categories	40	95	85
Normal	10	100	70
Diffuse Disease	20	95	85
C. hepatitis	10*	70	—
Cirrhosis/fibrosis	11*	82	—
Fatty infiltration	15*	80	—
Cirrhosis/fatty infiltration	17*	82	—
Tumor	10	90	97

Note: The number of patients in the diffuse disease subclasses (*) exceeds the total listed for diffuse disease because the patients for which *both* image analysis and subjective human observer evaluation was available was a subset of a larger group having image analysis only.

Adapted from Raeth, U. et al., *J. Clin. Ultras.*, 13, 87–99, 1985.

features led to sensitivities and specificities for detection of fatty liver in the 87 to 89% range and 76 to 78% for detection of cirrhosis. The group also noted that morphometric/chemical analysis of the samples yielded a better correlation to the image analysis findings than did simple histopathologic analysis.

Schuster et al.[37] have used image texture measures derived from 8-bit pixel data generated from a modified static B-scanner. Each pixel within a selected region of interest (ROI) is analyzed with respect to its neighbors according to one of 35 texture parameters that the user selects from five categories: textural edgeness, gray-scale run length, co-occurrence matrix, spread, and relative extrema density. The texture values are then pooled with those from another region of interest to which the first region is to be compared. The pooled texture values are split into those falling within a normal range and those which fall outside that range. The two ROI are then redisplayed assigning all pixels with texture values within the normal range as one gray level or color and all others as another.

The resultant images allow observers to distinguish between different tissues by looking for a nonrandom distribution of the two types of pixels in a side-by-side comparison of the two ROI. This method has been used to detect acute hepatitis, fatty liver, and cirrhosis with high accuracy in images from a group of 33 patients and normals (Table 3).[38] Observers were also able to distinguish between the various disease states with high accuracy although the number of abnormal patients was small (largest single group, 12 patients).

The method described has the advantage of allowing the observer to qualitatively distinguish between various diseases from images without per-

TABLE 3
Results of Local Texture Analysis

Task	Cases with clustered pixels	Accuracy (%)
Hepatitis vs. normal	14/14	100
Cirrhosis vs. normal	10/10	100
Fatty infiltration vs. normal	10/12	83
Overall (normal vs. disease)	34/36	94
Hepatitis vs. fatty liver	9/10	90
Cirrhosis vs. fatty liver	7/8	87
Hepatitis vs. Cirrhosis	8/8	100

forming numerical comparisons or manipulations. On the other hand, the method has not yet been validated on data from other more modern scanners which have 8-bit scan conversion. The authors resorted to a specialized image processor (IPSUS[37]) to produce 8-bit images for analysis in their reported work. Differing machines and scan conversion hardware may introduce considerable machine-dependent variation in the data sets. This variation could prevent the use of a single set of control ROI for comparison with data from different machines. A *new* set of control ROI (i.e., ROI collected from known normals and from patients with histologically proven disease) may be required for each scanner that the method is used on. This could represent a serious limitation to the method. All methods using image data face this potential problem and so far no group has tested image analysis methods derived using control data sets from one machine on patient data from another machine.

Insana et al.[39] have developed a different approach to statistical image analysis. Instead of first calculating a large number of image features and then evaluating them to determine which features allowed discrimination between normal and abnormal tissues, they began with a theoretical model of ultrasound scattering in tissues (see also Chapter 4). From this model, they developed several parameters describing the relative contributions to the image of the different types of scatterers in tissue. These parameters were then checked for their ability to separate normal from diseased liver tissue.

The model developed for scattering in soft tissues was prompted by the recognition that the textural statistics of many ultrasound images are similar to those that characterize texture in laser speckle.[40-42] Thus the granular texture of ultrasonic images has become known as ultrasonic speckle. The model is based on the observations that tissue scatterers vary in size and shape, and that different structures have varying degrees of spatial order. A simple biological scattering medium is unclotted blood where the scatterers are small and competely disordered. Ultrasonic backscatter from this type of material is known as Rayleigh scattering in which a histogram of pixel intensities (I_d) follows a characteristic distribution having a mean to standard deviation ratio (point signal-to-noise ratio) of 1.00 (1.91 for amplitudes). In addition to completely random scatterers, most biological tissues also have nonrandom

TABLE 5
Detection of Diffuse Liver Disease Using Four Features (β, $\bar{d}$, r, σ'_s)

Task (no. of patients in parentheses)	$A_z \pm$ S.D.
Normal (36) vs. chronic hepatitis (120)	.88 ± .03
Normal (36) vs. Gaucher's disease (68)	.94 ± .03
Hepatitis (120) vs. Gaucher's disease (68)	.84 ± .04
Normal (36) vs. glycogen storage disease (12)	.94 ± .05
Normal (36) vs. primary biliary cirrhosis (21)	.80 ± .10

TABLE 6
Human Performance Using Three Images

Task (no. of patients in parentheses)	$A_z \pm$ SD
Normal (29) vs. chronic hepatitis (32)	.53 ± .08
Normal (29) vs. Gaucher's disease (30)	.67 ± .08
Chronic hepatitis (32) vs. Gaucher's (30)	.71 ± .07

TABLE 7
Effect of Different Ultrasound Imagers on Feature Values

Feature	Machine 1	Machine 2
$\bar{d}$	1.16 ± .08	1.24 ± .07
r	.47 ± .12	.57 ± .12
σ'_s	.72 ± .11	.61 ± .08
Attenuation	.48 ± .16	.38 ± .09
A_{rms}	−2.80 ± 7.1	−7.61 ± 5.1

Note: Both machines were of the same model and manufacture. Both were used to acquire digitized radio frequency at 3.5 MHz. Average feature values ± SD are listed for 36 normal volunteers studied with machine 1 and 10 studied with machine 2.

cannot be readily transferred to another. Data was collected on a second scanner from the same manufacturer from a group of normal volunteers (several of which were included in the first data set). The calculated feature values were significantly different, even when the same transducer was used on each machine (Table 7). This difficulty has caused many to focus on features that can be calculated directly from radio frequency data and features that actually represent acoustic properties of the tissue, avoiding the nonlinearities of radio frequency detection and gray-scale mapping.

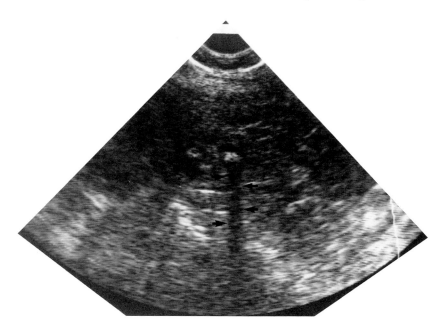

FIGURE 9.   Echogenic renal calculus with acoustic shadowing (small arrows).

C. ESTIMATION OF FEATURES DERIVED FROM ACOUSTIC PROPERTIES OF TISSUE

1. Attenuation Estimation

The simplest and most intuitive methods for attenuation estimation are the time domain methods where the actual decrease in backscatter intensity as a function of depth in tissue is used as an estimate of acoustic attenuation. Subjective attenuation estimates have been made by observers since the earliest clinical use of ultrasonic imaging. For focal lesions, attenuation within the lesion is based on the presence of ''acoustic enhancement'' or acoustic shadowing (Figures 4 and 9) in which increased or decreased echoes are seen posterior to a lesion relative to the adjacent background. This finding is due to the attenuation in the lesion being different from that of adjacent tissues. In clinical practice, one usually thinks of a lesion which shows acoustic enhancement as being fluid filled and one showing shadowing as being calcified or gas filled. Observers' inability to accurately quantify the degree of shadowing or enhancement (i.e., attenuation) makes it impossible to accurately diagnose many lesions where only slight enhancement or shadowing is present.

For entire organs, the estimation of attenuation subjectively is even more difficult. It depends upon the operator's subjective impression that higher or lower than usual gain and time gain compensation (TGC) must be used to penetrate the organ. An example of this is shown in Figure 10 where fatty

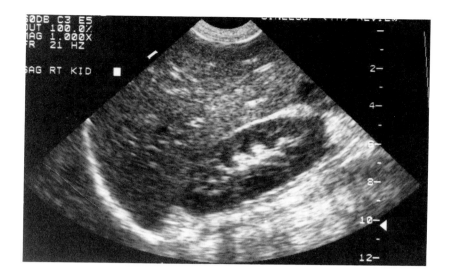

a

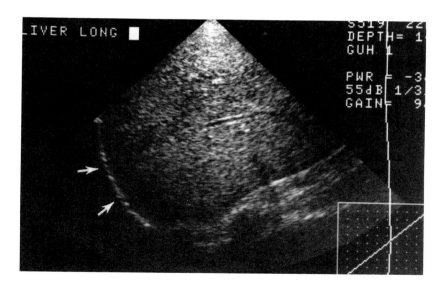

b

FIGURE 10. Longitudinal image (a) in a patient with Wiskott-Aldrich syndrome show a normal liver and kidney. Follow up scan after anti-fungal drug therapy (b) shows increased echogenicity anteriorly and a weak diaphragmatic echo (arrows, compare with (a)) suggesting increased attenuation due to fatty infiltration. Both scans were performed using a 5 MHz transducer.

infiltration of the liver has increased the attenuation of the liver and has made visualization of the diaphragm difficult. Gosink et al.[4] noted that the degree of beam penetration was the most difficult criterion to apply for diagnosis in 61 patients with various types of diffuse liver disease. Dewberry and Clark[48] studied 67 patients having cirrhosis and found effects due to attenuation on the B-scan images, but concluded that the observation was too subjective to be of value. Other factors in addition to the subjective nature of visual attenuation estimates include transducer focusing effects (increased echo intensity is usually seen at the transducer focus), and variable pre- and post-processing. Quantitative attenuation estimation holds the promise of eliminating several of these problems and has been the focus of most of the quantitative tissue characterization work done over the past 10 years.

Many methods are available for the quantitative estimation of attenuation, but only a few are suitable for *in vivo* measurements. Substitution methods in which the strength of the reflected ultrasound beam is measured with and without tissue interposed between an ultrasound transducer and a standard reflector are not suitable for *in vivo* applications. Transmission methods are only usable in peripheral organs such as the breast and testicles. The main methods used in the abdomen all involve analysis of backscattered echoes. Attenuation in soft tissue is commonly modeled using a power law function $\alpha = \beta f^n$, where α is the tissue attenuation in dB (or Nepers)/cm, β is known as the coefficient of frequency dependent attenuation (dB/cm/MHz^n) and n is the power law dependency (usually between 0.9 and 1.4). Clinical attenuation measurement methods have focused on estimating these three parameters with most techniques concentrating on β.

a. Time Domain Methods

There are two major approaches to *in vivo* attenuation estimation: time domain methods and frequency domain spectral estimation methods. Time domain (or amplitude based) methods are conceptually the simplest. In these methods, the loss in signal amplitude over depth at a single frequency or over a narrow band of frequencies is measured. The earliest attempts at attenuation estimation were time domain-amplitude based methods.[49] Since clinical ultrasound scanners measure amplitudes to generate images, amplitude based methods are easier to implement than spectral estimation methods.

Ophir et al.[50] described an interesting method for *in vivo* attenuation estimation in which a narrow-band ultrasonic pulse was used to estimate attenuation. The narrow-band pulse allows a simple amplitude relationship

$$A_d = A_o e^{-2d\alpha_f} \quad \text{and} \quad \frac{\log A_o - \log A_d}{2d} = \alpha_f$$

where A_d is the amplitude at depth d, A_o is the amplitude at the initial depth, f is the frequency of the ultrasound beam. Taking the difference of the log amplitudes gives the attenuation α at frequency f. By averaging many amplitude differences within a region of interest, the random variability of the measurement can be made very low. Transducer focusing effects on backscatter amplitude are eliminated by moving the transducer closer to the patient when data from deeper tissues are to be collected (a C-scan method). This allows all backscatter amplitudes to be measured at the same distance from the transducer. Since attenuation is estimated at only a single frequency, one must assume that attenuation is directly proportional to frequency (n = 1) for α to be used to estimate the coefficient of frequency-dependent attenuation β. As will be discussed later, this direct proportionality is approximately true for normal liver but does not hold as true for abnormal liver and for other organs. In later work,[51] the narrow-band method was extended to multiple frequencies eliminating the need for the assumption of direct proportionality and making possible the estimation of both β and n.

Time domain methods using broad-band ultrasound beams have also been used with some success. One approach recently described[52] calculated attenuation from the slope of the time gain compensator (TGC) curve after that curve had been set to give constant echo amplitude vs. depth in a region of interest. The attenuation in dB/cm is then divided by the average frequency in the ROI (measured by a zero crossing technique) to give the frequency dependent attenuation coefficient in dB/cm/MHz. The transducer was weakly focused to minimize error due to focusing effects. The method assumes an approximately Gaussian shaped spectrum for the US pulse and a linear frequency dependence of attenuation. The method gave lower variability in both patients and phantoms than did a spectral estimation (spectral shift) method.

Our own time domain method[53] is also a broad-band method based on the same assumptions. In our method root mean square (RMS) amplitudes are calculated at multiple depths within the ROI. TGC and transducer focusing effects are corrected for by using calibration phantom amplitudes obtained at the same time each patient is scanned. Backscatter amplitudes (relative to the phantom) corrected for attenuation are then calculated using different attenuation (β) values until the backscatter amplitudes remain constant over the depth range used. In this calculation the downward shift of the mean frequency of the backscattered ultrasound caused by attenuation (discussed later in spectral estimation methods) is accounted for in the computation. Attenuation values obtained by this method demonstrated lower variability than did those obtained by a spectral estimation method (spectral difference method) in phantoms and in patients.

A further reduction in the variance of attenuation estimates is possible if envelope peak (defined as the local maximum of the envelope) values are used instead of average amplitude or RMS amplitude.[54] This is largely due to the higher signal-to-noise ratio of the envelope peaks compared to the envelope signal as a whole (2.4 to 1.91).[55] The method was originally described for narrow bandwidth ultrasound but more recently has been modified

ATTENUATION ESTIMATION BY SPLIT
SPECTRUM PROCESSING

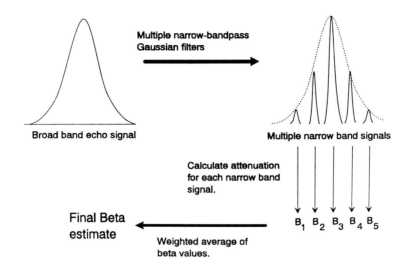

FIGURE 11. Schematic illustrating attenuation estimation using multiple narrow band filters.

for use with broad-band signals by splitting the broad-band ultrasonic echo into multiple narrow-band signals from which several attenuation estimates are made (Figure 11).[56] In the experiments described, Gaussian bandpass filters with a standard deviation of 0.16 MHz were sufficiently narrow to give good results.

In the method used by Parker et al.[57] a broad-band ultrasound beam (approximately 1 MHz) was also used. The returning signal was decomposed by discrete Fourier transformation to allow evaluation of amplitude decay at 15 discrete frequencies within the bandwidth. This method is therefore best thought of as a narrow-band amplitude method carried out at multiple frequencies. This method has the advantage of making no assumptions about the relationship of attenuation and frequency and allows the estimation of both β and n. Estimates of β and n obtained by this method have a relatively large uncertainty of approximately 20% but the normalized value of β at the center frequency using the power law fit of attenuation values over the entire bandwidth has a much lower error of approximately 3%. This level of uncertainty is similar to that of other amplitude based methods and is sufficient to distinguish some types of tissue abnormality.

b. Spectral Estimation Methods

The spectral difference method[58] was one of the first methods developed for attenuation estimation from backscattered ultrasound. The method assumes

SPECTRAL DIFFERENCE
METHOD

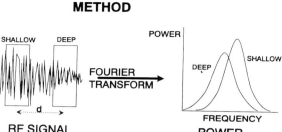

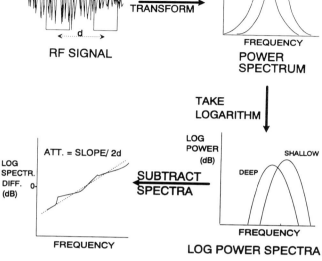

FIGURE 12.   Schematic illustrating the spectral difference method of attenuation estimation.

that attenuation increases linearly with frequency (n = 1) and attempts to compute the coefficient of frequency dependent attenuation (β) by pairwise comparison of backscatter spectra from different depths. The slope of a straight line fit to the log ratio of two spectra obtained at different depths yields β (Figure 12). Typically many pairwise spectral comparisons are made within a region of interest but the method has been shown to be an inefficient use of the available data[59] and thus requires large regions of interest for reliable results. This makes the method less useful for focal lesions and for organs other than the liver where smaller regions of interest must be used. Many of the early reports of *in vivo* liver acoustic attenuation used this method.

A second approach uses the same assumptions as the spectral difference method but attempts to estimate attenuation from the center frequency of the returning echo pulses (Figure 13). If attenuation is linearly dependent on frequency (n = 1) and the ultrasound pulse is Gaussian in shape, the increased attenuation of higher frequencies causes the center frequency of the pulse to decrease as the pulse traverses the tissue. The width of the pulse (represented by its standard deviation, σ) does *not* change.[60] The attenuation coefficient

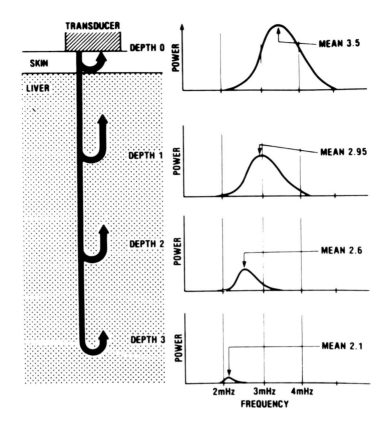

a

FIGURE 13. The spectral shift method. As the ultrasound beam traverses tissue, the beam intensity decreases and its center frequency shifts downward (a). The slope of a plot of center frequency vs. depth (b) may be used to calculate the attenuation coefficient β. (From Shawker, T. H., Garra, B. S., and Insana, M. F., in *Ultrasound Annual 1985,* Sanders, R. C. and Hill, M. C., Eds., Raven Press, New York, 1985, 112–113. With permission.)

(β) may be calculated from the slope of a plot of center frequency as a function of depth by using

$$\beta = \frac{\Delta \bar{f}}{\Delta d} \frac{1}{4\sigma^2}$$

In this equation $\Delta \bar{f}/\Delta d$ is the slope of a linear fit to the plot of center frequency vs. depth and σ is the standard deviation of the returning echo pulse.[61] The mean or center frequencies may be estimated from the power spectra of backscattered signals from different depths, or they may be estimated in the

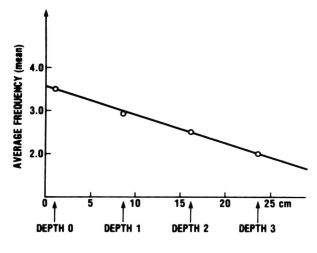

FIGURE 13b.

time domain by measuring the number of zero crossings per time period in the radio frequency signal from different depths.[62] The advantages of these methods, known as the spectral shift method and the zero crossing method, are their insensitivity to beam profile related amplitude artifacts and their ability to use data obtainable from current commercial sector scanners. These methods have been incorporated into prototype commercial sector scanners for clinical trials. Unfortunately, the clinical trials demonstrated large variations (approximately 100%) in the *in vivo* attenuation values.[63] Ophir et al.[64] then showed that when using the zero crossing technique, reducing variation in calculated β values to acceptable levels would require far more data than is feasible to collect in a clinical setting. This problem plus the effects of range dependent center frequency shifts exhibited by large aperture focused transducers[65] have damped the initial enthusiasm for this method.

Wilson et al.[66] used a variant spectral based technique in which average log spectra were calculated at several depths within a region of interest. Spectral slopes for each depth were calculated by fitting least squares linear regressions to the log power spectra and taking the slopes of the regression lines. The slope of a linear fit to a plot of spectral slope vs. depth was taken as the attenuation coefficient β. This method makes more efficient use of available data than does the spectral difference method. It also assumes that attenuation is directly proportional to frequency ($n = 1$). The authors found that this method gave results similar to a limited form of the method used by Parker et al.[57] where amplitude decay at 2.5 MHz was used to estimate the attenuation coefficient.

c. In Vivo *Attenuation Results in the Liver*

Table 8 summarizes the reported values[52,53,57,66,69-77] for acoustic attenuation in normal liver. The attenuation coefficient (β) of normal liver has

TABLE 8
In Vivo Liver Attenuation Results

Author	No. Patients	Attenuation (dB/cm/MHZ) (mean ± SD)	n	Method used
Kuc (1982)[69]	42	0.43 ± .07	*	Spectral diff.
Jones (1981)[70]	30	0.47 ± .07	0.9–1.1	Spectral diff.
Cooperberg (1984)[71]	10	0.68 ± .14	*	Zero crossing
Garra (1984)[53]	31	0.63 ± .13	*	Zero crossing
Kuc (1984)[72]	50	0.55	*	Spectral diff.
Maklad (1984)[73]	39	0.52 ± .03	*	Single-freq. amp
Parker (1984)[74]	11	0.59 ± .21	.81–1.5	Mult. freq. amp
Sommer (1984)[75]	3	0.71 ± .07	*	Spectral shift
Wilson (1984)[66]	12	0.53 ± .10	*	Spectral slope
Fredfeldt (1985)[76]	13	0.72 ± .14	*	Single-freq. amp
King (1985)[77]	10	0.79 ± .09	*	Spectral diff. (variant)
Taylor (1986)[52]	26	0.52 ± .12	*	Broad-band amp
	26	0.50 ± .20	*	Spectral shift
Garra (1986)[53]	18	0.49 ± .12	*	Broad-band amp
	18	0.62 ± .18	*	Spectral diff.
Parker (1988)[57]	15	0.47 ± .08	1.05 ± .25	Mult. freq. amp

generally been reported to be between 0.5 and 0.6 dB/cm/MHz. Many of the higher values for attenuation were reported in the early literature, with the more recent literature suggesting that attenuation in normal liver is approximately 0.5 dB/cm/MHz. Single-frequency and multifrequency time domain methods have become the most popular, possibly because they have shown less variation than the spectral estimation methods.[52,53,67] Variability in normal liver attenuation measurements is due to more than just variability in the attenuation estimation process. Tuthill et al.[68] have shown that there is a significant liver attenuation difference between well-fed and fasting individuals in animal experiments which appears to correlate with the glycogen content of the liver. Thus, to decrease variability in clinical attenuation estimates it may be necessary to somehow control amount of liver glycogen present by studying all patients either at a fixed time interval from their previous meal or while fasting. This could cause considerable scheduling difficulty in a typical clinical setting and has not yet been attempted.

Most attenuation measurements in diseased liver have been in diffuse diseases including cirrhosis (alcoholic, primary biliary, cardiac, and that due to hepatitis and hemochromatosis), chronic hepatitis, fatty infiltration of the liver, diffuse tumor infiltration, and Gaucher's disease. In most cases the diagnosis was confirmed either by liver function tests or by biopsy. Table 9 lists reported attenuation values for patients with cirrhosis.

Early results suggested that cirrhotics had elevated attenuation values. However, later studies of both alcoholic and nonalcoholic cirrhosis[52,53] did not confirm this. A more recent study[57] also confirmed elevated attenuation

TABLE 9
Attenuation in Cirrhotic Livers

Author	Cirrhosis type	No. patients	Attenuation (dB/cm/MHz) (mean ± SD) (range)	NL. range
Clinical proof				
Sommer (1984)[75]	Hepatitis	7	0.93 (.85–.98)	.65–.77
Wilson (1984)[66]	?	9	0.64 (.39–.87)	.38–.63
Maklad (1984)[73]	Cardiac	4	0.66	.48–.55
King (1985)[77]	Ethanol	8	1.07 ± .20	.79 ± .09
Histologic proof				
Maklad (1984)[73]	Ethanol	5	0.83 (.72–.92)	.48–.55
Fredfeldt (1985)[76]	Methotrexate	5	0.72 ± .12	.72 ± .14
Taylor (1986)[52]	Ethanol	23	0.57 ± .14	.52 ± .12
	(No fat)	16	0.53 ± .11	
	Hemochromatosis	2	0.43 ± .11	
Garra (1987)[53]	C. hepatitis	11	0.45 (.22–.57)	.29–.60
	Hemochromatosis	2	0.47 (.27–.68)	

Note: The range of values for attenuation is given whenever available, otherwise the SD is given.

in fatty livers and suggested that fibrosis could also produce elevated attenuation values. *In vitro* work by Lin et al.[78] demonstrated that both fibrosis and fatty infiltration produced increased attenuation, but that the effects of fat were much greater than those of fibrosis. These studies suggest that elevated attenuation in cirrhotics is primarily due to concomitant fatty infiltration with fibrosis playing a smaller (if any) role. This is unfortunate since fatty infiltration is reversible and of no real prognostic significance. Clinicians are primarily interested in the amount of fibrosis and inflammation rather than the amount of fatty infiltration.

Table 10 shows the reported results for patients with fatty infiltrated livers. Earlier studies gave somewhat variable attenuation values for fatty livers, but two of the most recent studies show that fatty livers have increased attenuation and that this increase correlates with the amount of triglyceride present. There is evidence that intrahepatic accumulations of other lipids (e.g., sphingomyelin in Niemann-Pick disease) may increase attenuation as well.[53]

Two studies[53,73] of patients with chronic hepatitis both showed attenuation values close to normal (.52 ± .04 and .50 ± .18 dB/cm/MHz, respectively). In each case, however, subgroups of the hepatitis patients with lower than normal attenuation values appeared to exist. These results have yet to be confirmed by larger studies in which close correlation with histology is available. Table 11 lists results reported for various other diseases including met-

TABLE 10
Attenuation in Fatty Infiltrated Livers

Author	Cause	No. patients	Attenuation (dB/cm/MHz) (mean ± SD) (range)	NL. range
Kuc (1980)	?	4	0.49 ± .09 (.41–.62)	.44 (one)
Maklad (1984)	?	5	0.49 ± .11 (.37–.66)	.48–.55
Wilson (1984)	?	3	0.87 (.47–1.34)	.38–.63
Fredfeldt (1985)	MTX[a]	9	0.91 ± .16	.72 ± .14
Taylor (1986)	MTX	7	0.77 ± .11	.52 ± .12
Garra (1986)	GLY[b]	6	0.85 ± .12 (.68–1.06)	.29–.60

[a] MTX = methotrexate therapy.
[b] GLY = type 1 glycogen storage disease.

astatic disease to the liver. Gaucher's disease (a glycolipid accumulation in the liver) and primary biliary cirrhosis both appear to have little effect on liver attenuation. The study of focal disease (metastases and hepatoma) is tantalizing, but the number of patients studied so far is too small to draw meaningful conclusions. The large variation in attenuation estimates of hepatomas (SD = .37) may be due to the small size of many of the lesions, and their inhomogeneity which violates the assumption of a homogeneous scattering required for reliable attenuation estimates. The small size of focal lesions is a major obstacle to obtaining accurate attenuation estimates.

In summary, refinements in attenuation estimation have made it possible to separate normal livers from certain disease states. Unfortunately, in the liver, elevated attenuation appears to be primarily related to fatty infiltration rather than the more important changes of inflammation and fibrosis. This severely limits the usefulness of attenuation estimation for the characterization of liver disease.

2. Sound Speed Estimation In Liver

A number of methods for estimation of sound speed in the liver have been tried. They may be grouped into methods which use two or more transducers and those which use only one transducer. Robinson et al.[79] proposed a two-transducer method of determining sound speed in which sound speed is determined from the apparent shift of a selected feature on B-scan images obtained from two transducers having a known relative geometry. The apparent shift of features was determined by cross-correlation of identically located regions of interest in the image from each transducer. The method was applied to data obtained from an Octoson immersion water bath scanner and was found to yield sound speed estimates with a variability of ±15 m/s (approximately 1%). Using the method, the authors were able to separate normal from fatty and cirrhotic liver.

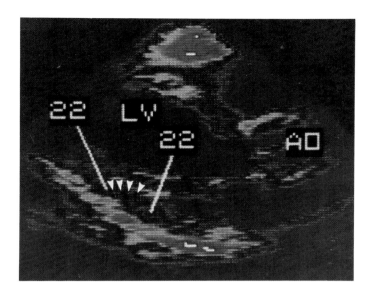

CHAPTER 10. PLATE 1. Color-encoded two-dimensional echocardiogram (parasternal long-axis view) from a patient with new inferior myocardial infarction. There is no change in color or pixel intensity (22) in the area of infarction (arrowheads) compared with adjacent normal muscle. AO = aortic root; LV = left ventricle. (From Chandraratna, P. A. N. et al., *Am. J. Cardiol.*, 56, 381, 1985. With permission.)

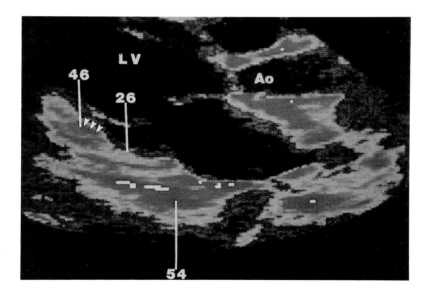

CHAPTER 10. PLATE 2. Color-encoded two-dimensional echocardiogram in a patient with old inferior myocardial infarction. There is a change in color and increase in pixel intensity (46) in the area of infarction (arrowheads) compared with adjacent normal muscle (26). The pixel intensity of the pericardium is 54. Ao = aortic root; LV = left ventricle. (From Chandraratna, P. A. N. et al., *Am. J. Cardiol.*, 56, 381, 1985. With permission.)

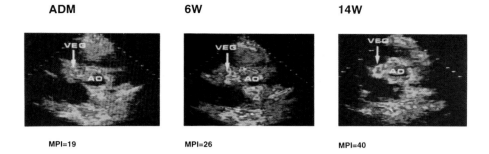

ADM 6W 14W

MPI=19 MPI=26 MPI=40

CHAPTER 10. PLATE 3. Example of a valvular vegetation (VEG) (in color) on the tricuspid valve (TV) showing the mean pixel intensity (MPI) at admission, at 6 weeks, and at 14 weeks. With bacteriologic cure, the mean pixel intensity of the VEG gradually increased from 19 to 40. (From Tak, T. et al., *Circulation*, 78, 116, 1988. With permission.)

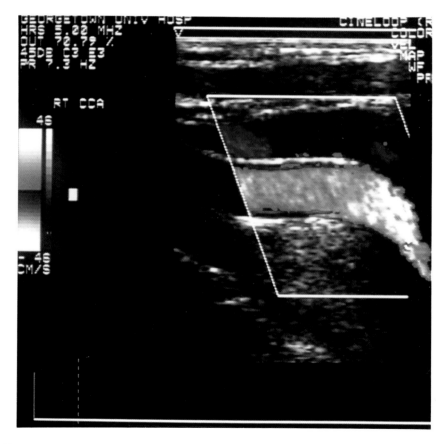

CHAPTER 11. PLATE 1. Color doppler image of a normal proximal common carotid artery (red) and internal jugular vein (partly stained blue). The colors are arbitrarily assigned to indicate flow towards or away from the transducer.

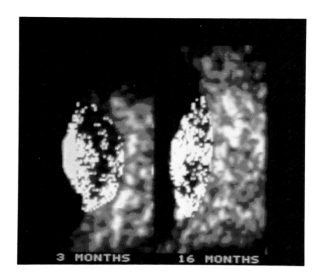

CHAPTER 12. PLATE 1.   Differential image. Highlighting shows regions of a small intraocular melanoma in which discriminant-function values have changed as a result of treatment. The left image shows the tumor 3 months after treatment; the right image shows the tumor 16 months after treatment. Although not indicated here, the changes in spectral parameters are consistent with an increase in scatterer size.

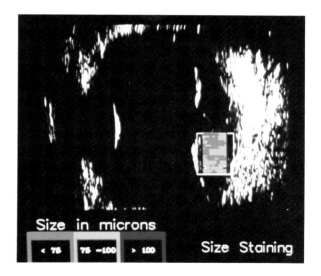

PLATE 2A

CHAPTER 12. PLATE 2.   Color-encoded size images in 3-D. A small posterior intraocular melanoma treated by cobalt plaque is color-encoded to depict computed scatterer size values. The encoding is shown in 2A, which depicts a central scan plane through the pre-treatment tumor. 2B shows the pre-treatment tumor in a cut 3-D block; the coding indicates the predominant scatterer size is less than 75 microns. 2C shows the post-treatment tumor; the encoding shows an increase in predominant scatterer size.

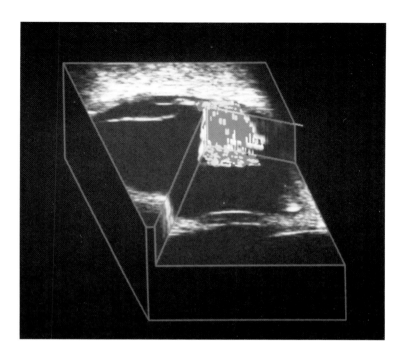

PLATE 2B

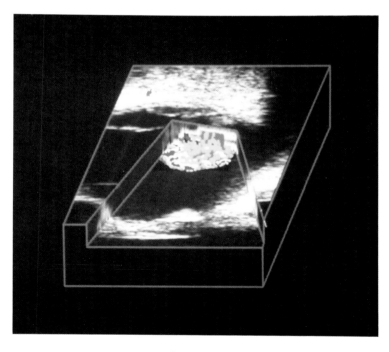

PLATE 2C

Garra et al.[53] estimated the average backscatter amplitude from regions of interest corrected for the overlying attenuation and found a strong correlation between increased attenuation and increased echogenicity in fatty livers. King et al.[77] calculated the y-intercept of a linear fit to an amplitude vs. frequency plot in a region of interest for use as a tissue characterization feature. This feature, known as the zero megaHertz amplitude (ZMA) is a measure of backscatter amplitude and was elevated in patients with severe alcoholic cirrhosis. Since the extent of fatty infiltration present in the alcoholics was unknown, fatty infiltration may have been the cause of the ZMA elevations. It is well known that the increased echogenicity of fatty liver infiltration is proportional to the lipid content of the tissue[91] and is pronounced enough to be readily visible on B-scans in many cases. On the other hand, Lyons and Parker[92] have shown that in many tissues absorption is the predominant contributor to attenuation. This does not preclude the possibility that a process such as fatty infiltration could cause an *increase* in attenuation primarily by increasing backscattering. Alternatively the disease process may increase both backscattering and attenuation by different mechanisms.

Lizzi et al.[93] have studied from a theoretical standpoint the relationship between backscatter amplitude and frequency used by King in earlier tissue characterization studies. By using spherical and Gaussian correlation models to describe backscatter, theoretical curves of the relationships between spectral slope (in dB/MHz), spectral intercept (ZMA) and effective scatterer size were generated (Figure 17). The spectral slope is dependent on the attenuation of the tissue (and overlying tissue) and on the effective scatterer size. In addition to being affected by the scatterer size, ZMA is also affected by system factors, the scatterer concentration, and the acoustic impedance differences between the scatterers and the medium. The relationships were reported as follows:

$$Y = Y_1 + 10 \log CQ^2$$

and

$$M = M_1 - 2\alpha R$$

where Y is the spectral intercept (ZMA), M is the spectral slope, α is the effective attenuation coefficient, and R is the distance to the transducer. Y_1 and M_1 depend on the scatterer diameter with

$$Y_1 \approx -38 - 3M_1$$

for scatterer diameters between 0.2 and 0.4 mm.[94] In the eye, where intervening attenuation is negligible, direct estimates of scatterer size and of the CQ^2 parameter could be made.[95] In the liver, where intervening attenuation is present, no absolute estimates of the scatterer size were made, although

Spectral Slope and Intercept
vs. Scatterer Size

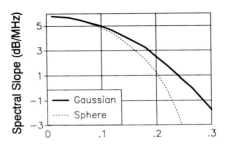

Scatterer Diameter (mm)

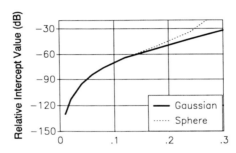

FIGURE 17. The theoretical relationship between scatterer diameter and two different spectral parameters for a model using spherical scatterers and for a model using Gaussian scatterers. (Adapted from Lizzi, F. L. et al., *IEEE Trans. Ultras. Ferroelectr. Frequency Control,* 34, 319–329, 1987.)

the authors were able to produce parametric images based on spectral slope, ZMA, and a third feature ($S_{(3)}$) dependent only on α and CQ^2.

Another method of investigating the frequency-dependent characteristics of backscatter is to perform narrow-band filtration on the radio frequency ultrasound signal.[96] Because backscatter is more strongly dependent on frequency for smaller scattering particles, the presence of larger scatterers will increase the relative amount of scattering at low frequencies. Narrow-band filtration can be used to examine the frequency-dependent backscatter of two different tissues at a frequency where the effects of different scatterer sizes produce a measurable difference in backscatter amplitude. In the method employed by Sommer et al.[97] on ten normal livers and ten cirrhotic livers, radio frequency data were acquired with a 3-MHz transducer to a standardized dynamic range from a region of interest 5 to 12 cm in depth and 30 to 50 A-

TABLE 13

Group	Number	Mean A_o	SD (range)
Healthy volunteers	12	0.375	0.087 (.289–.515)
Liver metastases	4	0.079	0.046 (.027–.119)

lines wide. The radio frequency data were then filtered with a narrow-band digital filter (center frequency 3.4 MHz, 800 KHz bandwidth), envelope detected and displayed as bistable images. With proper thresholding, the difference between normal liver and cirrhotic liver was readily apparent on the images. For normal liver $18 \pm 4.0\%$ of the pixel values were above the threshold (displayed as bright) while cirrhotic livers demonstrated a significantly higher percentage of pixel values above the threshold ($31 \pm 10\%$). Potential difficulties of the method are the lack of correction for overlying attenuation and lack of correction for focusing/diffraction effects. The authors used an unfocused transducer to minimize these effects but such an approach would be less feasible in a busy clinical setting where B-scan image quality would suffer. It is also unclear how much more complex the filtering method would become if it were required to both detect and distinguish between several types of liver disease.

4. Estimation of Physical Properties in Liver Tissue

Backscattered ultrasound has also been used to evaluate the "hardness" of tissues. Since many benign and malignant neoplasms are firmer than normal liver tissue, these tissues should exhibit different responses to externally applied pulsations than would normal liver parenchyma. A number of methods have been proposed to evaluate tissue motion. Many of these are briefly reviewed by Tristam et al.[30] These authors calculated the correlation coefficient (R') for segments of two temporally separated A-lines gathered from an M-mode display obtained from a region of interest in the liver. The time interval used was 1/10 of a cardiac cycle. In this analysis, plots of R' vs. time gave a complex pattern of peaks with the peaks in normal liver (12 healthy volunteers) being higher and more numerous than those obtained from liver metastases (four patients). The authors found the area under the correlation plots (A_o) to be a good separator of normal from abnormal liver tissue (Table 13).

In a later paper, the authors[31] performed a more detailed analysis of the correlation pattern using Fourier transforms. The plot of R' as a function of time was represented as a Fourier series

$$R'(t) = A_0 + \sum_{n=1}^{K} (A_n \cos nt + B_n \sin nt) + e(t)$$

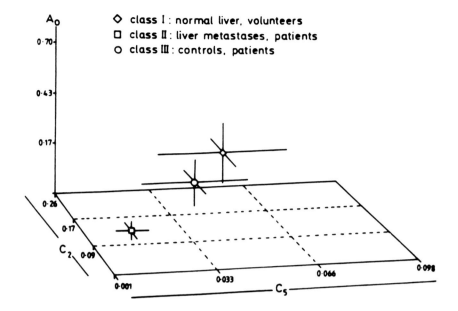

FIGURE 18.   Mean values and standard deviations of features A_0, C_2, and C_5 for normal liver, liver metastases, and controls. (From Tristam, M. et al., *Ultras. Med. Biol.*, 14, 695–707, 1988. With permission.)

where

$$A_0 = \frac{1}{2\pi} \int_0^{2\pi} R'(t) \, dt$$

$$A_n = \frac{1}{\pi} \int_0^{2\pi} R'(t) \cos nt \, dt$$

$$B_n = \frac{1}{\pi} \int_0^{2\pi} R'(t) \sin nt \, dt$$

and

$$C_n = \sqrt{A_n^2 + B_n^2}$$

is the amplitude of the nth Fourier harmonic. Fourier series with K = 20 were calculated resulting in 21 quantitative features describing each correlation pattern (A_0, C_1, C_2, . . . , C_{20}). The authors studied 10 healthy volunteers and 11 patients with liver metastases and were able to distinguish three types of correlation patterns: normal liver, metastasis, and uninvolved liver in patients with metastatic disease. Use of all 21 features was found to be unnecessary. Statistically significant separation of metastases from normal liver or uninvolved liver was achieved using only three features (Figure 18). They

90. **Ophir, J. and Yazdi, Y.,** A transaxial compression technique for localized pulse-echo estimation of sound speed in biological tissues, *Ultras. Imag.,* 12, 35–46, 1990.

91. **Freese, M. and Lyons, E. A.,** Ultrasonic backscatter from human liver tissue: its dependence on frequency and protein/lipid composition, *J. Clin. Ultras.,* 5, 307–312, 1977.

92. **Lyons, M. E. and Parker, K. J.,** Absorption and attenuation in soft tissues II-experimental results, *IEEE Trans. Ultras. Ferroelectr. Frequency Control,* 35, 511–521, 1988.

93. **Lizzi, F. L., Ostromogilsky, M., Feleppa, E. J., Rorke, M. C., and Yaremko, M. M.,** Relationship of ultrasonic spectral parameters to features of tissue microstructure, *IEEE Trans. Ultras. Ferroelectr. Frequency Control,* 34, 319–329, 1987.

94. **Lizzi, F. L., King, D. L., Rorke, M. C., Hui, J., Ostromogilsky, M., Yaremko, M. M., Feleppa, E. J., and Wai, P.,** Comparison of theoretical scattering results and ultrasonic data from clinical liver examinations, *Ultras. Med. Biol.,* 14, 377–385, 1988.

95. **Feleppa, E. J., Lizzi, F. L., Coleman, D. J., and Yaremko, M. M.,** Diagnostic spectrum analysis in ophthalmology: a physical perspective, *Ultras. Med. Biol.,* 12, 623–631, 1986.

96. **Sommer, F. G., Stern, R. A., Howes, P. J., and Young, H.,** Envelope amplitude analysis following narrow-band filtering: a technique for ultrasonic tissue characterization, *Med. Phys.,* 14, 627–632, 1987.

97. **Sommer, F. G., Stern, R., and Chen, H.,** Cirrhosis: US images with narrow band filtering, *Radiology,* 165, 425–430, 1987.

98. **Taylor, K. J. and Milan, J.,** Differential diagnosis of chronic splenomegaly by gray scale ultrasonography. Clinical observations and digital A-scan analysis, *Br. J. Radiol.,* 49, 519–525, 1976.

99. **Siler, J., Hunter, T. B., Weiss, J., and Haber, K.,** Increased echogenicity of the spleen in benign and malignant disease, *AJR,* 134, 1011–1014, 1980.

100. **Sommer, F. G., Hoppe, R. T., Fellingham, L., Carroll, B. A., Solomon, H., and Yousem, S.,** Spleen structure in Hodgkin disease: ultrasonic characterization, *Radiology,* 153, 219–222, 1984.

101. **Sommer, F. G., Joynt, L. F., Hayes, D. L., and Macovski, A.,** Stochastic frequency-domain tissue characterization: application to human spleens 'in vivo', *Ultrasonics,* 20, 82–86, 1982.

102. **Itoh, K., Yasuda, Y., Suzuki, O., Itoh, H., Itoh, T., Jing-Wen, T., Konishi, T., and Koyano, A.,** Studies on frequency-dependent attenuation in the normal liver and spleen and in liver diseases, using the spectral-shift zero-crossing method, *J. Clin. Ultras.,* 16, 553–562, 1988.

103. **Manoharan, A., Robinson, D. E., Wilson, L. S., Chen, C. F., and Griffiths, K. A.,** Ultrasonic characterisation of splenic tissue: A clinical study in patients with myelofibrosis, in *Proc. 4th Meet. WFUMB,* Gill, R. W. and Dadd, M. J., Eds., Pergamon Press, Sydney, 1985, 113, (abstr.).

104. **Nicholas, D.,** Evaluation of backscattering coefficients for excised human tissues: results, interpretation and associated measurements, *Ultras. Med. Biol.,* 8, 17–28, 1982.

105. **Friedman, P. A., Sommer, F. G., Chen, H. S., Rachlin, D. J., and Hoppe, R.,** Characterization of splenic structure in Hodgkin disease by using narrow-band filtration of backscattered ultrasound, *AJR,* 152, 1197–1203, 1989.

Chapter 12

IN VIVO OPHTHALMOLOGICAL TISSUE CHARACTERIZATION BY SCATTERING

Frederic L. Lizzi and Ernest J. Feleppa

TABLE OF CONTENTS

0-8493-6568-6/93/$0.00 + $.50

I. INTRODUCTION

A patient presenting with a small, solid subretinal mass may have one of a variety of neoplastic conditions including (1) benign nevi and hemangiomas, (2) primary malignant, choroidal melanomas of different potential for growth, invasion, and metastasis, and (3) several types of carcinomas metastatic to the eye from other sites. Some of the possible conditions are life threatening, and some forms of melanoma can become rapidly invasive. Each of these possible conditions requires a different response, and biopsy of intraocular masses is rarely considered possible. An urgent need exists for an accurate, noninvasive method of differentially diagnosing such masses. Furthermore, some cases may warrant treatment using ionizing radiation or ultrasonic hyperthermia. In such cases, a reliable, quantitative means for monitoring the effect of therapy may greatly increase its efficacy. Ultrasonic tissue characterization based on spectrum analysis of backscattered ultrasound has shown the potential to provide the information needed for effective diagnosis and monitoring.

This chapter provides an overview of the evolution of our ophthalmological tissue-characterization studies, describes data-acquisition and data-processing system features, discusses the underlying theoretical framework, and presents a summary of clinical applications.

II. BACKGROUND AND OVERVIEW

In the early 1970s, our laboratories initiated our collaborative investigations of ophthalmic techniques with Dr. D. Jackson Coleman and colleagues (now at the Cornell University Medical College). These early studies were motivated by the work of Sigelmann and Reid on basic scattering phenomena related to tissues,[1] of Waag, Lerner, and Gramiak on the manifestations of tissue differences in power spectra,[2] of Lele and Namery on computer analysis of ultrasonic echo signals,[3] and of Purnell et al. on ways of displaying the results of spectrum analysis in color-coded images.[4] Our very earliest studies utilized a dithering technique during scanning to provide necessary averaging, an analog spectrum analyzer to compute power spectra of tissue and calibration targets, a manual digitization of output spectral plots, and a mainframe computer for spectral normalization and computation of spectral parameters.[5-10] Our data were obtained using a 10-MHz, sector-scanning, water-immersion instrument. Although we currently employ real-time contact scanners, an upgraded version of the original, water-immersion configuration is frequently utilized for various special studies, such as three-dimensional (3-D) scanning, in our present investigations.

Following these early studies, we implemented our tissue-characterization methods using minicomputer technology and high-speed, transient-waveform digitizers. The general configuration of this sytem is shown in Figure 1.

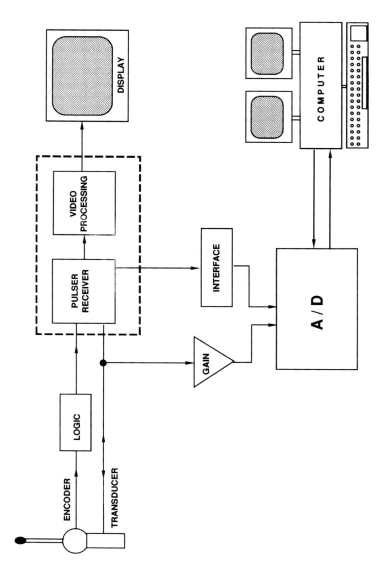

FIGURE 1. Importance of constant depth water path. (A) Uniform echo amplitude levels in the phantom are present when water path depth is constant. (B) Echo amplitude levels increase with increasing water path depth because of TGC effects.

Typically, these studies utilized data acquired with 6-bit samples acquired at 100 MHz. Each scan consisted of 100 scan lines and 1024 samples per line; each digitized scan line spanned approximately 7.5 mm; the scan extended over an arc of approximately 22 mm. Since the 100-MHz sample rate provided ample oversampling, radio frequency data were averaged to obtain an additional bit of signal resolution. After scanning, the computer generated B-mode, gray-scale images, to verify proper data acquisition, and to guide data processing. The analyst then examined this image and defined a rectangular region of interest for analysis. Analysis started with computation of normalized power spectra and other methods described below.

Our current methods are significantly improved over the earlier versions; however, they are based on the same basic principles and operations. Today, typical ophthalmic data acquisition utilizes 2048 samples per scan line and 8-bit sampling at a frequency of 50 MHz. These parameters acquire data over the full depth of the eye, i.e., over an area of approximately 30×30 mm. A typical image analysis region and computed spectrum are illustrated in Figure 2.

The primary objective of our ophthalmic studies was to improve the diagnosis of malignant ocular disease. We found that spectrum-analysis methods applied to intraocular tumors are capable not only of distinguishing malignant from benign disease, but also metastatic from primary tumors, and spindle from mixed or epithelioid primary melanomas within the eye. A database of over 2000 confirmed cases has been developed and provides an objective basis for assigning probabilities that an unknown tumor is one of the following four disease categories: metastatic tumor, spindle melanoma, mixed/epithelioid melanoma, and hemangioma. In addition, tumors can be followed over long periods of time (e.g., months or years) and compared quantitatively to a baseline scan to evaluate changes associated with disease progression or therapeutic efficacy; in such cases, changes can be detected and evaluated even though the lesion image features (such as echogenicity, size, or texture) shown by conventional methods remain unaltered.[11-14]

Our studies also have developed a theoretical framework capable of relating the spectra computed from acquired data to the scattering properties of tissue and the diffraction properties of the ultrasound beam.[15] Effective scatterer size and a property termed "acoustic impedance" (the product of scatterer concentration and the square of relative scatterer impedance) are two scattering properties that can be evaluated using this framework. The framework has been fruitfully applied to studies of liver parenchyma and tumor microstructure, and to diagnosis, treatment selection, and treatment monitoring in ophthalmology.[16,17]

III. DATA ACQUISITION AND PROCESSING

As shown in Figure 1, the acquisition system consists of a conventional, high-quality ophthalmic scanner; interface components for position encoding,

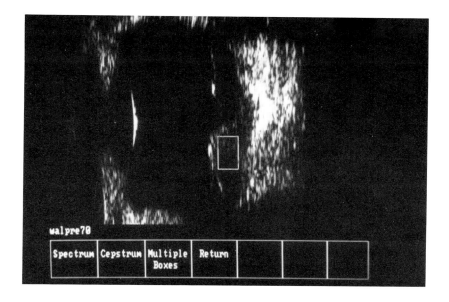

A

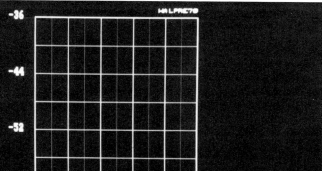

B

FIGURE 2. Spectrum of an ophthalmic tumor. (A) The computer-generated image on which the analysis region (rectangular box) is specified; the dimensions of the area portrayed in the image are approximately 30 × 30 mm. (B) The output spectrum with the associated straight-line approximation.

triggering and signal amplification; a high-speed transient recorder for digitization; and an AT-compatible, desk-top computer system with associated monitors. The processing system requires only the computer system and monitors.

Data acquisition is performed in the course of standard clinical scanning. The conventional ultrasonic scanner is utilized to identify volumes of interest (e.g., a tumor) within the eye or orbit. If a single scan plane is to be examined, the transducer is placed in the position of the first scan line, the data-acquisition process is enabled using the computer keyboard, the plane of interest is scanned, and the radio frequency echo-signal of each scan line is digitized and immediately transferred to extended memory within the computer.

Radio frequency data contain information on a fine scale that is lost during the process of envelope detection used to produce video images. In addition, the use of radio frequency data enables us to compensate for system properties that otherwise make ultrasonic scanning results heavily system dependent; by allowing correction for system factors including instrument settings, radio frequency data make results obtained using spectrum analysis independent of the operator and the instrument.

As stated above, our current configuration acquires data at a rate of 50 MHz; covers a 30-mm segment of each scan line; and each digitized scanline segment includes 2048, 8-bit samples, i.e., 2 KBytes of data are acquired on each scan line. Each scan plane includes 128 lines, so that 256 KBytes are acquired for each scan. The computer immediately generates a B-mode image from the digitized radio frequency data for each acquired scan to provide verification of data quality. In addition to the standard operating mode, we have implemented a 3-D scanning capability utilizing computer-controlled stepper motors to linearly scan a set of parallel, closely spaced (e.g., separated by 0.5 mm) planes. This 3-D capability is applied when full-volume data and a spatial perspective are required.

Data-acquisition software, termed DIGITIZE, controls data acquisition. DIGITIZE is menu driven; incorporates fields for file-header data (such as patient-identification number, calibration file name, distance to the first sample, etc.); sets parameters for the waveform digitizer (such as sampling frequency and delay time to the first sample); creates data files; and displays images of acquired data. It also can display A-mode traces of radio frequency echo signals (magnified by selectable factors of 1, 2, or 4) prior to digitization.

After data are acquired, a second software package, termed GENSPEC, generates higher-quality B-mode images and performs interactive analysis of acquired data. Upon activation, GENSPEC displays a menu containing a list of file numbers (filed by patient-identification number); using a mouse, the analyst selects a file of interest. This software first generates a set of well-defined gray-scale images of all acquired scans for the selected patient, and displays them as a montage. Again using a mouse, the analyst selects an image of interest, and GENSPEC displays it in an enlarged rectilinear form. The procedures used for data analysis are described below.

IV. THEORETICAL BASIS AND ANALYTIC PROCEDURES

As stated previously, the theoretical basis of tissue characterization by scattering is described in several references.[15-17] This paper summarizes the theoretical framework of scattering by tissue, then describes the manner in which this framework is applied to ophthalmological tissue characterization.

A. SUMMARY OF THEORETICAL FRAMEWORK

As shown in our original theoretical framework,[15] the use of focused transducers generates echo signals within or beyond the focal zone, resulting from scattering of ultrasound by tissue at sites where changes in acoustical impedance occur. Scattering results from large, extended structures such as the interface between the retina and the vitreous, or from microscopic structures such as blood vessels or melanin-filled cells in the choroid, i.e., from entities having sizes on the order of the wavelength of the incident ultrasound. Changes in acoustic impedance result from changes in mass densities and compressibility. In soft tissue, these impedance differences are small, and scattering is termed "weak", i.e., echo signals derived from ultrasonic energy scattered back to the transducer arise from single, rather than multiple, scattering events. As described in the references, the "calibrated" or "normalized" spectrum of ultrasound backscattered under these conditions is dependent on three spatial, autocorrelation functions: $R_Q(\Delta x)$, $R_D(\Delta y, \Delta z)$, and $R_G(\Delta x)$. $R_Q(\Delta x)$ is the spatial autocorrelation function of the relative acoustic impedance, Q; $R_D(\Delta y, \Delta z)$ has the form of an autocorrelation function representing the two-way directivity function, $F^2(y,z)$, of the ultrasonic beam; and $R_G(\Delta x)$ is similarly described as the autocorrelation function of the gating function, g(x), used to window the echo signals prior to analysis. The coordinates are Δx, which represents the set of three-dimensional, lagged coordinates $(\Delta x, \Delta y, \Delta z)$; here x represents the range (direction of propagation) dimension; y and z represent the cross-range (perpendicular to the propagation axis) dimensions. The relative acoustic impedance, Q, is defined as the acoustic impedance, Z, of the scatterers divided by the mean acoustic impedance, Z_o, of the tissue in the vicinity of the scatterer. $F^2(y,z)$ describes the beam directivity function in the focal zone where the acoustical wavefront can be approximated as a plane wave, which is a valid approximation in ophthalmology.

Consequently, the spectrum of ultrasound backscattered from tissue is a function of three parameters: R_Q, which is dependent on tissue properties, specifically, the spatial distribution of acoustical impedance; R_D, which is dependent on beam properties that can be measured or approximated; and R_G, which can be calculated from the gating function (e.g., Hamming window) used in processing. (A Hamming window effectively multiplies the signal by a squared cosine function with its maximum value, 1, in the center of the

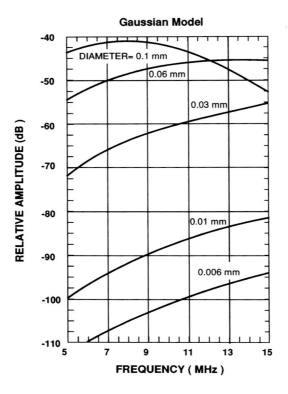

FIGURE 3. Theoretically predicted spectra. A set of theoretically predicted spectra, for scatterer sizes ranging upward from 6 μm, shows an increase of spectral amplitude with increasing size. However, at some point, slope decreases with further increases in size, and while intercept continues to increase, spectral amplitude at the high-frequency portion of the band may in fact decrease below the spectral amplitude of smaller scatterers.

windowed region falling smoothly to minima at the start and end of the window.)

Figure 3 shows predicted spectra for a Hamming window and a typical ophthalmic transducer (e.g., 35-mm focal length and 10-mm aperture); a Gaussian autocorrelation function is assumed for R_Q, where the effective scatterer size is the diameter at which the value of the autocorrelation function is 1/e. Spectral amplitude clearly increases with increasing scatterer size; however, it also is linearly proportional to CQ^2, or "acoustic concentration", where C is scatterer volume concentration and Q (as defined above) is the relative acoustic impedance of the scatterers.

Although the spectra predicted by theory are curved, they can be approximated by straight lines over the 10-MHz bandwidth typically used in a high-quality ophthalmic system. The straight-line approximating the curved spectrum is computed by least-squares regression, and the line is defined in

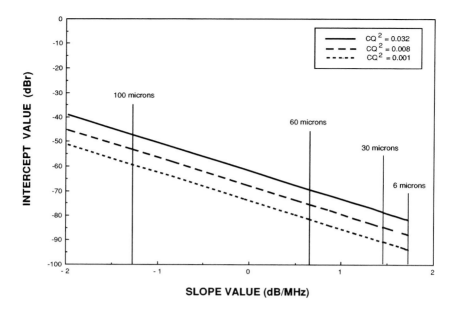

FIGURE 4. Theoretical dependence of slope and intercept on size and acoustic impedance. Intercept value (dBr) is plotted vs. slope (dB/MHz) for CO^2 values of 0.001, 0.008, and 0.032 mm^{-3}; four points are indicated for scatterer sizes of 6, 30, 60 and 100 μm. These curves are derived for a high-quality ophthalmic transducer of 35-mm focal length, 10-mm aperture, 10-MHz center frequency, and 10-MHz bandwidth.

terms of its slope, and intercept values, m and Y_O, respectively; the slope and intercept have proven to be very valuable in characterizing ophthalmic tissues. The curve of Figure 4 shows the predicted dependence of these two parameters on size and acoustic impedance for the frequency range employed in ophthalmology. These curves are derived for a typical ophthalmic transducer having a focal length of 35 mm, an aperture of 10 mm, and a center frequency of 10 MHz.[17] Acoustic concentration is expressed as scatterers per volume; its units are mm^{-3}. The curves are shown corresponding to CQ^2 values of 0.001, 0.008, and 0.032 mm^{-3}. Size is expressed as the diameter of the corresponding Gaussian autocorrelation function; its units are microns. Four values of size 6, 30, 60, and 100 μm are noted on each curve. As scatterer size increases, intercept value increases and slope value decreases. As acoustic-concentration value increases, intercept value increases. Accordingly, spectra computed from digitized radio frequency echo-signal data can be defined in terms of the values of the slope, m, and intercept, Y_O, parameters of the associated straight-line approximation, and these in turn provide a basis for estimating scatterer size and acoustic concentration. However, since attenuation also affects slope, slope can be used to estimate scatterer size only in the absence of intervening tissue, or if it is present, then after correction is made for known attenuation. These relationships are shown in the equation below, which was derived for the typical ophthalmic transducer described above (35-mm focal length, 10-mm aperture, 10-MHz effective bandwidth, and 10-MHz center frequency).

$$10 \log(CQ^2) = Y_0 + 11.5m + 47$$

$$d = 24(10 - 5.75m)^{1/2}$$

In these equations, CQ^2 is expressed in dB, and the effective scatterer diameter, d, is expressed in microns.[17]

B. PROCEDURES FOR TISSUE ANALYSIS

As stated above, the echo-signal voltages generated by the transducer from tissue backscattering are amplified by known amounts, digitized, and stored on hard disk (with tape backup). Data acquisition on each scan line is delayed to allow time for signals to return from the focal zone of the transducer before acquiring the first sample. (Beam inhomogeneities proximal to the focal zone preclude uniform tissue insonification and prevent meaningful signal analysis.) In addition to acquiring echo signal data from tissue, echo signals are digitized from a reference target (such as an optically flat glass plate) placed in the focal zone of the transducer. These signals serve to calibrate or normalize the tissue spectrum by providing the transfer function of the system. DIGITIZE controls data acquisition and generates echo-signal data files and their headers, which contain scan parameter values and comments.

Interactive processing includes the following steps:

1. Run GENSPEC. Upon selection, the program lists a menu of patient-scan files in the disk directory.
2. Select a case of interest from the GENSPEC menu. The program displays a montage of images of the scans for that patient.
3. Select a scan plane of interest from the displayed image montage. The program displays an enlarged, high-quality image of the selected scan with a menu of processing-mode options below the image.
4. Select a processing mode from the menu displayed below the image; this menu includes a variety of methodologies, including power-spectrum analysis and A-scan display. The program displays a cursor superimposed on the image for the purpose of defining a region of interest (ROI).
5. Define the scan-line segments to be analyzed by specifying the size and location of the ROI on the image. The program displays a rectangle showing the ROI from which stored radio frequency data will be selected for processing; it also displays numbers indicating region size and location coordinates.

are the current norm in medicine, but the implementation of 3-D scanning, data acquisition, and processing for tissue characterization permits presentation of information in 3-D. Color or gray-scale encoding provides this information in easy-to-interpret form enabling the clinician to visualize the internal properties as well as their anatomy, i.e., the 3-D extent of lesions and their 3-D spatial relationships with other normal and disease entities. For example, 3-D parameter images of an unknown lesion may, by virtue of clear depiction of boundary and shape features as well as parameter-value distribution, aid in differential diagnosis and treatment selection. Subsequently, the treated lesion may show no change in tumor volume, shape, or other macroscopic qualities, but may present significant changes in internal properties indicative of successful therapy.

VI. CONCLUSION

This chapter has presented an overview and summary of spectrum analysis of backscattered echo signals as a basis for ultrasonic tissue characterization in ophthalmology. A theoretical framework and extensive experimental and statistical effort have fostered the development of clinically useful methods for use in differential diagnosis of ophthalmic disease, primarily intraocular neoplasms, and in monitoring observed as well as treated lesions. However, much work remains to be done in improving clinical methods, refining computer analyses, and developing new data-presentation modes if the full potential of these techniques is to be realized.

ACKNOWLEDGMENTS

The studies described in this chapter were supported by National Institutes of Health Grants EY01212, EY03183, and RR05853. Furthermore, these studies were made possible by the collaborative participation of D. Jackson Coleman and colleagues, particularly Ronald Silverman and Mark Rondeau, in the Ophthalmology Research Department at Cornell University Medical College. We also thank our colleagues at Riverside Research Institute for their on-going participation and contributions to the research, and specifically thank Mary Rorke and Joan Sokil-Melgar for their help and suggestions regarding this manuscript.

REFERENCES

1. **Sigelmann, R. and Reid, J.**, Analysis and measurement of ultrasound backscattering from an ensemble of scatterers excited by sine-wave bursts, *J. Acoust. Soc. Am.*, 5, 13351–13355, 1973.
2. **Waag, R. C., Lerner, R. M., and Gramiak, R.**, Swept-frequency ultrasonic determination of tissue macrostructure, in Ultrasonic Tissue Characterization, Linzer, M., Ed., U.S. Government Printing Office, Washington, D.C., 1976.
3. **Lele, P. and Namery, J.**, A computer-based ultrasonic system for the detection and mapping of myocardial infarcts, *Proc. San Diego Biomed. Symp.*, 13, 121–132, 1974.
4. **Purnell, E. W., Sokollu, A., Holasek, E., and Cappeart, W.**, Clinical spectra-color ultrasonography, *J. Clin. Ultras.*, 3, 187–189, 1975.
5. **Coleman, D. J. and Lizzi, F. L.**, Computer-processed acoustic spectral analysis of ophthalmic tissues, *Trans. Am. Acad. Ophthalmol. Otolaryngol.*, 83, 725–730, 1977.
6. **Lizzi, F. L., Laviola, M. A., and Coleman, D. J.**, Ultrasonic tissue characterization utilizing spectrum analysis, *Proc. Soc. Photo-Optical Instrument. Eng.*, 96, 322–328, 1976.
7. **Lizzi, F. L., Laviola, M. A., and Coleman, D. J.**, Tissue signature characterization using frequency-domain analysis, *1976 IEEE Ultras. Symp. Proc.*, 76, 714–719, 1976.
8. **Lizzi, F. L., Saint Lewis, L., and Coleman, D. J.**, Applications of spectral analysis in medical ultrasonography, *Ultrasonics*, 14, 77–80, 1976.
9. **Coleman, D. J. and Lizzi, F. L.**, Computerized ultrasonic tissue characterization of ocular tumors, *Am. J. Ophthalmol.*, 96, 165–175, 1983.
10. **Lizzi, F. L., Feleppa, E. J., and Coleman, D. J.**, Ultrasonic tissue characterization, in *Characterization of Tissue with Ultrasound*, Greenleaf, J., Ed., CRC Press, Boca Raton, FL, 1986, 41–46.
11. **Silverman, R. H., Coleman, D. J., Lizzi, F. L., Topey, J. H., Driller, J., Iwamoto, T., Burgess, S. E. P., and Rosado, A.**, Ultrasonic tissue characterization and histopathology in tumor xenografts following ultrasonically induced hyperthermia, *J. Ultras. Med. Biol.*, 12, 639–645, 1986.
12. **Feleppa, E. J. and Yaremko, M. M.**, Ultrasonic tissue characterization for diagnosis and monitoring, *IEEE Eng. Med. Biol. Mag.*, 6, 18–26, 1987.
13. **Coleman, D. J., Rondeau, R. H., Silverman, R. H., and Lizzi, F. L.**, Computerized ultrasonic biometry and imaging of intraocular tumors for the monitoring of therapy, *Trans. Am. Ophthalmol. Soc.*, 85, 49–81, 1987.
14. **Feleppa, E. J., Lizzi, F. L., and Coleman, D. J.**, Ultrasonic analysis for ocular tumor characterization and therapy assessment, *News Physiol. Sci.*, 3, 193–197, 1988.
15. **Lizzi, F. L., Greenebaum, M., Feleppa, E. J., and Elbaum, M.**, Theoretical framework for spectrum analysis in ultrasonic tissue characterization, *J. Acoust. Soc. Am.*, 73(4), 1366–1373, 1983.
16. **Feleppa, E. J., Lizzi, F. L., Coleman, D. J., and Yaremko, M. M.**, Diagnostic spectrum analysis in ophthalmology: a physical perspective, *Ultras. Med. Biol.*, 62(8), 623–631, 1986.
17. **Lizzi, F. L., Ostromogilsky, M. O., Feleppa, E. J., Rorke, M. C., and Yaremko, M. M.**, Relationship of ultrasonic spectral parameters to features of tissue microstructure, *IEEE Trans. Ultras. Ferroelectr. Frequency Control*, UFFC-34(3), 319–329, 1987.
18. **Coleman, D. J., Silverman, R. H., Rondeau, M. J., Lizzi, F. L., McClean, I. W., and Jakobiec, F. A.**, Correlations of acoustic tissue typing of malignant melanoma and histopathological features as a predictor of death, *Am. J. Ophthalmol.*, 110, 380–388, 1990.

Chapter 13

FETAL LUNG TISSUE CHARACTERIZATION BY SCATTERING

Gary A. Thieme, Paul L. Carson, Charles R. Meyer, and Richard Bowerman

TABLE OF CONTENTS

I. INTRODUCTION

In 1986, Carson et al.[1] presented a review of ultrasound tissue characterization for determining fetal lung maturity in a CRC Press book entitled *Tissue Characterization with Ultrasound*. The present chapter is an up-to-date review of this subject, including some unpublished results.

II. BACKGROUND

As a consequence of obstetric problems leading to preterm delivery, about 50,000 newborns are at risk for complications of prematurity in the U.S. each year. One of the most common life-threatening problems encountered is respiratory distress syndrome (RDS). A newborn with this condition has physiologically immature lungs which cannot support adequate gas exchange without medical intervention. The developmental aspects of the human lung are summarized as follows.[2-7]

The pathophysiologic complex known as respiratory distress syndrome occurs when surface-active compounds are not present in sufficient amounts for alveoli to remain open at the end of expiration.[1] The lung collapses and can only be opened for further gas exchange by the application of high positive pressure. Normal lung remains open at the end of expiration because surfactants (surface-active compounds) lower surface tension on the alveolar surfaces and allow residual air to remain in the individual alveoli. For normal lung, only a small increase in pressure is needed to expand the alveoli during inspiration. Type II cells lining the alveoli of normal newborn lung must synthesize surface active compounds at a high rate. The enzyme systems of immature lung cannot produce surfactant fast enough to prevent atelectasis (collapse of alveoli).

The biochemical component of fetal lung maturation is surfactant production. The anatomic component of fetal lung maturation is the development of airways and alveoli with appropriate fibroelastic components. Structural development of the lung progresses through three stages.[3] During the *glandular* stage (first 16 weeks) the lobes of the lungs become well demarcated and bronchi and bronchiole airway divisions from 20 to 32 generations (branchings) develop. The cells lining the airways are thick and columnar proximally and change to cuboidal peripherally. During the *canalicular* stage from 16 to 24 weeks, distal airway development occurs in the form of respiratory bronchiole branching and vascular proliferation at the ends of airways. The cells in these distal airways change from cuboidal proximally to thinner flattened epithelial cells distally. The lungs are not yet capable of respiratory function. During the *alveolar* stage from 24 weeks to term, respiratory tissue begins to appear at the ends of respiratory bronchioles as alveolar sacs and eventually small alveoli. Flat type I epithelial cells form the lining of the alveoli walls and are thin enough to allow gas exchange by diffusion between the air-filled alveolus space and the capillary blood vessel network in the wall. In the fetus, though, the alveoli are filled with fluid. During this stage, respiration can occur in a premature newborn if surfactant production by type II cells is sufficient to lower surface tension and maintain open airspaces. A term newborn (37 to 40 weeks) will have about 24 million alveoli each with a diameter of about 50 μm and will have a mature enzyme system for the production of surfactant.

Anatomic development of fetal lung seems to be closely related to gestational age, while biochemical maturity can occur as early as 28 weeks or as late as term. Accelerated maturation can occur under *in utero* circumstances where the fetus is stressed (e.g., intrauterine growth retardation). Delayed lung maturation is a recognized risk in maternal diabetes mellitus. Both accelerated and delayed maturation seem to be directly related to biochemical maturity and only indirectly linked to anatomic development. It appears that a baseline anatomic maturity of at least 26 weeks is necessary for respiratory function, however.

The variable relationship between lung maturity and gestational age can be related to the two enzyme systems responsible for the production of surface active compounds.[2] The methyl transferase enzyme system begins to function at about 24 weeks and makes it possible for a prematurely born infant to sustain respiration. However, this system is not very robust and is adversely affected by conditions associated with prematurity, such as acidosis, hypothermia, and hypoxia. The functions of modern neonatal intensive care units attempt to prevent these complications and to assist the enzyme system. At about 35 weeks the more important phosphocholine transferase enzyme system appears. Its development parallels the maturation of the respiratory portion of the lung and would seem to explain why RDS rarely occurs in newborns whose gestational age is greater than 36 weeks.

Prediction of fetal lung maturity becomes important when potential premature delivery is a clinical problem. Basically, if the fetal lungs are both structurally and biochemically sufficiently mature to sustain the newborn with minimal or no respiratory support, then the clinical benefits and economic costs of prolonging the pregnancy (e.g., in a premature labor situation) are no longer justified. However, if the fetal lungs are immature, then the risks and costs of prolonging the pregnancy can be justified in terms of reduced newborn complications and neonatal intensive care unit costs. Thus, the clinical management strategy can be altered to reduce the incidence, risks, and complications of RDS in the newborn, when the fetal lung maturity status is known.

Methods for determining fetal lung maturity include physical measurements for fetal size, accurate determination of gestational age, placenta condition, and biochemical tests on amniotic fluid. Physical measurements of the fetus, such as biparietal diameter, head circumference, abdominal circumference, and femur length, can be used to estimate fetal weight. Unfortunately, these parameters are poor predictors of fetal lung maturity.[8,9] Mature lungs are usually associated with a grade III placenta pattern. However, exceptions occur, and this sign is frequently not present when lungs are known to be mature. Thus, this is just a loose association of probably coincidental events; there is no direct physiologic link between lung maturity and placenta pattern.[10,11] For pregnancies where the gestational age is accurately known from a crown-rump length measurement made during a first-trimester ultrasound examination between 8 and 12 weeks, a known gestational age of 37 weeks or greater virtually assures lung maturity. However, accurate gestational age knowledge is a poor predictor of lung maturity for situations of threatened premature delivery at less than 37 weeks. The incidence of RDS is about 20% at 35 weeks, 50% at 32 weeks, and greater than 70% below 30 weeks.[12]

Biochemical tests on amniotic fluid samples have been used to predict lung maturity state for nearly 20 years.[6,7,13] These tests assume that surfactants found in amniotic fluid accurately reflect the developmental and functional state of type II cells in fetal lung. The lecithin-sphingomyelin (L/S) ratio measures the rise in relative concentration with gestational age of a surface active compound lecithin with respect to the more constant sphingomyelin, which is not related to lung maturation. Gluck and Kulovich's original work showed that a ratio of 2.0 or greater indicates lung maturity since hyaline membrane disease did not occur in newborns with ratios greater than 2.[13] Subsequent studies have shown that a mature L/S ratio predicts the absence of RDS in 98% of newborns. About 50% of newborns with a ratio between 1.5 and 1.9 will develop RDS. Below 1.5 the risk of RDS increases to 73%. Obviously, many newborns with an immature L/S ratio will not develop RDS.[7]

Presence of phosphatidylglycerol (PG), another surface active agent, predicts lung maturity with virtually 100% reliability but does not appear until 35 weeks gestation. However, lungs are usually mature when the L/S ratio is mature and PG is absent. Thus, absence of PG in amniotic fluid does not mean that the lungs are immature. Another simpler measure of lung maturity, the shake test, can be done at the bedside by observing the amount of foam produced after physically shaking an aminiotic fluid sample mixed with ethanol in a tube. A positive test virtually excludes development of RDS in the newborn; however, normal lungs occur frequently when the test is negative.[7]

To obtain the benefits of biochemical determination of lung maturity state, a sample of amniotic fluid must be obtained by inserting a needle into the amniotic cavity through the maternal abdominal wall. This procedure is called amniocentesis and can usually be performed with minimal risk to the mother and fetus. However, an adequate pocket of fluid in an accessible location may not always be present. Under these circumstances, amniocentesis cannot be safely performed. In addition, amniocentesis may need to be performed repeatedly until an amniotic fluid sample indicating biochemically mature lung is obtained.

Despite the benefits of amniocentesis, this invasive test is less than ideal. A positive test is highly predictive of mature lungs; however, a negative test is frequently obtained when lungs are mature (rather than immature). A non-invasive method of determining fetal lung maturity state could eliminate the risks associated with amniocentesis, could be performed even when amniotic fluid volume is low, and could be done repeatedly without risk until lung maturity is established. Diagnostic ultrasound imaging might fulfill this role if sonographically detectable fetal lung features change from the immature state to the mature state and if the observed morphologic structural change is in synchrony with measurable biochemical changes.

Ultrasound cannot measure any of the biochemical parameters of fetal lung maturity, nor can ultrasound provide direct histologic information about fetal lung development. However, it is reasonable to assume that both morphologic changes and possibly biochemical changes associated with the transition to fetal lung maturity would alter the diffuse scattering and other propagation properties of fetal lung. These changes might be detectable by conventional or specially modified pulse-echo ultrasound systems.

As the fetal lung matures, the developing alveolus can be viewed as a sphere of tissue with a central cavity. During the transition from thick cuboidal epithelium to flat type I and type II epithelium, the wall of the sphere becomes thinner. The initially solid tissue volume changes to a fluid-filled space, whose diameter is about 50 μm or 0.05 mm. Since the wavelength of clinically useful diagnostic ultrasound varies from about 1.5 mm at 1 MHz to about 0.15 mm at 10 MHz, these 24 million fluid-filled aveloli should act as Rayleigh scatterers. The diffuse scattering properties should change as the fluid-content to tissue-content ratio changes. The macroaggregate effect of these morphologic changes should produce detectable changes in echogenicity and/or attenuation of fetal lung.[14] Fetal liver might be used as a comparison reference standard, since its histology should not change appreciably during gestation and since no changes in diffuse scattering or attenuation are anticipated. Support for these assumptions is based upon histologic and physiologic studies of human fetal lung and lamb fetal lung (similar developmental features). Thus, the expectation that ultrasound imaging may indirectly detect the morphologic and biochemical changes of fetal lung maturation is a reasonable one.

III. FETAL LAMB STUDIES

In 1978, at the University of Colorado Health Sciences Center, Michael Johnson suggested to the ultrasound fellow, Lawrence Mack, that ultrasound might be able to detect fetal lung maturity. Intermittent observation of the sonographic appearance of fetal lung over the next 2 years showed no consistent changes. During this time Carson observed from information on fetal lung morphology in the literature that alveolar bud structures might be large enough to be resolved by frequency dependent scattering at the highest diagnostically useful frequencies and that maturational changes might result in sonographic appearance changes.[1]

Experiments to observe the sonographic features of fetal lung during maturation were planned, and the study was initiated in 1980.[14] The fetal lamb was chosen as an experimental model, since it had been used extensively in pulmonary physiology research and since lamb lung progresses through the same stages of development as human lung. Also, the fetal lamb is similar in size to the human fetus at various stages of development.

To sonographically examine fetal lamb lung under optimal and reproducible conditions, the ewe was anesthetized, and the living lamb was delivered by caesarean section into a water bath at 37°C while still attached by its long umbilical cord to the ewe. The hairy skin was surgically removed from the right lateral chest and upper abdomen so that the right lung and liver could be scanned through the intact chest wall at a constant water path depth under relatively artifact-free conditions. A 3.5 or 5.0 MHz 19 mm diameter focused transducer at the end of the articulated scanning arm of a modified Rohnar Model 5580 B-scanner (Phillips Ultrasound, Santa Ana, CA) was attached to a ball-bearing slide on a steel scanning rod positioned parallel to the chest wall so that lung and liver were contained within the focal zone of the transducer. After obtaining both sonographic images and digital data of lung and liver, the ewe and lamb were euthanized so that the lamb lungs could be removed for correlative physiologic function tests (surface tension and pressure volume measurements).

Physiologic function tests performed on the excised lungs were regarded as the standard for categorizing lung as mature or immature. Ability to fully inflate the lung at 30 cm water pressure indicated maturity. No significant inflation of the lung indicated immaturity. Surface tension measurements on minces of lung were correlated with inflatability.

By using pen-bred ewes, gestational age was accurately known to within 4 days. Complete sets of data were obtained from 13 of 19 sheep experiments. Three age groups were examined. The *term* group (142 to 147 days) should have physiologically mature lungs capable of sustaining life with minimum surface tension (MST) less than 7 dynes/cm. The *middle* group (120 to 130 days) should have physiologically immature lungs near the transition region (between canalicular and terminal sac stages) with MST greater than 10 dynes/cm. The *early* group (90 to 110 days) should have physiologically immature lungs (canalicular stage) with MST greater than 15 dynes/cm.

A. QUALITATIVE VISUAL ANALYSIS

By scanning in both the longitudinal plane (across the fetal ribs) and the transverse plane (parallel to the fetal ribs), both lung and liver appear on the same image cross-section for comparison. Maintaining a constant fluid path depth between the transducer and the fetus is important so that echo amplitude levels would not be altered artifically by time-gain-compensation (TGC) effects. This principle is demonstrated using a phantom (Figure 1). The Philips scanner has carefully calibrated ten-turn slope and gain dials, carefully shaped exponential time-gain-compensation, probe compensation for transducer fo-

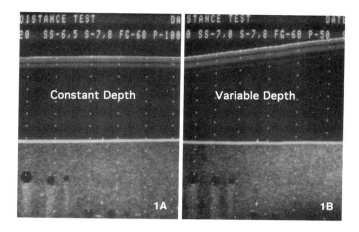

FIGURE 1.   Importance of constant depth water path. (A) Uniform echo amplitude levels in the phantom are present when water path depth is constant. (B) Echo amplitude levels increase with increasing water path depth because of TGC effects.

cusing, and narrow-band filters in the receiver. This set-up permits semi-quantitative evaluation of attenuation in dB cm^{-1} MHz^{-1} and scattering level differences in dB between adjacent tissues.

The following sonographic parameters were assessed:

1. *Attenuation* — The slope dial was adjusted until uniform brightness throughout the depth of lung tissue was observed. The process was repeated for liver. The attenuation in dB/cm was read from the calibration chart and then divided by the nominal center frequency of the transducer to obtain the attenuation coefficient in dB cm^{-1} MHz^{-1}.

2. *Relative scattering amplitude* — To assess differences between lung and liver backscattering (amplitude brightness), the calibrated system gain in dB was decreased until echoes from lung or liver could no longer be seen. Three categories were defined.

   Equal: No dB difference between lung and liver; visually the same brightness

   Greater: Lung at least 2 dB greater than liver; easily visible difference in brightness

   Intermediate: Lung 1 or 2 dB greater than liver; subtle difference in brightness

3. *Texture* — Although texture assessment must be interpreted cautiously because of transducer depth and image-processing-dependent features, comparison at constant depth for adjacent lung and liver within the focal zone should be valid, though subjective.

Coarse: Big bright scatterers in a nonuniform pattern
Fine: Low-level scatterers in a homogeneous pattern
Intermediate: Features between coarse and fine

The sonographic data (5 MHz) and physiologic data from these experiments have been reproduced in Tables 1 and 2. All lambs appeared healthy and appropriate for gestational age except lamb L. This lamb was small for gestational age, growth retarded, and meconium stained; these factors may explain why the lungs of this term fetus were not inflatable.

Sonographic classification was based upon brightness and texture of lung with respect liver. The following patterns were observed:

Mature: Lung brightness less than or equal to liver; fine lung texture equal to liver
Immature: Lung brightness greater than liver; lung texture coarser than liver
Transitional: Intermediate brightness and texture features

Mature lung pattern in a term fetal lamb is shown in Figure 2. Equal brightness amplitude and equal fine texture for lung compared to liver are more easily observed in the transverse view than in the coronal view where rib shadows are present. Immature pattern in a fetal lamb at 124 days gestation is shown in Figure 3. The boundary between lung and liver is easily observed in both transverse and coronal views because the lung brightness is greater and lung texture is coarser than adjacent liver. The higher attenuation of liver compared to lung for one mature term lamb is evident in Figure 4, where the liver brightness decreases as a function of depth because the slope dial has been set to compensate for the lower attenuation of lung (uniform brightness with depth). A comparison of echo amplitude brightness can only be valid at a level nearest the transducer, before attenuation effects alter echo amplitude relationships. The profile plots were recently obtained by digitizing the original film image and confirm the attenuation difference, which can be seen visually.

The data from the fetal lamb experiments show some interesting trends.

1. An immature sonographic pattern indicated an immature lung state (five of five lambs). But, physiologically immature lungs did not always have an immature sonographic pattern (five of eight lambs). An immature sonographic pattern was not observed when lungs were physiologically mature.
2. A mature sonographic pattern was associated with mature lung state in three of five cases. However, a significant false positive occurred for one lamb (N) with mature sonographic pattern but immature lung state.
3. A transitional sonographic pattern was not helpful, since it was seen in two of five mature states and two of eight immature states.

TABLE 1
Sonographic Characteristics of Lung and Liver for the Fetal Lamb 5 MHz Transducer Data

| Sheep | Age (days) | Attenuation coefficient dB/cm/MHz | | Brightness (lung/liver) | Texture (lung/liver) | Sonographic classification[b] | Physiologic function tests | Minimum surface tension (dynes/cm) | |
		Lung	Liver				Inflatability	Upper lobe	Lower lobe
G	145	0.30	0.55	Less	Fine, equal	M	Yes	3.5	3.0
J	145	0.3	0.3	Equal	Fine, equal	M	Yes	7.5	10.5
K	145	0.25	0.35	Equal	Fine, equal	M	Yes	5.0	4.0
L[a]	145	0.35	0.45	Intermediate	Intermediate	T	No	12.0	15.0
A	143	0.37	0.37	Intermediate	Intermediate	T	Yes	7.0	3.0
B	143	<0.5	0.5	Intermediate	Intermediate	T	Yes	2.0	7.0
H	126	0.28	0.28	Greater	Coarser	I	No	23	13
I	125	0.3	0.3	Greater	Coarser	I	No	12	13
E	124	0.3	0.3	Greater	Coarser	I	No	25	16.5
N	123	0.35	0.35	Equal	Fine, equal	M	No	11	12.5
M	122	0.30	0.30	Intermediate	Intermediate	T	No	10	17
F	103	0.3	0.3	Greater	Coarser	I	No	18	18
O	98	0.32	0.32	Greater	Coarser	I	No	18	15

[a] Fetal lamb was small for gestational age, growth retarded, and meconium stained.

[b] M = mature; T = transition; I = immature.

Adapted from Thieme, G. A. et al., *Invest. Radiol.*, 18, 18–26, 1983. With permission.

TABLE 2
Summary of Sonographic and
Physiologic Results

Sonographic	Status of lung	
classification	Mature	Immature
Mature	3 (K,J,G)	1 (N)
Transitional	2 (A,B)	2 (L,M)
Immature	0	5 (H,I,E,F,D)

Adapted from Thieme, G. A. et al., *Invest.*
Radiol., 18, 18–26, 1983. With permission.

4.   Texture patterns correlated with brightness patterns but were too subjective to apply as diagnostic criteria.
5.   As a measurement of diffuse scattering, brightness of fetal lung decreased as a function of gestational age. This assumed that diffuse
scattering from fetal liver remained constant.
6.   Attenuation differences between lung and liver were apparent in some
term lambs. Visual estimation of the attenuation coefficient suggested
that lung attenuation did not change and that liver attenuation increased.
Because of attenuation differences between lung and liver, brightness
comparisons were valid only at the most superficial depth; comparison
at deeper depths gave the false impression of lung brightness greater
than liver.

Another important observation from a related set of experiments is helpful
in correlating ultrasound appearance of fetal lung and anatomic development.
For several lambs, images of lung and liver were obtained both before and
after removal of the chest wall (Figure 5). At 126 days gestation immature
lamb lung does not collapse, and the brightness amplitude of lung remains
much greater than liver. Fetal lung in this canalicular stage consists of predominantly columnar and cuboidal epithelium, and fluid-filled alveolar sacs
have not yet developed. The lung is solid and stiff. At 143 days gestation
mature lamb lung brightness is equal to liver before removal of the chest wall.
After removal of the chest wall, the lung collapses (concave boundary), and
lung brightness is much greater than liver. Fetal lung in this alveolar stage
consists of predominantly tiny fluid-filled alveolar sacs lined with flat epithelial cells. Egress of lung fluid via the airways results in loss of lung volume.
Thus, diffuse scattering properties of fetal lung are significantly altered by
the presence of fluid in the alveolar sacs, as predicted by the morphologic
model. Furthermore, the diffuse scattering level, as reflected by brightness
amplitude, appears to decrease as gestational age increases. This is consistent
with observations previously stated.

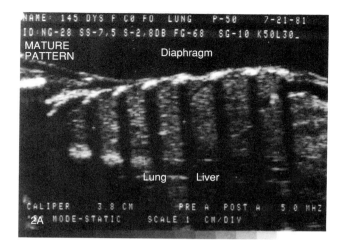

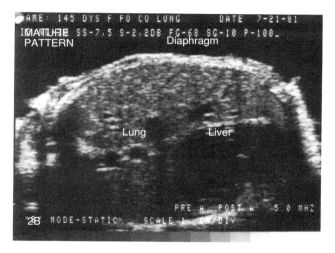

FIGURE 2. Mature fetal lung pattern examples. (A) Sagittal view shows indistinguishable boundary between lung and liver. Liver attenuation is slightly greater than lung attenuation. (B) Transverse view shows equal lung and liver sonographic features.

B. QUANTITATIVE DIGITAL DATA ANALYSIS

The details of data acquisition and analysis were discussed by Meyer et al.[15] To summarize, a Philips B-mode scanner was modified by the vendor to allow switch-selectable bypass of existing filter traps in the receiver so that broad-band radio frequency signal could be obtained after application of time gain compensation (TGC) but before any other signal processing. This post-TGC radio frequency output signal was digitized at a sampling rate of 25 MHz with full 8-bit significance. An adjustable gating signal allowed

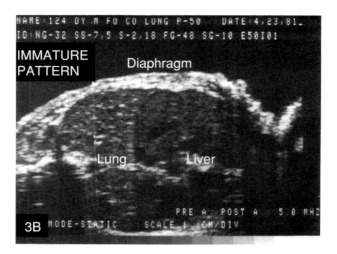

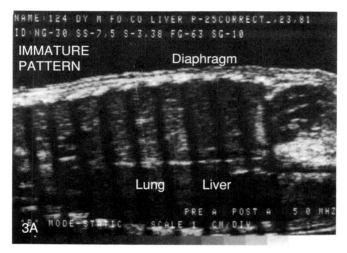

FIGURE 3.   Immature fetal lung pattern. (A) Sagittal view shows lung brightness greater than liver. (B) Similar features are seen in the transverse view.

digitization of only the A-line segment incorporating the lung or liver; the path length was 4.8 cm. Sequentially spaced digitized A-lines obtained at millimeter intervals during single sweep scanning were stored on disk for later off-line processing.

Two quantitative parameters that can be computed from backscattered signals and correlated with fetal lung maturity are (1) the slope of the attenuation coefficient with respect to frequency and (2) the mode of the backscattered power cepstrum. Calculation of both parameters is dependent upon the shapes of the backscattered spectra and independent of absolute signal

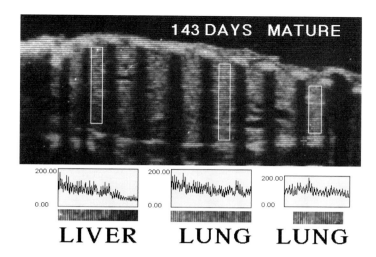

LIVER   LUNG   LUNG

FIGURE 4. Lung liver attenuation difference. Constant lung brightness with respect to depth is observed because the slope is set to compensate for lung attenuation. Liver brightness decreases indicating a higher attenuation coefficient. Comparison of brightness is valid only nearest the transducer. The plots display lung brightness with respect to depth for each region of interest.

amplitude (system gain and TGC setting dependent). Sufficient digital data was obtained for 8 of the 13 sheep. Data sets were sorted so that rib shadows, chest wall, and other nonlung structures were excluded and so that only radio frequency data from lung were analyzed (Figure 6).

Each digitized line of lung data was divided into overlapping 128 point segments where the overlap was 64 points (Figure 7). The slope of the attenuation coefficient with respect to frequency was derived for each lamb lung by using the technique described by Kuc[16] of averaging the difference of log spectra from overlapping data segments. The backscattered power cepstrum provides a measure of average backscatterer size within the lung and assumes that tissues are composed of a random collection of small rigid spheres. Due to the stochastic nature of the process, a large data volume consisting of many independent A-lines from lung must be recorded in order to achieve useful statistical significance in the attenuation and cepstral results.

The attenuation and cepstral results are summarized in Table 3. Attenuation as a function of minimum surface tension is graphically displayed in Figure 8. An increase in the attenuation coefficient of about 25% is evident as minimum surface tension declines during the transition from immature to mature lung state. However, this is presently just a statistically observed trend and is not clinically applicable. The amount of independent data (nearly 5 MByte per sheep of unsorted radio frequency data) required to obtain an attenuation estimate whose 95% confidence limits were less than 0.2 dB cm^{-1} MHz^{-1} is large even by current data processing capabilities (nearly 10 years

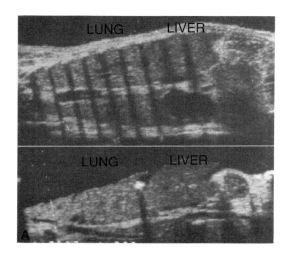

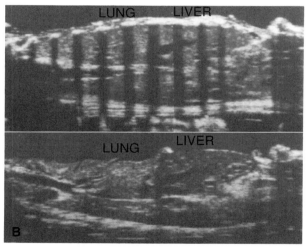

FIGURE 5.   Fetal lung stiffness. (A) At 126 days immature lung is rigid and does not collapse after removal of the chest wall. (B) At 143 days mature lung collapses after removal of the chest wall and brightness increases.

later). The digital processing technique does demonstrate either that information is lost through signal processing stages used to produce the B-scan image or that semiquantitative visual analysis using a calibrated TGC dial is not sensitive enough to detect this relatively small increase in attenuation with increasing lung maturity state.

One would expect that cepstral results might reflect anatomic development of the lung since the cepstral peak is a measure of the structural size of scatterers. Indeed, the correlation between cepstral peak and sonographic

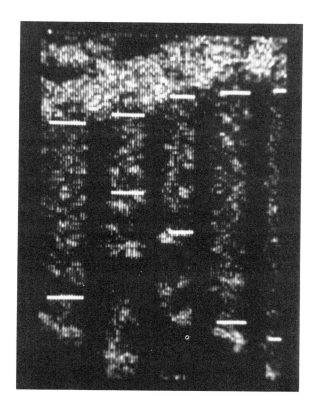

FIGURE 6. B-Mode echogram computed via envelope detection of digitized radio frequency data. Lines of sorted radio frequency data lie between bright pixels. (From Meyer, C. R. et al., *Ultras. Imag.*, 6, 13–23, 1984. With permission.)

visual results is striking for this small sample. Cepstral peaks above 100 μm are associated with transitional or mature sonographic patterns (reduced lung brightness); and, cepstral peaks significantly below 100 μm are associated with immature sonographic patterns (increased lung brightness). However, there is no apparent correlation with gestational age, lung inflatability, or minimum surface tension for this small sample of eight sheep.

In summary, the digital data analysis of the radio frequency signal detected a small statistical increase in fetal lung attenuation from the immature state to the mature state. This change could not be appreciated by semiquantitative visual estimation of the attenuation coefficient. Thus, digital data processing of the radio frequency signal may prove rewarding if the acquisition process becomes practical in the clinical setting. Cepstral results may reflect anatomic development of the lung. A large clinical study would be necessary to investigate any potential relationships between results from radio frequency signal analysis and fetal lung maturity.

ESTIMATION OF ATTENUATION AND BACKSCATTERER SIZE

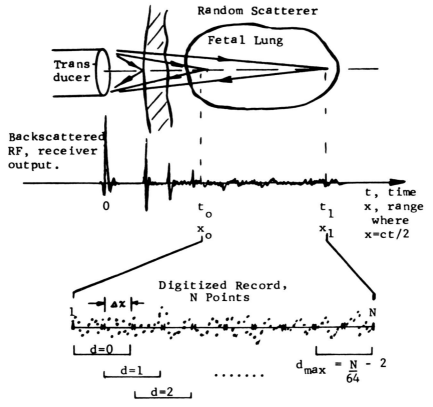

FIGURE 7. Relationship of overlapping data segments to physical setting of experiment. (From Meyer, C. R. et al., *Ultras. Imag.*, 6, 13–23, 1984. With permission.)

IV. HUMAN STUDIES

Sonographic imaging of the human fetal lung in the normal *in utero* environment is usually technically difficult because of acoustic shadowing from the fetal ribs. Visualizing fetal lung and liver in the same image plane is dependent upon fetal position, technical skill, and suitable imaging equipment. Control over the depth of amniotic fluid between the transducer and the fetus is limited. Furthermore, the fetus must be imaged through the maternal abdominal wall which may significantly degrade image quality for obese

TABLE 3
Comparative Findings (Ranking by Attenuation Coefficient)

| Sheep (Id) | Independent variables | | | | Experimental results | |
| | Functional maturity | | | | Quantitative results | |
	Minimum surface tension (dynes/cm)	Opened at 30 cm of water pressure	Gestational age (days)	Visual results[a]	Atten + 2 SEM (dB/cm MHz)	Abscissa of cepstral peak (μm)
K	3.0	Yes	145	M	0.67 + 0.03	110
G	3.0	Yes	145	M	0.66 + 0.12	110
J	7.5	Yes	145	M	0.62 + 0.04	115
L[b]	12.0	No	145	T	0.62 + 0.05	100
M	10.0	No	122	T	0.58 + 0.04	160
N	11.0	No	123	M	0.53 + 0.09	145
I	12.0	No	125	I	0.49 + 0.04	55
H	13.0	No	126	I	0.48 + 0.12	25

[a] M = mature; T = transitional; I = immature.
[b] Meconium stained, growth retarded.

From Meyer, C. R. et al., Ultrason. Imag., 6, 13–23, 1984. With permission.

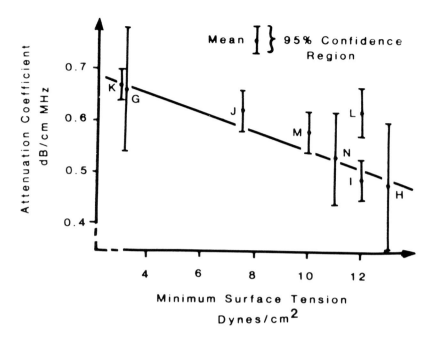

FIGURE 8.   Linear regression of fetal lung attenuation coefficient estimates with respect to minimum surface tension for eight sheep. (From Meyer, C. R. et al., *Ultras. Imag.*, 6, 13–23, 1984. With permission.)

individuals. Consequently, the ideal imaging conditions used for the fetal lamb studies cannot be duplicated in the clinical setting. In addition, biochemical tests on amniotic fluid do have significant false negative results (previously discussed). Perhaps the best "gold standard comparison" is the newborn outcome, where the criteria is normal respiration vs. development of respiratory distress syndrome due to hyaline membrane disease. These factors must be considered when clinical studies are reviewed.

A. QUALITATIVE VISUAL STUDIES USING COMMERCIAL SYSTEMS

The first recorded reference in 1980 regarding the sonographic appearance of human fetal lung stated that the reflectivity of lung is equal to or less than that of the liver throughout most of pregnancy but that this relationship reverses in late gestation.[17] A second reference to human studies in 1983 reported a strong relationship between a high lung-to-liver echogenicity ratio and subsequent absence of hyaline membrane disease in the newborn, in agreement with the first report.[18] A third reference to human studies in 1984 confirmed these results in a group of 48 patients where 32 had complicating factors such as diabetes, hypertension, and intrauterine growth retardation.[19]

TABLE 4
Human Fetal Lung Study 1982

Case	L/S ratio	PG	Age (weeks)	SONO (lung/liver)	Delivery interval	Newborn outcome
HO	5.1		33	Equal		
TA	5.5	Absent	37	Equal		
JO	6.2	Present	37	Equal	1 day	Normal
KI	4.6	Absent	35	Equal		
MA	3.7	Present	37	Equal		
RY			38	Equal	Same day	Normal
EN	4.0	Present	37	Equal		
HU	6.0	Absent	35	Equal	1 day	Normal
SM	5.3	Present	36	Equal		
WI	3.0	Absent	37	Greater		Maternal diabetes
IN			26	Greater	2 days	HMD
DE			29	Greater		
84			28	Equal		
DA			31	Greater		
BL			31	Greater		

L/S ratio	Probability of HMD (%)	Laboratory L/S test
<2.0	57	Basis was a study of 100
2.0–3.5	20	patients delivering within
>3.5	<5	72 h of amniocentesis

In contrast, the fetal lamb results show an opposite pattern.[14] Lung reflectivity is greater than liver reflectivity during mid-gestation and is equal to liver reflectivity at term. There are no apparent interspecies differences to account for these contradictory observations. Unpublished pilot study data by Thieme in human pregnancies are in agreement with the fetal lamb data.

In Table 4 (pilot study by Thieme) the lung-liver reflectivity relationship is compared to L/S and PG values (amniocentesis on the same day as the ultrasound) or to newborn outcome (delivery within 48 h after the ultrasound) or both. For this laboratory, the probability of HMD is less than 5% for an L/S ratio greater than 3.5, 20% for a ratio between 2.0 and 3.5, and 57% for a ratio less than 2.0. The presence of PG is reassuring of lung maturity. Absence of PG when the L/S ratio is greater than 3.5 at this laboratory is associated with lung maturity. Performance of the ultrasound exam within 2 days of delivery was achieved in only 4 of 15 cases, due to practical limitations. Lung and liver reflectivity were equal for nine of nine patients having either a mature L/S ratio (greater than 3.5) or normal newborn outcome or both conditions. Typical mature lung/liver images are shown in Figure 9. For patient WI with maternal diabetes, the fetus at 37 weeks has lung reflectivity much greater than liver and an L/S ratio indicating greater than 20% probability

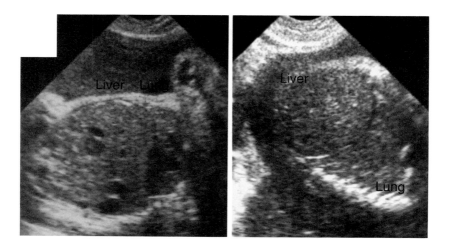

FIGURE 9. Mature human fetal lung pattern. Echo amplitude levels from lung and liver are equal for each fetus with mature lungs. Center frequency is 3.5 MHz. (A) Case SM in Table 4. (B) Case HU in Table 4.

of HMD. This is an immature sonographic pattern according to the fetal lamb work; delayed maturation of fetal lung is a known problem in maternal diabetes. Patient IN was the only case below 32 weeks where delivery occurred within 48 h of the ultrasound examination. Lung reflectivity greater than liver correlated with the development of HMD in the newborn, as expected at 26 weeks. Four of the five fetuses examined at less than 32 weeks had lung reflectivity greater than liver; one fetus had lung reflectivity equal to liver. From the prior discussion of the incidence of RDS with respect to gestational age, this is a reasonable ratio of immature to mature patterns at less than 32 weeks. The variability of the immature sonographic patterns is apparent in Figure 10.

In Table 5 (pilot study by Thieme) the results for a small group of patients between 31 and 36 weeks are summarized. All six had prolonged premature labor problems. Newborns at less than 32 weeks have a greater than 50% probability of developing HMD; and, newborns at greater than 36 weeks have a low probability of developing HMD. Thus, this group is in a transition zone where lung maturity is frequently achieved during the days between the onset of premature labor and the time of delivery. The importance of serial examinations is demonstrated by patient WI, where the transition from lung reflectivity greater than liver to lung reflectivity equal to liver is documented (Figure 11). In addition to establishing the direction of change in fetal lung reflectivity, this case also shows that a change in attenuation of lung or liver has occurred over the 10-day interval between examinations, since two different slope settings are necessary to achieve uniform brightness through the

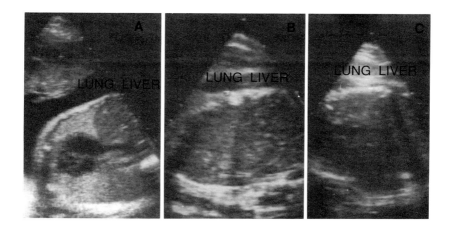

FIGURE 10. Immature human fetal lung patterns. Echo amplitude levels from lung are greater than liver. Considerable variability in the brightness ratio is evident for cases SC (A), FE (B), and GO (C) in Table 5.

TABLE 5
Human Fetal Lung Study 1982 (Thieme)

Case	L/S ratio	PG	Age (weeks)	Sono (lung/liver)	Deliver interval (days)	Newborn outcome
WI			33.5	Greater	11	
			35	Equal	1	Normal
SC	2.4	Absent	34	Greater	2	TTN vs. mild HMD on RA within 28 hours
FE			34	Greater	5	Mild respiratory distress on RA within 56 h
CR			34	Greater	2	Normal
GO	1.2	Absent	33	Greater	2	Mild HMD
WH			31	>>or =	2	Mild HMD betamethasone

depth of tissue. An important consequence of this observation is that lung and liver reflectivity can be meaningfully compared only at the depth nearest to the transducer; attenuation effects alter echo level relationships at greater depths. Mild respiratory distress symptoms were experienced by three of four newborns (SC, FE, CR, GO) where fetal lung reflectivity was greater than liver. Patient WH illustrated the technical difficulties of imaging; in some images lung and liver were equal and in others lung greater than liver (Figure 12).

These two small human pilot studies provide general confirmation of the fetal lamb study results.

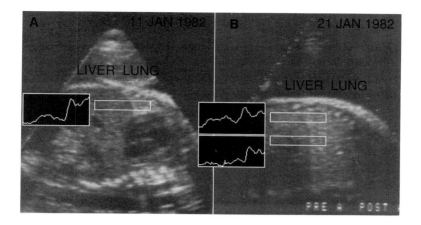

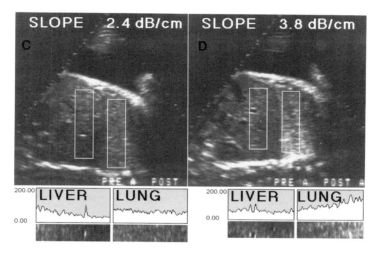

FIGURE 11. Transition from immature to mature pattern for case WI in Table 5. (A) The first scan clearly demonstrates lung echogenicity greater than liver. (B) Ten days later, lung echogenicity is equal to liver echogenicity near the transducer. Liver echogenicity decreases with increasing distance from the transducer implying higher attenuation of sound by liver than lung. (C) and (D) The slope dial setting has been changed from 2.4 dB/cm to 3.8 dB/cm. At 2.4 dB/cm compensation for lung attenuation of sound is correct; liver brightness decreases because of higher attenuation than lung. At 3.8 dB/cm compensation for liver attenuation is correct; lung brightness increases because of lower attenuation than liver. Comparison profiles have been plotted using digitized data from the original images.

1. Lung and liver echogenicity are equal for most mature lungs.
2. Lung greater than liver echogenicity is seen for most immature lungs.
3. The trend over time appears to be for the lung/liver reflectivity ratio to change from significantly greater than one during the early third trimester to near unity at term.

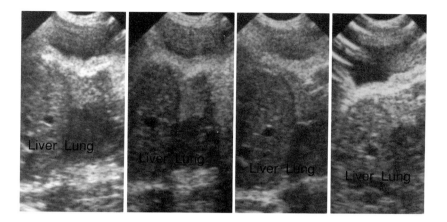

FIGURE 12. Unexplained technical variance. Four different lung-liver echogenicity relationships were obtained in a brief period of scanning for this 38 week fetus. PG was present, indicating mature lungs. Images at 3.5 MHz.

4. The studies are technically demanding and cannot be accomplished for some fetal positions and maternal conditions. Problems with overlying structures, particularly ribs, can generate conflicting images for the same patient (Figure 13). Variable fluid path can also alter echo amplitude relationships. Attenuation differences between lung and liver must be recognized in order to interpret images properly.
5. The best gold standard is newborn outcome. A mature L/S ratio and presence of PG are considered the "gold standards" for determining lung maturity prior to delivery. However, lung maturity may be present even if test results are negative. The last ultrasound study should be within 24 h of delivery so that the ultrasound results can be compared to the newborn outcome (presence or absence of RDS).

A study reported by Cayea in 1985 compared fetal lung/liver images obtained at the time of aminocentesis to the lecithin/sphingomyelin (L/S) and phosphatydylcholine (PC) values of the amniotic fluid.[20] Apparently, great care was taken to eliminate artifact effects from ribs and other overlying structures during imaging. Three different commercial ultrasound systems were used; but no comparison of results from each system was presented to determine if differences exist between equipment manufacturers. An L/S ratio greater than or equal to 2 was defined as mature; and, presence of PC was defined as mature. Three ultrasound parameters were examined: (1) lung/liver echogenicity comparison, (2) lung texture, and (3) sound transmission through lung. Of an original 81 cases, complete data sets were obtained for 59 cases; diabetic pregnancies were excluded because of known effects on L/S and PC values. The data in Table 6 for lung echogenicity and L/S ratio

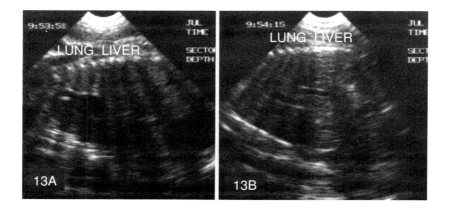

FIGURE 13. Technical pitfalls. The focal zone of this 3 MHz long focus transducer is too deep such that the beam diameter is too large to pass through rib interspaces without significant attenuation. Lung echogenicity appears greater than liver in one view (A) and less than liver in another view (B). Imaging technique is crucial to interpretation. The time interval between images is less than 30 s. The transducer position has been shifted inferiorly.

TABLE 6
Fetal Lung Study

| | Lecithin/sphingomyelin ratio | | | |
| | Mature | | Immature | |
Echogenicity	# Case	Percent	# Case	Percent
Lung = Liver	13	26	3	22
Lung > Liver	32	74	11	78
Totals	45	100	14	100

Data derived from Cayea, P. D. et al., *Radiology,* 155, 473–475, 1985.

were derived from the paper and are representative of the results for the other parameters. The authors concluded that there is no statistically significant correlation between the sonographic features and biochemical fetal lung maturity indices. Contrary to some prior reports, no cases of lung echogenicity less than liver were observed. Another important observation was an overall increase in fetal lung brightness with respect to liver and apparent coarsening of lung texture with increasing gestational age. The authors concluded that distinguishing mature from immature fetal tissues with current clinical equipment was not possible.

A study reported by Fried in 1985 compared fetal lung/liver images to gestational age for 185 cases and to amniotic fluid L/S ratio and PG for 37 cases.[21] Commercial clinical ultrasound equipment was used to obtain images between 15 weeks and term. Lung echogenicity was described as hypodense, isodense, slightly hyperdense, and hyperdense with respect to lung. Fetal growth retardation and macrosomia cases were excluded. The authors could not demonstrate any correlation between lung echogenicity and gestational age. There was no statistically significant correlation between lung echogenicity and L/S and PG determinations either. These authors concluded, similar to Cayea, that sonographic features of fetal lung are of no predictive value with regard to fetal lung maturity.

A study reported by Feingold in 1987 also compared sonographic fetal lung/liver reflectivity to amniotic fluid biochemical measures of lung maturity (L/S ratio, PG, and OD650).[22] Only images showing liver, diaphragm, and lung free of acoustic artifacts from overlying structures were included for the 20 cases between 34 and 38 weeks. A single sonographic system was used. Quantitative values for lung and liver brightness were obtained by using a densitometer to measure the film optical density over a 3 mm diameter aperture. Three measurements at increasing depths were obtained for both lung and liver. No statistically significant associations between lung/liver reflectivity relationships and biochemical parameters were demonstrated in this paper either.

Carson et al.[1,23] presented an evaluation of the 19 patients (45 examinations) in which lung/liver echogenicity was evaluated serially according to the following visual scale.

0. Nondiagnostic study
1. Lung echogenicity less than liver
2. Lung echogenicity equal to liver
3. Lung equivocally or slightly of greater echogenicity than liver
4. Lung definitely but minimally greater echogenicity than liver
5. Lung significantly greater echogenicity than liver
6. Lung much greater echogenicity than liver

A trend was noted in these patients who had multiple ultrasound examinations. To reveal this trend and to avoid variance effects across patients, serial relative echogenicity assessments for each patient are plotted vs. time as shown in Figure 14, with data points for each patient joined by straight lines. The first data point (denoted by concentric circles) for each patient is plotted at the estimated gestational age at the time of the first examination. Subsequent data points are plotted relative to the actual elapsed time between successive examinations. Gestational age was estimated via ultrasonic biparietal diameter measurements and other clinical data such as date of last menses. This method of plotting exact times between examinations is more appropriate than using the gestational age estimates at each ultrasound examination. Note that although data from many patients are plotted on the same graph, this need not be interpreted as pooled data; instead, each case

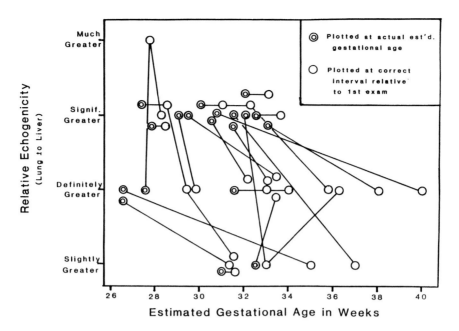

FIGURE 14.   Relative lung-to-liver echogenicity (scale defined in text) as a function of the best estimate of gestational age.

is used as its own control, and echogenicity is always compared to an assessment at an earlier examination.

The plot shows that in most cases echogenicity decreases over time between successive examinations as would be expected. Where there are only two examinations, there is, of course, only one interval between the examinations over which the echogenicity change can be considered. However, in cases having, for example, three examinations, there are three intervals: first to second, second to third, and first to third. In this manner, from the 44 examinations performed on 18 patients represented on the plot, there are 34 intervals to consider. For examination intervals greater than 21 days, echogenicity *always* decreases over time. There are 18 intervals that are greater than 8 days; echogenicity decreases with respect to time over 15 of these intervals (or 83%), increases over one of the intervals and stays the same over two of the intervals. There are 16 intervals that are 8 or fewer days in duration. In over 31% of these intervals, echogenicity decreased, remained the same in over 50% of them, and increased over the remaining 19%. For examinations closer together than 8 days, echogenicity assessments appear to be rather random.

Some of the noise in the data might be explained by the fact that different lobes of the lung were imaged in the different examinations. The lobes of the fetal lung mature at different rates, with the result that at any time, the

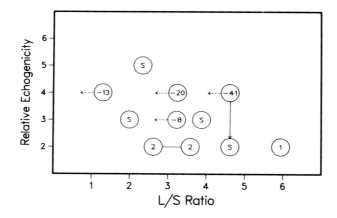

FIGURE 15. Relative lung-to-liver echogenicity (scale defined in text) vs. L/S ratio. Temporal relationship of ultrasound in days before (−), after (no sign), or the same (S) as amniocentesis noted in circle. Interconnecting bar and arrow indicate same patient. Broken arrows indicate reduction in L/S ratio assumed if amniocentesis had been performed nearer the time of sonography, when lungs were presumably less mature. (From Carson, P. L. et al., in *Tissue Characterization with Ultrasound*, Greenleaf, J. F., Ed., CRC Press, Boca Raton, FL. With permission.)

lung contains regions having different maturities and therefore, under a hypothesis of the study, different ultrasonic properties.

Within this group of 19 patients,[1,23] six patients (seven examinations) had scans within 2 days of amniocentesis to determine the L/S ratio. Four patients (four examinations) had scans 8 to 41 days prior to amniocentesis. The relative lung/liver echogenicity score vs. the L/S ratio is plotted in Figure 15. Lung echogenicity equal (or nearly equal) to liver is clearly associated with high (mature) L/S ratios. Lung echogenicity significantly greater than liver echogenicity is likely to have been associated with a lower L/S ratio if the amniocentesis and ultrasound had been done at the same time. The one patient with a repeat examination also shows decreasing lung echogenicity with increasing gestational age; if amniocentesis had been performed at the time of the examination 41 days earlier (6 weeks), one would reasonably assume that the L/S ratio would have been much lower since this fetus almost certainly had immature lungs then.

Even though the study design is not ideal and the population is small, two trends are apparent in the data presented by Carson et al.[1,23] Lung echogenicity equal to liver echogenicity is seen in association with mature L/S ratios; by extrapolation, lung echogenicity greater than liver echogenicity may be associated with immature lung state. Lung echogenicity with respect to liver decreases with increasing gestational age for the patients with repeat examinations.

The different results from five research groups are summarized in Table 7. The conflicting results may seem irreconcilable initially. All groups have

TABLE 7
Summary of Results by Five Research Groups of Visual Qualitative Analysis of Lung/Liver Echogenicity

Criteria	Thieme/Carson (Lamb)	Thieme (Human)	Carson (Human)	Cayea (Human)	Fried (Human)	Feingold (Human)
Lung echogenicity with increasing age	Decreases	Decreases	Decreases	Increases	No trend	No comment
Lung echogenicity less than liver	One exception	Never seen	Never seen	Never seen	17% of cases	35% of cases
Correlation with lung maturity tests	Good trend	Good trend	Good trend	None	None	None
Study size	Small	Small	Small	Medium	Large	Medium

recognized the meticulous detail required for obtaining high-quality fetal lung images and display technically good examples in their presentations. Differences in patient population characteristics would seem to be an unlikely explanation. Observer bias hopefully has been eliminated but remains a potential reason for the different results. The results from Thieme and Carson may not reflect the true situation because of the small sample sizes. Perhaps the best explanation is the differences in image interpretation since visual assessments are subjective and subtle.

For example, an attenuation difference between liver and lung is evident in Figure 1 by Cayea et al.[20] Lung attenuation is less than liver since lung brightness stays the same with respect to depth while liver brightness decreases. Near the transducer, lung and liver echogenicity would be interpreted as equal; but far from the transducer, lung brightness is clearly greater than liver because of the lower attenuation of sound in lung with respect to liver. One must assume that the authors did not recognize the effect of attenuation differences on reflectivity relationships between tissues. This could easily lead to the misclassification of cases since the image shown could be be misinterpreted as lung echogenicity greater than liver. This attenuation difference has been observed by both Thieme and Carson since the time of the fetal lamb work; and Carson[24] describes it as a potential pitfall. Since liver attenuation greater than lung is thought to be present in some near-term fetuses due to the increased glycogen storage in the liver,[25] a misclassification of such cases could be misinterpreted as increasing lung echogenicity with increasing age. Similarly, the attenuation difference pitfall apparently was not recognized by Feingold et al.[22] The image of lung and liver shows liver brightness remaining the same throughout the depth of tissue and lung brightness increasing with depth from the transducer. Densitometry measurements from the three regions in liver would be expected to provide similar values. The measurements from lung would increase with depth because the TGC slope setting overcompensates for the lower lung attenuation. The average of the lung measurements would be greater than liver and would lead to the false conclusion of lung echogenicity greater than liver. When compared at the depth nearest the transducer before attenuation difference effects occur, lung and liver echogenicity are equal. Obviously, if this condition occurs with any significant frequency, then significant misclassification of cases will occur. Therefore, conclusions from the Feingold study may be skewed for similar reasons.

The images obtained by Fried et al.[21] are obviously high-quality sonograms; and one must assume that all study images met this standard. The authors do describe some of the technical pitfalls. However, the assignment of the images with respect to the classification scheme raises some questions about reader interpretation. The hypodense lung example can be explained by rib shadow artifact due to sound attenuation. The isodense example appears appropriate. The slightly hyperdense example shows lung which is equal in brightness to liver near the transducer and slightly hyperechoic to liver farther from the transducer. This represents an attenuation difference effect between lung and liver. The case should be classified as isodense at the depth nearest to the transducer. The hyperdense example has an unexplained hypodense band in the liver just inferior to the diaphragm; this may represent a reduction in sound transmission from the diaphragm (possibly a refraction effect due to the curvature of diaphragm). Comparison of lung to this portion of the liver leads to the misclassification as hyperdense lung; the lung is really isodense with respect to liver since lung brightness is similar to the bulk of the liver caudal to the artifact band. These classification problems with the reference images for each category raise concerns about the results and conclusions of this study.

This discussion serves to point out the subtlety of image interpretation and the complexity of sound interactions. Carson[24,26] presents some of the difficulties of assessing lung maturity from ultrasonic backscattering. Thus, it seems reasonable to suggest that image interpretation difficulties are the most likely explanation for the variance in results among the researchers. A list of known pitfalls follows:

1. Attenuation differences between lung and liver dictate that echogenicity comparisons be made at a depth nearest the transducer (Figures 4 and 11).

2. Not all rib artifacts resulting in attenuation of sound are easily recognized (Figure 13).

3. Subtle refraction effects which alter sound transmission may not be readily apparent, even to experienced observers.

4. The soft tissue and fluid paths between the transducer and the fetus should be made as equal as possible since differences in sound transmission may affect the echogenicity relationships of lung and liver (Figure 1).

5. Lung-to-liver contrast is greater at 5 MHz than at 3 MHz (Figure 16). Attenuation effects are also greater at higher frequencies. The highest frequency possible should be used.

6. The beam intensity profile of the transducer might produce unrecognized variations in echogenicity of tissues as a function of distance from the transducer. The focal depth of the transducer should pass through the lung and liver. Echogenicity comparisons between lung and liver should be made at equal distances from the transducer whenever possible.

7. Segmented TGC and slope TGC settings must be carefully adjusted to provide proper attenuation compensation. Gain settings may alter echogenicity relationships in subtle ways (Figure 17). A standard for system control settings is important.

8. Dynamic range, pre-processing, and post-processing functions may have subtle effects on image interpretation (Figure 17). Establishing a standard which optimizes image contrast is important. The number of gray shades displayed would not seem to be a factor for modern systems with 128 or 256 digital levels.

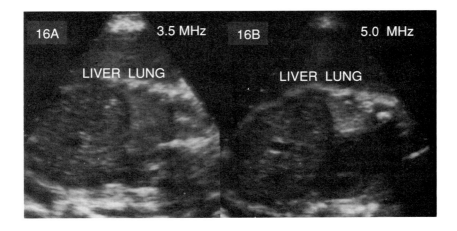

FIGURE 16. Effect of transducer frequency. Backscatter amplitude from diffuse scattering media is frequency dependent. At 3.5 MHz (A) the lung-to-liver echogenicity ratio is less than at 5.0 MHz (B). Image texture is coarser at 3.5 MHz than at 5.0 MHz. These images are from CR in Table 5.

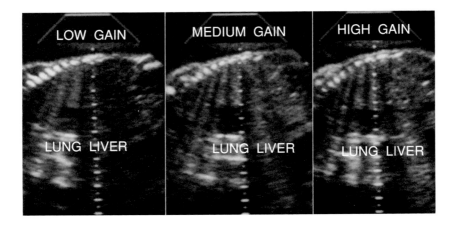

FIGURE 17. Effect of system gain and image processing. These images were obtained from an antique scanner with effectively eight levels of gray. At low gain lung echogenicity is much greater than liver. At medium and high gain, they are indistinguishable. Only low gain properly displayed contrast differences between lesions for this machine. These images are from case N in Table 4.

9. If more than one machine is used, it is important to do comparisons of equipment performance since differences might affect results.
10. Fetal position and maternal body habits may result in significant technical limitations.

In addition to technical and interpretation pitfalls, choosing comparison standards for fetal lung maturity is clearly critical. Newborn outcome with respect to absence or presence of RDS appears to be the best standard for determining lung maturity state. However, one must differentiate between transient tachypnea of the newborn (TTN) and true hyaline membrane disease (HMD) as established by clinical and radiographic criteria. Also, newborns with nonpulmonary problems probably should be excluded because of potential adverse effects upon respiratory status.

Lung maturity standards, such as L/S ratio and PG, derived from amniotic fluid samples have significant false negative results; fortunately, false positive results are few.[6,7,13] These factors will have a significant impact on classification, especially for the group assigned as immature.

Timing of ultrasound studies is critical since the transition from immature lung state to mature lung state can occur within 72 hours.[13] Therefore, the ultrasound study should be done within 48 h of delivery (preferably 24 h) if newborn outcome is the comparison standard. When L/S ratio and PG are the comparison standard, an ultrasound examination at the time of amniocentesis would seem to be ideal. However, there is even some doubt about this since both the transit time of surfactant from lungs to amniotic fluid and the hysteresis of change in surfactant amniotic fluid concentration are not absolutely known. To establish the direction of change of lung echogenicity with respect to liver and the occurrence of attenuation differences between lung and liver, serial examinations on each patient must be done. Both the optimum number of examinations and the timing of examinations has not been clearly established.

Attempting to relate individual single examination data to gestational age will not be rewarding because of the wide range of ages over which lungs can become mature. However, establishing a distribution of age of onset of lung maturity by serial examinations seems important. The magnitude and distribution of the transition time between immature and mature sonographic patterns would be interesting as well.

One must remember that a clear linkage between sonographic features of fetal lung and the morphologic and biochemical development of fetal lung has not been established. Visual assessment of standard sonographic images appears to be technically difficult from the standpoint of both the generation and the interpretation of the images. More objective methods and criteria need to be developed. Application as a clinical tool for routine practice has not been realized.

B. OTHER PERSPECTIVES AND TECHNIQUES

Visual assessment of the lung/liver echogenicity relationship as a means of determining fetal lung maturity state has not proven clinically useful to date. Technical limitations and subjectivity of image interpretation are significant problems impeding progress. Quantitative, rather than qualitative, methods for analysis of sonographic digital data might prove to be a more fruitful approach and would provide more objectivity in the assessment of the magnitudes of backscattered sound energy.

Carson et al.[27] have examined a quantitative method to measure relative echogenicity between lung and liver, which corrects for digitally estimated differences in attenuation coefficients of two tissues with presumably identical overlying tissues. This measure of relative echogenicity was studied on fetal lungs and livers in a sample of 172 digital images from eight third trimester fetuses.

The raw radio frequency signal from a region of interest (ROI), which included fetal lung and liver, was digitized to 8 bits at 25 MHz during obstetrical sonograms obtained using a 5-MHz transducer. The stored radio frequency digital data from an ROI for a particular patient was displayed on a CRT. Each radio frequency data line within the ROI was envelope-detected using the Hilbert-Transform method. Smaller ROIs separating lung and liver were outlined; and the envelope-detected lines from each tissue type (fetal lung or liver) were averaged to generate a mean envelope representing the magnitude of backscatter for that particular tissue.

A crucial simplifying assumption for this measurement of relative echogenicity (RE) is that the path lengths through both amniotic fluid and soft tissues are the same for both of the smaller ROIs representing lung and liver. Practically, this requires that the fetal body wall be normal to the ultrasound beams as illustrated in Figure 18. Transmission/reflection paths for liver (1) and lung (2) can be described in terms of the initial echo amplitude A at the surface of lung or liver and the path distance r in the intervening tissues. The backscatter magnitude difference D between lung and liver as a function of tissue depth z can be graphically displayed. The Y-intercept represents the echogenicity difference in dB between lung and liver, if the intervening tissues are identical. A correction factor can be applied for the initial echo amplitude difference, if the intervening tissues are not identical, as will be demonstrated.

The average Hilbert transformed backscatter function B(x) at a distance x from the transducer in imaging medium i is

$$B(x) = \frac{b_i A_i T(x) e^{-2\alpha_i(x-r)}}{T(r)x}$$

b_i = backscatter transfer function at frequency = f_0
A_i = initial amplitude of wavefront at x = r
α_i = attenuation coefficient at frequency = f_0

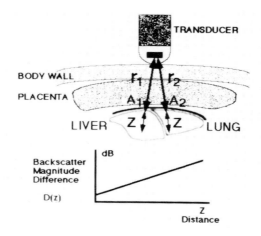

FIGURE 18.   Diagram of spatial relationships for the transducer, fetus, and intervening tissues and accompanying graphical display of backscatter magnitude difference for lung and liver.

x   =   propagation range
T(x)   =   aperture diffraction effect (including focusing)
r   =   range to beinning of ROI within medium

Taking the log difference (times 20 to express it in dB) of the backscatter function for two media $B_1(x)$ and $B_2(x)$ cancels out the aperture diffraction effect $T(x)$. A new range variable

$$z = x - r$$

describes the propagation distance within the media. The log difference $D(z)$ in dB between the two media is reduced to

$$D(z) = 20 \log_{10}((b_1 A_1)/(b_2 A_2)) + 17.4(\alpha_2 - \alpha_1)z$$

Thus, a graph of the difference in dB between the backscatter magnitudes of the two media as a function of distance z yields a regression line where the

$$\text{slope} = 17.4(\alpha_2 - \alpha_1) \quad \text{and the}$$

$$\text{intercept} = 20 \log_{10}(b_1/b_2) + 20 \log_{10}(A_1/A_2)$$

The intercept (z = o) represents the dB difference between the two media before attenuation differences alter the relative echogenicity relationship at distance z > o. Two conditions exist:

1. If intervening tissues between the transducer and the lung/liver tissues are the same, then $A_1 = A_2$; and the RE relationship is simply the intercept

$$D(O) = 20 \log_{10}(b_1/b_2)$$

2. If the intervening tissues are not of identical attenuation, then an initial amplitude difference at the depth r of the start of the lung/liver tissues may exist; and an error of $20 \log_{10}(A_1/A_2)dB$ may be added to the true RE relationship. Obviously, if both A_1 and A_2 can be measured, then a correction to $D(O)$ may be applied to yield the true RE relationship.

The relative liver/lung attenuation coefficients, proportional to the slope of the regression line, will be zero when the attenuation coefficients of lung and liver are equal, negative if lung is greater than liver, and positive if liver is greater than lung. Absolute attenuation coefficient values cannot be derived from this method.

This derivation is confirmation of the intuitive concepts presented in the prior section covering qualitative methods of describing lung/liver echogenicity:

1. If the tissues and path lengths between the transducer and lung and between the transducer and liver are identical, then visual comparison of lung/liver relative echogenicity is valid.
2. If attenuation coefficients for lung and liver are different, then echogenicity comparisons are only valid at the distance where the lung/liver is nearest the transducer. At greater distances, echogenicity comparisons are not valid.

This method of comparison of fetal lung/liver echogenicity provides a quantitative means for eliminating the subjectivity of simple visual inspection of images. Good agreement between digital and visual relative echogenicity was demonstrated for the eight patients (Figure 19). The somewhat higher precision and expected higher objectivity of the digital measure suggests possible value in repeating the generally marginal visual studies by us and others of relative lung-to-liver echogenicity as a possible indicator of fetal lung maturity. In addition, the use of an attempted absolute backscatter measure should at least provide a more accurate relative measure of lung/liver backscatter and may do much more.[28]

While this text concentrates on sonographic diagnosis based upon diffuse scattering interactions, the attenuation of sound energy also contributes important diagnostic information and is the aggregate effect of energy losses due to both scattering and absorption processes. Unlike echogenicity relationships, attenuation differences between tissues are not usually detectable

Digital and Visual Relative Echogenicity
mean per case

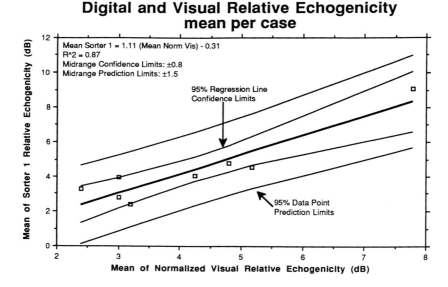

FIGURE 19.   Mean relative echogenicity difference between lung and liver. Comparison of digital measurement and visual assessment.

by simple visual assessment of images. The anatomic relationship described previously where fetal lung and liver are directly adjacent and easily compared is a rather unique circumstance. Estimation of the attenuation coefficient of a tissue is difficult in the spatial domain. Special hardware and sophisticated software signal processing techniques in the frequency domain are desirable.

Carson and associates digitized radio frequency signals from the lungs of 228 human fetuses from 26 to 40 weeks gestation and processed the data using frequency domain techniques.[29] The data (Figure 20) show a 21% reduction in the attenuation coefficient with advancing gestational age. Even though lung maturity state was not known for each fetus and maturity varies greatly with gestational age, the data strongly suggest that a statistically measurable reduction of lung attenuation from immature to mature state may occur.

Carson et al.[30] also digitized radio frequency signal from the livers of 178 fetuses from 26 to 40 weeks that were processed using frequency domain techniques to calculate the liver attenuation coefficient for each fetus. The data (Figure 21) show a 26% increase in the attenuation coefficient with advancing gestational age. The authors suggest that increased glycogen storage in the liver before birth may be responsible for this change in the attenuation of sound energy. Since healthy, well-nourished fetuses accumulate glycogen in the liver prior to birth and since nutrition-deprived fetuses have depleted glycogen stores, this observation might help to identify fetuses affected by intrauterine growth retardation. While the data were quite variable, particu-

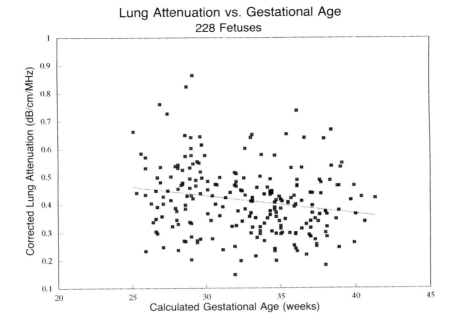

FIGURE 20. Lung attenuation vs. gestational age for 228 fetuses.

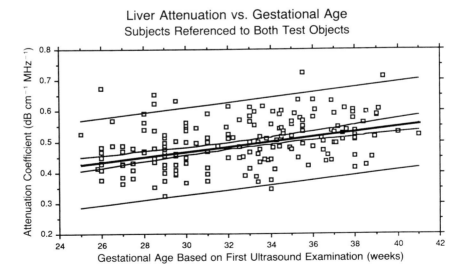

FIGURE 21. Liver attenuation versus gestational age for 178 fetuses. (From Carson, P. L. et al., *Ultras. Med. Biol.*, 16(4), 399–407, 1990. With permission.)

larly for the fetal lung, the observation that the lung attenuation coefficient changed in the opposite direction from that of the liver as a function of gestational age reduces the concern that the changes are artifacts of changes in overlying maternal tissues or fetal depth.

Both of these quantitative studies by Carson et al.[29,30] demonstrate important statistical trends which correlate with visual observations of fetal lung and liver attenuation relationships described previously. Mean values for lung fall from 0.44 dB cm^{-1} MHz^{-1} at 26 weeks to 0.36 dB cm^{-1} MHz^{-1} at 40 weeks. Mean values for liver rise from 0.43 dB/cm/MHz at 26 weeks to 0.55 dB cm^{-1} MHz^{-1} at 40 weeks. This quantitative data would support the qualitative visual observation of no significant difference in attenuation at early gestational ages and of liver attenuation greater than lung attenuation at late gestational ages. However, no link has yet been established with lung maturity state. Also, these statistical trends may not necessarily be applicable to individual cases.

A most promising approach achieving noninvasive indications of fetal lung maturity is to combine physically independent, quantitative measures. In an as yet unpublished multivariate analysis, even the visual echogenicity measure improved slightly the correlation of lung attenuation and gestational age.[29]

Linear regression on a group of 172 subjects yielded the following relations between, gestational age, G, visually determined relative echogenicity, E, and lung attenuation coefficient α_f. In the single variate linear regression with lung attenuation coefficient,

$$\alpha_f = \underset{\pm 0.0025}{-0.005} \times G + \underset{\pm 0.13}{0.65}$$

and with relative echogenicity,

$$G = 7.6(\pm 2.6) \times E + 30$$

In this particular subject group, the correlation coefficient (R^2) was only 0.024 for the correlation with lung attenuation coefficient and 0.04 for the correlation with relative echogenicity. However, in the bivariate analysis,

$$\alpha_f = \underset{\pm 2.2}{-4.5} \times \alpha_f + \underset{\pm 2.7}{7.8} \times E + \underset{\pm 3.8}{32.0}$$

and R^2 was 0.07, an almost doubling of R^2 for the same attenuation-only analysis. The F probability that the relationship was due to chance was F $\leq .002$.

The wide differences in relative lung to liver echogenicity and lung attenuation coefficient between fetuses and the rather poor overall correlation coefficients appear to be due in part to control of the acoustic measures by

Chapter 14

QUANTITATIVE BACKSCATTER IMAGING

James A. Zagzebski, Lin X. Yao, Evan J. Boote, and Zheng Feng Lu

TABLE OF CONTENTS

I. INTRODUCTION

B-Mode scanning, the most frequently used diagnostic ultrasound modality, produces gray-scale images from echo signals resulting when pulsed ultrasound beams propagate through soft tissues. These images are related to the attenuation and scattering properties of the tissues scanned. However, they also are highly operator and instrument dependent so that current ultrasonography is a qualitative, or at best "semiquantitative", imaging modality.

There is evidence that quantitative ultrasound imaging systems might provide more sensitive diagnosis of some abnormalities. For example, cirrhotic livers and livers with chronic active hepatitis with fibrosis are known to be more echogenic than normal livers.[1] Also for the liver, Garra et al.,[2,3] find that echogenicity and attenuation vary from normal in patients with Gaucher's disease, and that patients with fatty infiltration have higher liver attenuation and echogenicity. Although present ultrasound equipment allows easy detection of advanced liver disease, it appears to be more limited for detecting diffuse disease in early stages. This was pointed out by Sandford et al.,[4] who studied correlations between liver biopsy results and diagnostic image features (e.g., texture of foci, overall echogenicity, and qualitative impressions of attenuation) scored from B-mode images. When the tissue structure was at an advanced degree of abnormality, strong correlations were found between fat, as well as fibrosis, and these diagnostic signs. However, the ability to detect these changes was found to be much more limited for early-stage disease.[4]

In other areas, echo signals from the enlarged spleen are weaker than those from normal organs,[1] suggesting that quantitative imaging could play a role in diagnosis of abnormalities in this organ. Lamminen et al.[5] have shown that scattering from muscle tissue is affected by disease processes. Muscle parenchyma is weakly echogenic when normal, but appears highly echogenic in patients with diffuse muscular disease. Helguera et al.[6] found that the integrated backscatter levels from muscle tissue of patients with Duchenne's muscular dystrophy are 15 dB higher than those of normal individuals.

In addition to overall scattering and attenuation levels, the frequency dependence of scattering and attenuation,[7-10] as well as tissue structural properties computed from frequency-dependent scattering parameters,[11-14] may be useful diagnostic indicators.

The parameter commonly used for quantifying acoustic scattering in soft tissue is the backscatter coefficient, discussed extensively in other chapters of this book. A number of researchers have measured and reported on backscatter coefficients for normal and abnormal tissues.[15-22] Backscatter coefficient imaging, initially proposed by O'Donnell[23] and O'Donnell and Reilly,[24] and more recently studied by our group,[25-28] is the spatial mapping to gray-scale images of acoustic backscatter coefficients measured over a region.

Such quantitative images, either in the form of frequency-dependent "backscatter estimators" (see below) or accurately measured, local backscatter coefficients might be more sensitive to subtle tissue changes than current ultrasound systems, particularly for detection of early disease and for monitoring changes due to therapy.

In this chapter we will outline progress at producing quantitative backscatter images in our laboratory. Initial work incorporated an absolute method of data reduction to construct backscatter coefficient images. This data reduction method involves use of planar reflector as well as beam modeling to compute instrument-independent backscatter estimators and backscatter images (see Chapter 7). More recent work has been carried out using a well-characterized reference phantom to account for instrument and transmission path factors. Both methods have yielded accurate backscatter images of phantoms; however, with current clinical imaging technology, the latter has been easier to implement for patient studies.

II. QUANTITATIVE IMAGING BASED ON ABSOLUTE DATA REDUCTION METHOD

A. IMAGING ALGORITHMS

In this method of quantitative backscatter imaging, the data acquisition and reduction methods are extensions of those described in earlier reports.[29,30] Previously, these methods were applied for measuring backscatter coefficients of small test samples. In the present application the scattering object consists of a large volume of tissue or phantom material extending beyond the face of the transducer, rather than an isolated sample. The transducer is scanned systematically over the sample and the spatial coordinates of the transducer are retained during the echo data acquisition. The data reduction is applied to echo signals from throughout the sample.

Figure 1 illustrates a typical backscatter imaging configuration. An ultrasound transducer is translated perpendicularly to its beam axis over the sample. For discrete positions of the transducer (specified by beam line m and scanning plane n) an ultrasound pulse is emitted and a long duration echo signal wavetrain is recorded. This wavetrain is divided into contiguous segments, each segment identified with index ℓ. The echo signal voltage at the transducer for the ℓth segment is designated $V_s(t;\ell,m,n)$, where t is the time following emission of the pulse.

The analysis consists first of calculating "backscatter estimators"[26] for signals obtained from throughout the scanned volume. Backscatter estimators are the square moduli of the Fourier transforms of the echo voltages, divided by a term which depends on the acoustic path, the transducer beam as well as instrumental factors. Backscatter coefficients at locations in this volume are obtained by averaging estimators from neighboring points.

Backscatter estimators are calculated using

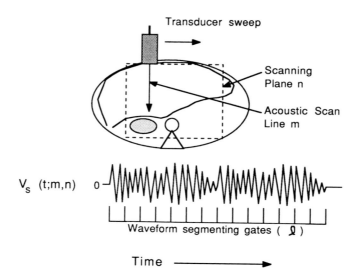

FIGURE 1. A backscatter coefficient imaging system. The transducer is scanned in a rectilinear fashion over the sample. The indices 1, m, and n refer to the echo signal voltage segment at depth increment 1, acoustic beam line m and scanning plane n.

$$BSE(\omega_o, \, \ell, \, m, \, n) \; = \; \frac{|V_s(\omega_o; \, \ell, \, m, \, n)|^2}{a_\ell(\omega_o)} \qquad (1)$$

where $V_s(\omega_o;\ell,m,n)$ is the Fourier transform of $V_s(t;\ell,m,n)$ at frequency ω_o. The denominator in Equation 1 accounts for instrumental and acoustic path dependencies on the echo signal. It is evaluated for each segment using[29-31]

$$a_\ell(\omega_o) \; = \; \left(\frac{\tau}{2\pi}\right)^2 \iiint_\Omega d\vec{r} \; \left| \int_{-\infty}^{+\infty} d\omega \; T(\omega) B_o(\omega) \; \frac{g(\omega)}{g(\omega_o)} \right.$$

$$\times \; \left. \mathrm{sinc}\left(\frac{(\omega - \omega_o)\tau}{2\pi}\right) [A_o(\vec{r}, \, \omega)]^2 \right|^2 \qquad (2)$$

where τ is the duration of a time gate used to select the ℓth segment of the signal; the *sinc* function is the representation of this temporal rectangular gate in frequency space; $T(\omega)B_o(\omega)$ represents the product of the frequency content of the emitted acoustic pulse and a transducer-receiver "force-to-voltage transfer function". The term $g(\omega)$ represents the frequency dependence of the backscatter coefficient and $g(\omega_o)$ is the value of this function at the analysis frequency. The volume Ω includes all scatterers that contribute to the signal over τ.

The parameter $A_o(\vec{r},\omega)$ accounts for the beam shape of the transducer and is proportional to the acoustic pressure at field point $\vec{r}$ and frequency ω. It is obtained by computing the integral

$$A_o(\vec{r}, \omega) = \iint\limits_S ds \, \frac{e^{ik|\vec{r} - \vec{r}'|}}{|\vec{r} - \vec{r}'|} \tag{3}$$

where S is the area of the transducer, $\vec{r}'$ ends on area element ds, and k is the complex wavenumber for the medium. k is calculated using

$$k = \frac{\omega}{c(\omega)} + i\alpha(\omega) \tag{4}$$

where $c(\omega)$ is the speed of sound and $\alpha(\omega)$ is the attenuation coefficient.

The backscatter estimators themselves may be used to construct quantitative images of the volume scanned. For uniform regions these images may be used to estimate the backscatter coefficient by averaging estimators over this region. The backscatter estimator images are subject to statistical fluctuations because of interference effects, similar to the texture in standard B-mode images. Backscatter coefficient images are obtained by averaging estimators over a local, three-dimensional volume and displaying these values for a lattice of points. Thus, the backscatter coefficient at angular frequency ω_o, corresponding to a lattice position ℓ', m', n', is given by[26]

$$\eta(\omega_o)_{\ell',m',n'} \simeq \frac{1}{(\Delta L + 1)(\Delta M + 1)(\Delta N + 1)}$$

$$\times \sum_{\ell = \ell' - (\Delta L/2)}^{\ell = \ell' + (\Delta L/2)} \sum_{m = m' - (\Delta M/2)}^{m = m' + (\Delta M/2)} \sum_{n = n' - (\Delta N/2)}^{n = n' + (\Delta N/2)} BSE(\omega_o; \ell, m, n) \tag{5}$$

The values ΔL, ΔM, and ΔN are the number of discrete steps spanned by the summation.

Although the development just outlined is for a linear scanning arrangement, analogous expressions may be written for sector scanning probes.

B. DETERMINING THE TRANSDUCER GEOMETRY

For single-element transducers, computation of the quantity, $A_o(\vec{r},\omega)$ in Equation 3 requires knowledge of the element diameter and radius of curvature. These may be obtained by exciting the transducer with a narrow-band burst with the beam propagating in water and probing the beam with a miniature hydrophone. The hydrophone is scanned perpendicular to the beam axis, searching for the "radius of curvature" of the transducer element. This can be identified from beam plots because at the radius of curvature, the resultant "lateral beam profile" is identical to the far field directivity function of an unfocused circular disk.[32] Thus the radius of curvature is taken as the axial distance to the plane in which the best fit to the far field directivity pattern is obtained. The projected diameter of the radiating element is then obtained from the angular width of the main lobe of the beam.

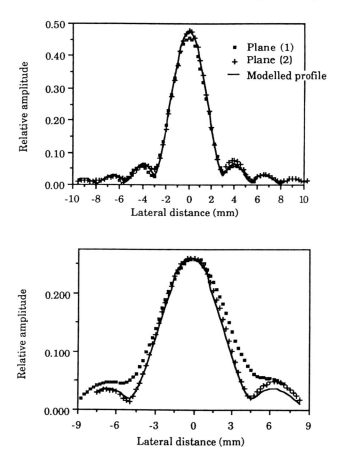

FIGURE 2. Comparison of computed and measured lateral beam profiles for a 5 MHz, 9 mm diameter, 85 mm radius of curvature transducer in a mechanical sector scanner transducer assembly at a depth of 7 cm in water (top); and at a depth of 12 cm (bottom). In both cases "plane 1" refers to a beam profile measured in the scanning plane and "plane 2" to a plot perpendicular to the scan plane. The solid line shows the computed profile. (From Boote et al., *Med. Phys.*, 19, 1145–1152, 1992. With permission.)

This method of transducer characterization has led to excellent agreement between predicted and measured beam profiles for single element probes.[32,33] It also provides very good agreement between computed and measured profiles for single element transducers in a real-time clinical ultrasound imaging system. Figure 2 shows comparisons of computed and measured lateral beam plots at depths of 7 and 12 cm from a mechanical scan head (ATL Mark III) transducer. In each diagram one experimental plot is in the scan plane of the scan head, while the other is perpendicular to the scan plane. At both depths, the in-plane profile is in excellent agreement with the profile computed using the geometric parameters determined as above. The measured profile perpendicular to the scane plane also is in agreement at depths less than the radius of curvature; beyond the radius of curvature agreement was seen for most of the main lobe, but slight disagreement was observed for points far off axis. This may be due to the shape of the window in the scan head, distorting the beam somewhat in the slice thickness direction. These distortions did not contribute a major amount of error in data reduction because the scan head was applied only for depths up to 85 mm.[28]

C. DETERMINATION OF $T(\omega)B_o(\omega)$

The acoustic pulse spectrum emitted by the transducer-pulser system and the sensitivity of the transducer-receiver system are represented by the factors $T(\omega)B_o(\omega)$ in Equation 2. $B_o(\omega)$ is used in the representation of the pulsed field from the ultrasound transducer, where the pressure is represented as a superposition of continuous wave beams varying sinusoidally in time. $B_o(\omega)$ is a complex superposition coefficient in this representation.[32,33] The scattered wave from a point in the field, incident on the surface of the transducer produces a net force on the probe, and $T(\omega)$ is the force-to-voltage transfer function for the system, in this case referred to the amplifier input. The product $T(\omega)B_o(\omega)$ can be obtained from the echo signal from a planar reflector[29] using

$$T(\omega)B_o(\omega) = \frac{V_r(\omega)}{R \iint\limits_{S_{mirror}} ds \, A_o(\vec{r}, \omega)} \tag{6}$$

where $V_r(\omega)$ is the Fourier transform of the echo signal, R is the amplitude reflection coefficient of the reflector and the integral is over the surface of a mirror image transducer at two times the transducer to reflector distance. Note, this is not a simple normalization of the echo data to the signal from a planar reflector, but allows absolute determinations of the backscatter coefficient of the material scanned.

D. TESTS IN PHANTOMS

The accuracy of the imaging methods just outlined has been evaluated using ultrasonically tissue mimicking phantoms. In one series of tests[26,27] a laboratory apparatus utilizing single element, fixed focused transducers, scanning across the phantom surface in a rectilinear fashion, was employed. Three transducers were each driven at their center frequencies (2.5, 3.5, and 5 MHz) using 3-μsec tone bursts. During each scan the probe was translated 1 mm between successive pulse-echo sequences, and the distance between parallel scanning planes also was 1 mm. For each transducer location echo signals were digitized at 50 MHz by a Le Croy TR8828C transient recorder and

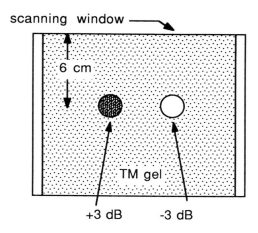

FIGURE 3.   Phantom used to test the quantitative accuracy of backscatter coefficient imaging algorithms.

stored on disk for analysis. Gate durations of 3 μsec were used in segmenting the echo signal waveforms. By using overlapping segments new BSEs were computed every 1.3 μsec, corresponding to axial increments of 1 mm.

Figure 3 presents a diagram of one test phantom used in these studies. It contains a background region that has tissue-like attenuation and scattering. The phantom also contains two cylindrical objects that exhibit backscatter contrast with respect to the background. The attenuation coefficient of the phantom was measured using a narrow-band substitution technique applied to test samples poured during construction.[34] These acoustic properties are listed in Table 4 below, where the phantom is identified as "B-A".

Figure 4 is a backscatter coefficient image of this phantom, taken with the 3.5-MHz transducer. The image was obtained by averaging BSEs (see Equation 5) over a cube having 4-mm sides; thus, each pixel in the image corresponds to 4 × 4 × 4, or 64 BSEs. The cube was shifted by one row or column for calculating a new pixel value.

Mean backscatter coefficients at all three frequencies obtained for different sections of the phantom are presented in Table 1. In column 2 of this table are shown results of calculated backscatter coefficients for the background material. These were obtained using the theory of Faran[35] applied to the glass bead scatterers in the gel matrix. The backscatter coefficients measured from the image (column 3) are within 6% of the computed values. In columns 4 and 5 are shown backscatter coefficients, expressed in dB relative to that of the background, for the lower and higher scattering cylinders. These values were determined by selecting a circular region of interest in the cylinders seen in the image and computing the mean backscatter coefficient for that region. All measured values are within $1/2$ dB of the expected scatter levels.

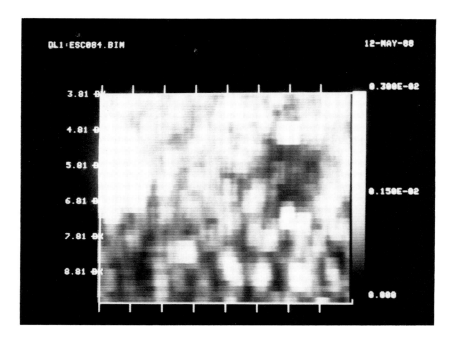

FIGURE 4. Backscatter coefficient image, obtained at a 3.5 MHz frequency of the test phantom shown in Figure 3. The ± 3 dB regions with backscatter contrast are seen. (From Boote et al., *Ultras. Imag.*, 10, 121–138, 1988. With permission.)

TABLE 1
Scattering Results Measured from Narrow-Band Backscatter Coefficient Images of the Phantom in Figure 3

Frequency ω_0 (MHz)	Backscatter coefficient of background (cm^{-1} sr^{-1})		Backscatter coefficients in cylindrical volumes in dB relative to background	
	Predicted	Experiment	−3 dB	+3 dB
2.5	2.73×10^{-4}	2.56×10^{-4}	−2.9	+3.1
3.5	9.33×10^{-4}	9.80×10^{-4}	−2.6	+3.4
5.0	3.03×10^{-3}	2.94×10^{-3}	−2.7	+3.5

Measurements of the backscatter coefficient require simultaneous determinations of attenuation coefficients in order to obtain meaningful results. In the present tests the attenuation coefficients measured using the narrow-band substitution technique applied to test samples were used in Equation 4 when determining the wavenumber for evaluating the pressure field of the transducer during the data analysis. When the attenuation coefficient is unknown, it must be estimated; this can be done as shown by previous workers,[2,10,13] for regions where both the attenuation and backscatter coefficients are constant. (See also Section III.) When the attenuation coefficient must be measured simultaneously with the backscatter coefficient, this contributes to the uncertainty in estimates of backscatter coefficients in practical situations.[24,46]

E. IMPROVED RESOLUTION IMAGES

Data displayed in Figure 3 and presented in Table 1 are accurate, absolute measurements of the backscatter coefficients of the volume scanned. However, the image texture in Figure 3 is coarse and the spatial resolution low because of the narrow-bandwidth pulses and relatively long duration time gates in the data acquisition and analysis. When short duration, broad-bandwidth pulses are applied, it becomes necessary to account for the frequency dependence of scattering in the data reduction.[27] This is introduced by the ratio $g(\omega)/g(\omega_o)$ in Equation 2, where ω_o is the analysis frequency. Under narrow-bandwidth conditions, this ratio is not expected to change significantly over the frequency range contributing to the integral in Equation 2. Consequently, the ratio has been set to unity for most previous tests of the data reduction method. For broad-bandwidth conditions, Madsen et al.[29] have suggested an iterative data analysis technique. A modified iterative technique[36,37] in which the backscatter coefficient is represented as a truncated power series in ω has worked effectively for determining the backscatter coefficient in a phantom having low attenuation and reasonably effectively in one with tissue-like attenuation.

Alternatively, accurate results with short-duration pulses and time gates have been obtained without introducing the scattering frequency dependence in the data reduction when (1) the echo spectrum is symmetric about its center frequency and (2) the analysis is restricted to the pulse center frequency.[36] These positive results have led to the utilization of broad-band pulses and short duration time gates in backscatter coefficient imaging.[27] The analysis frequency is restricted, however, to the center frequency of the echo signal spectrum.

For example, Figure 5 presents a backscatter coefficient image of the phantom in Figure 3, obtained when a 3.5-MHz transducer was excited with single cycle pulses at its center frequency. Other scan parameters, including the distance between acoustic beam lines and scan planes, were identical to those used for Figure 4. In the present case, data analysis was done for 0.5-μsec duration segments, rather than the 3 μsec segments used previously. Evaluation of Equation 2 required an integral over the entire bandwidth of the pulse. However, the analysis frequency, ω_o, was at the center frequency of the echo signal spectrum; thus, the frequency dependence of scattering was not introduced during this evaluation.

The same phantom was scanned using 2.5- and 5-MHz center frequency transducers driven with single-cycle pulses. Table 2 shows results of region

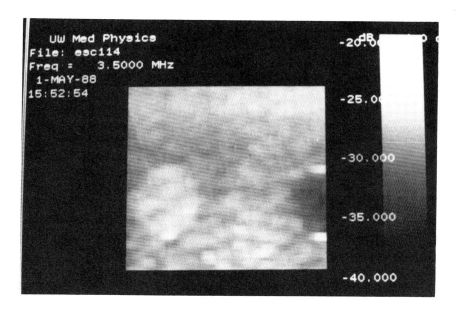

FIGURE 5. Backscatter coefficient image obtained with the same equipment as in the previous figure, only here the transducer was driven with a single cycle pulse at 3.5 MHz and the duration of the signal segmenting time gates used in the data analysis was 0.5 μsec. (From Boote et al., *Ultras. Imag.*, 13, 347–352, 1991. With permission.)

of interest analyses for each experiment. For the 3.5-MHz image, the average backscatter coefficient measured from the background region in the phantom is $8.9 \times 10^{-4} \, cm^{-1} \, sr^{-1}$. The regions containing ± 3 dB backscatter contrast exhibited a -3.6 dB and $+2.9$ dB backscatter level compared to the background. These values are in quite good agreement with predicted properties also listed in Table 2. Results for 2.5 and 5 MHz also are in excellent agreement with the expected backscatter values in the phantom. Thus, this method also yields accurate backscatter coefficient images.

Besides the apparent improvement in spatial resolution another advantage in using broad-band pulses and short duration segments in this application is the potential for reducing speckle variations through averaging. The number of statistically independent backscatter estimators increases roughly by a factor of 6 when the segment duration is decreased from 3 to 0.5 μsec.

F. *IN VIVO* BACKSCATTER COEFFICIENT IMAGES

Methods for obtaining *in vivo* backscatter coefficient images in our lab were initially developed by applying the absolute data reduction technique just outlined to echo signals acquired using a real-time clinical ultrasound scanner.[25,28] The experimental configuration is shown in Figure 6. An ATL Mark III mechanical sector scanner equipped with both a 3- and a 5-MHz

TABLE 2
Scattering Results Measured from Broad-Band Backscatter
Coefficient Images of the Phantom in Figure 3

Frequency ω_o (MHz)	Backscatter coefficient of background $(cm^{-1} sr^{-1})$		Backscatter coefficients in cylindrical volumes in dB relative to background	
	Predicted	Experiment	-3 dB	$+3$ dB
2.5	2.73×10^{-4}	3.1×10^{-4}	-2.7	$+3.5$
3.5	9.33×10^{-4}	8.9×10^{-4}	-3.6	$+2.9$
5.0	3.03×10^{-3}	2.7×10^{-3}	-3.0	$+2.8$

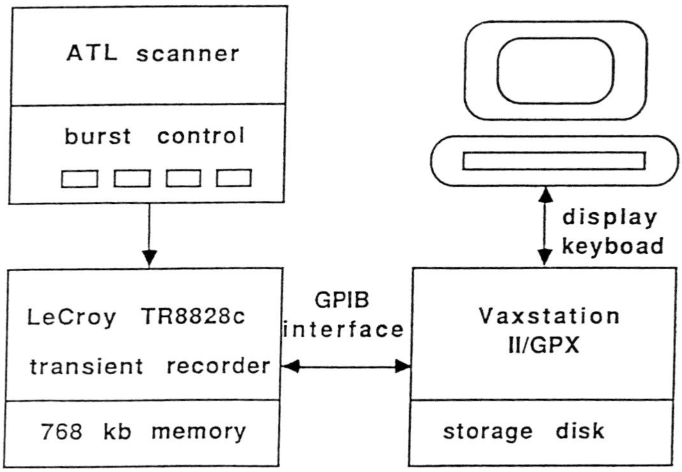

FIGURE 6. System used for *in vivo* backscatter coefficient imaging. (From Yao et al., *Ultras. Imag.*, 12, 58–70, 1990. With permission.)

probe was used. During scanning the transducer was driven with the same short-duration pulses that are applied for B-mode imaging. Radio frequency echo signals from the scanner's preamplifier were digitized in the LeCroy TR8828C transient recorder, operated in a "burst" mode. In this mode digital data are stored in the transient recorder's memory only when the store function is activated externally. A specially constructed "burst control" circuit initiated the transient recorder storage when echo signals from a user selected region of interest were present. The burst control uses frame synchronization and transmit-pulse trigger signals from the scanner, along with the beginning and ending scan lines for each frame and a time window within each scan line to activate the recorder. The scan lines and time window are set by the user. A 768-kbyte buffer memory provides storage of all echo data for each sweep of the transducer across the scan plane. Data analysis and image display was done off-line.

Because of the depth ranges involved in imaging a large organ such as the liver, it is necessary to use time gain compensation (TGC) to maintain adequate signals within the 8-bit range of the digitizer for the entire region of interest. The TGC system in the scanner was calibrated using an acoustically coupled burst generator (Nuclear Associates, Carle Place, NY), placed in contact with the transducer assembly. In response to the transducer transmit pulse, the burst generator emits simulated echo signals, the amplitude of which are controlled by externally adjusted attenuators. By monitoring the peak-to-peak signals from the preamplifier as a function of time following the transmit pulse, the amplification vs. time, $g(t)$ was determined. Look up tables were constructed so that $g(t)$ could be calculated using the ''near gain'', ''slope'', and ''far gain'' control settings. These control settings were recorded by the experimenter during each scan.

Following digitization and storage in the computer, the radio frequency data were corrected for the gain applied using

$$V_s(t; \ell, m, n) = \frac{V'_s(t; \ell, m, n)}{g(t)} \tag{7}$$

where $V'_s(t;\ell,m,n)$ is the recorded signal voltage. $V_s(t;\ell,m,n)$ is used for determining the backscatter estimators in Equation 1.

In vivo backscatter coefficient images were obtained in five individuals. The subjects for this study were normal, healthy adult volunteers, four males and one female between the ages of 24 and 27. Before data were taken using the ATL system, the subjects were scanned using an Acuson Model 128 ultrasound scanner and Kitecko© (3M Co.) standoff pads. This was done in order to estimate the body wall thickness at the site where subsequent backscatter coefficient images would be obtained. A 5 MHz Acuson transducer (L538) was aimed in either a transverse or longitudinal direction below the rib cage approximately 8 cm to the right of the midline. Measurements of the thickness of the subcutaneous fat layer and the abdominal muscles were made from the interfaces between each layer using the calibrated cursors on the Acuson scanner. These thickness measurements were recorded for subsequent use in the backscatter coefficient image data analysis.

The subject was then scanned using the ATL system. Echo signals were recorded from the right anterior lobe of the liver through the same subcostal acoustic window from which the previous thickness measurements were made. Data were acquired for four parallel image sectors, each approximately 1 mm apart, while the subject held his breath. Following data acquisition from the liver the transducer was mounted above a planar reflector in a water tank, the instrument was placed in M-mode and the echo signal from the reflector was recorded. This signal was used to determine $T(\omega)B_o(\omega)$ as discussed above.

The analysis frequency, ω_o, applied to the 3-MHz transducer data was 2.75 MHz; this coincided closely with the peak of the function $T(\omega)B_o(\omega)$ determined with this transducer. Data acquired with the 5-MHz transducer were reduced using a 4-MHz analysis frequency, since most of the echo signal was centered about this frequency. The homogeneous tissue mimicking phantom diagrammed in Figure 3 was used to test accuracy for both transducers. At 2.75 MHz a backscatter coefficient of 4.55×10^{-4} cm^{-1} sr^{-1} was measured from the quantitative image of this phantom; this compares to a predicted backscatter coefficient of 4.0×10^{-4} cm^{-1} sr^{-1}. At 4.0 MHz, the experimental result was 1.73×10^{-3} cm^{-1} sr^{-1}, compared to the predicted value of 1.5×10^{-3} cm^{-1} sr^{-1}. Results from the regions of backscatter contrast were in agreement to within ± 0.5 dB of the actual backscatter coefficients.

For analysis of the human data, fat and muscle layers were taken into account when computing the transducer pressure fields for use in Equation 2. The approach followed was to estimate an effective speed of sound, $c(\omega)_{eff}$, and attenuation coefficient, $\alpha(\omega)_{eff}$ and use these in Equation 4 for the wave number. With the transducer in contact with the tissue, and at depth z along the beam axis, these parameters are estimated using[28]

$$c(\omega)_{eff} \simeq \frac{1}{z} [c(\omega)_{fat} z_{fat} + c(\omega)_{muscle} z_{muscle} + c(\omega)_{tissue} \qquad (8)$$

$$(z - z_{fat} - z_{muscle})]$$

and   (9)

$$\alpha(\omega)_{eff} \simeq \frac{1}{z} [\alpha(\omega)_{fat} z_{fat} + \alpha(\omega)_{muscle} z_{muscle} + \alpha(\omega)_{tissue}$$

$$(z - z_{fat} - z_{muscle})]$$

where $z_i, c_i(\omega)$, and $\alpha_i(\omega)$ are the thicknesses of these layers, their speeds of sound, and their attenuation coefficients. The speeds of sound and attenuation coefficients used were 1460 m s^{-1} and 0.6 dB cm^{-1} MHz^{-1} for fat; and 1600 m s^{-1} and 1.3 dB cm^{-1} MHz^{-1} for muscle.[38] The attenuation and speed of sound in normal liver were assumed to be 0.5 dB cm^{-1} MHz^{-1} and 1570 m s^{-1}, respectively. The *in vivo* speed of sound in liver is from a very recent clinical study of 21 normal, healthy volunteers.[39]

Figure 7a is a backscatter coefficient image from one subject. This is a longitudinal image of the liver, obtained by averaging backscatter estimators from all four scan planes. The right kidney is visible at the lower right of this image, while the remaining part of the image is the right lobe of the liver. For comparison, Figure 7b is the corresponding B-mode image, taken

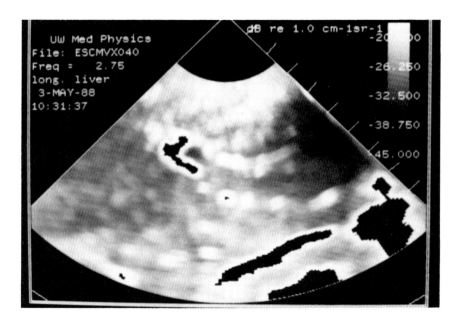

a

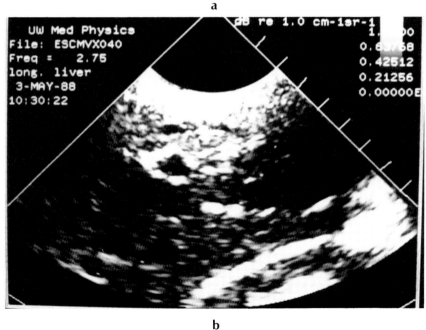

b

FIGURE 7. (a) Backscatter coefficient image obtained at 2.75 MHz, of the right lobe of the liver in a normal adult. (b) B-mode image of one of the scanning planes used in forming the backscatter coefficient image. (From Boote et al., *Med. Phys.*, 19, 1145–1152, 1992. With permission.)

from a central plane in the scan series. The backscatter coefficient image displays a rather uniform shade of gray over the liver, indicating a consistent backscatter coefficient. The attenuation estimate used in the data analysis (0.5 dB cm^{-1} MHz^{-1}) seems reasonable considering the relatively uniform brightness with respect to depth. Larger vessels are visible on the backscatter coefficient image, even though this image is smeared due to the averaging process. These large vessel or organ interfaces should be avoided in any region of interest analysis, or else artificially high backscatter results would be obtained.

Following formation of each backscatter coefficient image, a region of interest analysis was performed on the liver parenchyma. The analysis was aimed at regions where a fairly uniform gray level was seen, with no significant infiltration of large reflecting surfaces. Table 3 is a summary of body wall thickness and backscatter coefficient measurements for the five subjects in this study. The overall mean backscatter coefficient at 2.75 MHz was 7.9 × 10^{-4} cm^{-1} sr^{-1}, with a standard deviation of 2.3 × 10^{-4} cm^{-1} sr^{-1}. At 4.0 MHz, the mean backscatter coefficient was 3.1 × 10^{-3} cm^{-1} sr^{-1}, with a standard deviation of 1.7 × 10^{-3} cm^{-1} sr^{-1}.

The results at 2.75 MHz are in fairly good agreement with *in vitro* results reported by Nicholas.[20] However, results at 4 MHz are somewhat higher than the Nicholas' results (1.5 × 10^{-3} cm^{-1} sr^{-1}). This may have been related to differences between *in vivo* and *in vitro* results; differences in materials and/or experimental subjects; or experimental errors introduced by electronic noise, choice of the attenuation coefficient assumed in the data reduction and the model for estimating body wall losses.

III. QUANTITATIVE IMAGING USING A REFERENCE PHANTOM

A. DETERMINING ATTENUATION AND BACKSCATTER COEFFICIENTS

The accuracy of the previously described method for determining backscatter coefficients is related to the fact that detailed transducer pressure fields are included and the temporal nature of the measurement process is maintained throughout the data reduction. However, an important consideration in that method is the time needed for computations and the need for accurate specification of the size and shape of the radiating aperture.

Hall[40] has shown that accurate backscatter results may be obtained over a sizeable range of the acoustic field using beam approximations rather than the extensive modeling as in Equation 3. Nevertheless, for complex transducer systems, such as arrays, where the transmitting and receiving apertures often are different, apodization is present, and dynamic focusing is used, accurate, three-dimensional field specification may still be a difficult task. In addition, the need to use time varied gain further complicates the task of quantifying

TABLE 3
Body Wall Thicknesses and Backscatter Coefficients from Five Normal Adults

Subject no.	Body wall thickness (mm)		Backscatter coefficients measured *in vivo*	
	Fat	Muscle	2.75 MHz	4.0 MHz
1	4	9	7.2×10^{-4}	1.1×10^{-3}
2	7	12.5	5.7×10^{-4}	5.4×10^{-3}
3	4.5	14.5	9.6×10^{-4}	4.1×10^{-3}
4	7	13	1.1×10^{-3}	2.8×10^{-3}
5	4	11	6.0×10^{-4}	1.9×10^{-3}
Mean ± SD			$7.9 \pm 2.3 \times 10^{-4}$	$3.1 \pm 1.7 \times 10^{-3}$

backscatter. Although future scanners could easily have absolute calibrations of TGC settings built into the instrument, present scanners still require detailed experiments such as described in Section II to achieve the necessary calibration of the gain controls and pulser-receiver characteristics.

An alternative processing method[41] that does not require explicit knowledge of the transducer beam pattern or the transmission and reception properties of the pulse-echo instrument will now be outlined. This method involves comparison of echo data from the sample with data recorded from a reference phantom, for which the backscatter and attenuation coefficients are known. Echo signals are first obtained from the area of interest in the patient or sample after optimizing system sensitivity controls, TGC, etc. Then with the same equipment settings, data are recorded from the reference phantom. Ratios of the square of the echo amplitude from the sample and the reference phantom yield the backscatter and attenuation coefficients of the sample.

In most cases the reference phantom has uniform acoustic properties. We assume for the moment the same situation exists for the sample. For both data sets, the analysis consists of narrow-band filtration and computation of the square of the amplitude of the echo signal vs. time, $i(\omega_o, t)$, where ω_o is the analysis frequency. These are used to calculate a mean square echo signal, $I(\omega_o, t)$ for the sample and a corresponding $I'(\omega_o, t)$ for the reference phantom.

Let $\eta(\omega_o)$ and $\alpha(\omega_o)$ represent the backscatter and attenuation coefficients at angular frequency ω_o for the sample and $\eta'(\omega_o)$ and $\alpha'(\omega_o)$ the corresponding values for the reference phantom. It is shown in Reference 41 that the depth-dependent ratio of the filtered, amplitude-squared echo signal is

$$\frac{I(\omega_o, t)}{I'(\omega_o, t)} = \frac{\eta(\omega_o)e^{-4\alpha(\omega_o)z}}{\eta'(\omega_o)e^{-4\alpha'(\omega_o)z}} = \frac{I(\omega_o, z)}{I'(\omega_o, z)} \qquad (10)$$

Here we have mapped the signal at time t to the signal at range $z = ct/2$, where c is the speed of sound. Also we have assumed the same speed of sound for both media. The ratio of the $I(\omega_o, z)$'s for the unknown sample and

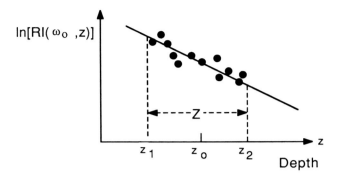

FIGURE 8.   Typical appearance of ln(RI(ω_o,z)), that is the ratio of the filtered average amplitude squared signa, vs. depth.

the reference phantom at each depth and frequency contains the backscatter coefficient and attenuation coefficient of the sample.

Let $RI(\omega_o,z)$ be the ratio of the average squared amplitude signal from the sample to that from the reference phantom for depth z. Also, let $RB(\omega_o)$ be the ratio of the backscatter coefficients of the sample and the reference phantom, and let $\Delta\alpha(\omega_o) = \alpha(\omega_o) - \alpha'(\omega_o)$ be the difference in attenuation coefficients of the media. Equation 10 then becomes

$$RI(\omega_o, z) = RB(\omega_o)e^{-4\Delta\alpha(\omega_o)z} \tag{11}$$

This is the basic equation used to compute the attenuation and backscatter coefficients of the sample. Define

$$X(\omega_o, z) \equiv \ln[RI(\omega_o, z)] \tag{12}$$

Taking logarithms of both sides of Equation 11, we have

$$X(\omega_o, z) = \ln[RB(\omega_o)] - 4\Delta\alpha(\omega_o)z \tag{13}$$

A least squares analysis is used to fit the function $X(\omega_o,z)$ vs. depth to a straight line (Figure 8). The slope of this line yields $\Delta\alpha(\omega_o)$ and, hence, the attenuation coefficient for the region of interest in the sample. Once $\Delta\alpha(\omega_o)$ is known, the backscatter coefficient ratio can be determined. The backscatter coefficient of the sample is then computed using

$$\eta(\omega_o) = \eta'(\omega_o)\overline{RI(\omega_o, z)e^{4\Delta\alpha(\omega_o)z}} \tag{14}$$

where the bar implies an average over a depth interval for which the phantom is assumed to have uniform properties.

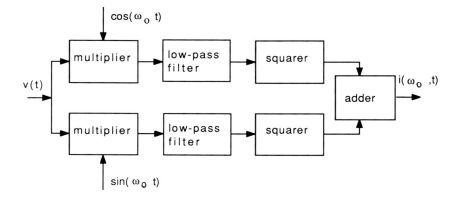

FIGURE 9. Signal processing used to form the filtered amplitude squared data in the reference phantom method.

B. SIGNAL PROCESSING

The signal processing for obtaining the filtered, amplitude-squared echo data is shown in Figure 9. $V(\omega_o,t)$, the amplitude of the signal at frequency ω_o and depth $z = ct/2$ is derived from the echo signal $v(t)$ using

$$V(\omega_o,\ t)\ =\ \int_0^\infty v(t')p(t\ -\ t')e^{-i\omega_o t'}\ dt' \tag{15}$$

$$=\ \int_0^\infty v(t')\cos(\omega_o t')p(t\ -\ t')t'\ dt'$$

$$+\ i\int_0^\infty v(t')\ \sin(\omega_o t')p(t\ -\ t')\ dt' \tag{16}$$

where $p(t)$ is a time window. Thus, the echo signal is fed into two channels and multiplied by orthogonal sinusoidal waves with frequency ω_O. The voltages output from the low-pass filters are the real and imaginary parts of $V(\omega_o,t)$ weighted by the window area (see below). These are squared and added, forming the squared amplitude data, $i(\omega_o,t)$ for each beam line. The analysis frequency can be flexibly changed by changing the frequency of the reference sinusoidal waves. Thus, the method can be easily implemented in hardware, conceivably in real time. For the tests of the method, all processing was done in software using routines written in Fortran.

The low-pass filter $p(t)$ in the signal processing is a 3-term Blackman-Harris window,[42] expressed as

$$p(t)\ =\ a_o\ -\ a_1\cos(bt)\ +\ a_2\cos(2bt) \tag{17}$$

TABLE 4
Properties of Phantoms Used in Tests of the Reference Phantom Method

Phantom	Attenuation coefficient α_1 dB/cm/MHz	α_2 dB/cm/MHz^2	Speed of sound (m/s)	Density (g/cm)	Scatterer diameter (μm)
H-2	0.04592	0.0127	1532	1.00	58.8
H-3	0.14050	0.0252	1587	1.02	88.6
H-4	0.44420	0.0198	1580	1.05	88.6
B-A	0.48286	0.0185	1525	1.04	60.5
TM-Liver	0.52900	-0.0020	1540	1.05	—

Here $b = 2\,\pi\tau$, where τ is the duration of the time window, $a_o = 0.42323$, $a_1 = 0.49755$, and $a_2 = 0.07922$. The spectrum of this window has very low side lobes, the highest being 67 dB below the main lobe.[42]

The duration of the window determines the width of the frequency band of the bandpass filter used to obtain $i(\omega_o,t)$ from the broad-band echo signal. In applications using pulses from a clinical transducer the resultant frequency spectrum of the filtered signal is the product of the power spectrum of the echo signal and the transfer function of the bandpass filters. The average frequency of the filtered spectrum tends to shift towards the frequency at which the power spectrum of the echo signal has the greatest value. Thus, a narrow frequency band is desired to guarantee a small frequency shift. However, this also means a long duration time window, which would deteriorate the axial resolution.

To find an optimal window size,[43] radio frequency echo data from two separate phantoms were acquired using the ATL scanner and 5 MHz center frequency transducer. The phantoms are labeled H-3 and H-4 in Table 4. They have identical backscatter coefficients, the level being controlled by the presence of 88.6 μm diameter glass beads. They differ, however in that H-4 has tissue-like attenuation, while H-3 has much lower attenuation. Echo signals from both phantoms were analyzed at frequencies of 4.25, 4.5, 4.75, 5.0, 5.25, 5.5, and 5.75 MHz. This analysis was done for a range of window durations from 0.25 μs to 32 μs.

$RI(\omega_o,z)$ the ratios of the average amplitude-squared from the H-4 to that from the H-3 phantom are plotted vs. the window duration in Figure 10. This ratio is less than one at all frequencies because of the higher attenuation in the H-4 phantom. Since attenuation is frequency dependent, the ratio should be lower for higher frequency components of the echo spectrum. As expected, successful detection of this feature depends on the window size. For short duration windows, the ratios at high frequencies tend to have the same values as those at lower frequencies. In these situations the filtering function has a very wide spectral range, and the output of the bandpass filters includes nearly

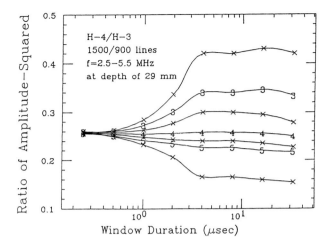

FIGURE 10. RI(ω_o,z) vs. window duration for echo data acquired from two tissue mimicking phantoms that have different attenuation coefficients. Analysis frequencies ranging from 3.0 MHz (top) to 5.5 MHz (bottom) were applied to echo signals from a 5-MHz center frequency pulse. For the frequency ranges used in the tests of the reference phantom method, window durations of 4 μsec or longer produce stable, accurate results.

the entire echo signal spectrum no matter what the analysis frequency, ω_o, is. Thus, the amplitude-squared ratios at different frequencies are practically identical. However, for a window duration of 4 μs and greater, the results appear to be stable, i.e., they do not change with increasing window duration. Similar results were obtained for the attenuation coefficients measured from the data.[43] In all subsequent applications of this method discussed in this chapter 4 μs time windows have been used.

C. TESTS OF THE REFERENCE PHANTOM METHOD

Phantoms whose properties are summarized in Table 4 were used to evaluate the accuracy of the method. The scattering medium in all but one (TM-liver) consists of microscopic glass spheres randomly distributed in a gel matrix. The backscatter coefficients for these samples can be calculated using the theory of Faran[35] and measured using the absolute method. Phantoms H-4 and B-A also have very finely powdered graphite uniformly distributed throughout the medium, which raises the ultrasonic absorption to that of soft tissues.[44] The graphite particles are small enough that their contribution to the backscatter coefficient is negligible compared to that of the glass beads.

TM-Liver is of different construction from the others.[45] This is a complex phantom with agar spheres surrounded by a matrix material consisting of animal hide gel laced with graphite particles. The agar spheres contain no graphite and have a lower speed of sound and density than the animal hide gel matrix. The diameter of these spheres ranges from about 0.5 to about 2.8

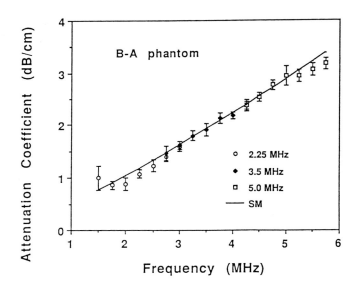

FIGURE 11. Attenuation coefficient vs. frequency for the B-A phantom, where phantom H-2 was used as the reference. Three different transducers, each with a different center frequency, were used to span the frequency range. The solid line is from independent, substitution measurements of attenuation in the sample. (From Yao et al., *Ultras. Imag.*, 12, 58–70, 1990. With permission.)

mm with a peak in the distribution at about 1.5 mm. The matrix gel has graphite particles ranging in size from 0.1 to 90 μm. The combination of scatterers has been shown to provide a backscatter coefficient frequency dependence mimicking that of normal *in vitro* liver tissues.[45]

Figures 11 and 12 show measured attenuation and backscatter coefficients for the B-A phantom, where the H-2 phantom was used as the reference. Three different transducers, with center frequencies of 2.25, 3.5, and 5.0 MHz, were used to extend the measurements over the range shown. The error bars represent ±1 standard deviation for ten such measurements. The continuous curve in Figure 11 is calculated using $\alpha(f) = (0.48286f + 0.0185f^2)$ dB/cm, where f is the frequency in MHz. These parameters were obtained by curve fitting the attenuation data from narrow-band substitution measurements[45] applied to test cylinders containing the same material as in the B-A phantom. Results obtained with the reference phantom method are shown by data points.

In Figure 12, the solid curve is the result of curve fitting backscatter coefficients measured using the absolute method;[29] the data points are results of measurements using the reference phantom. Except for the 28% error at higher frequencies, the reference phantom method did quite well in yielding the backscatter coefficient.

The H-4 phantom was used as a reference to measure the backscatter coefficients of TM-Liver. The results are shown in Figure 13. Also shown

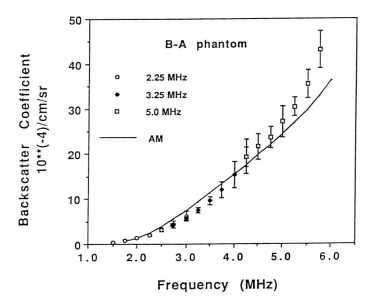

FIGURE 12. Backscatter coefficients vs. frequency for the B-A phantom. Data points are from the reference phantom method while the solid line is a result of curve fitting data obtained using the absolute data reduction method. (From Yao et al., *Ultras. Imag.*, 12, 58–70, 1990. With permission.)

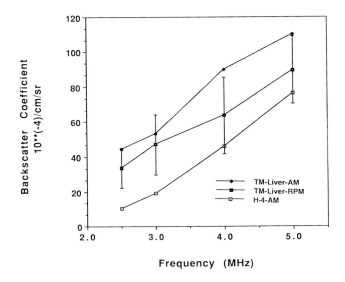

FIGURE 13. Backscatter coefficient vs. frequency in TM-Liver, measured using the reference phantom method (RPM). Results for the absolute data reduction method (AM) are shown for comparison, as are backscatter coefficients in the reference phantom (H-4-AM). (Adapted from Yao et al., *Ultras. Imag.*, 12, 58–70, 1990.)

are the results from the absolute method applied to small samples of TM-Liver. The frequency dependence of the backscatter coefficient of TM-Liver is quite different from that of the H-4 Rayleigh scattering phantom. Still, the reference phantom method appears to yield backscatter coefficient results that are in agreement with those of the absolute method.

The reference phantom method appears to work well for determining the frequency dependent attenuation and backscatter coefficient in volumes where these parameters can be considered uniform. The method yielded results that are in agreement with those obtained using an absolute data reduction method, even when the frequency dependence of scattering in the sample was different from that of the reference phantom. This finding has important implications since the TM-liver phantom used in this study has a frequency dependence of scattering that mimics scattering from human liver.

D. *IN VIVO* QUANTITATIVE BACKSCATTER IMAGES

Quantitative images of livers of normal individuals have also been obtained using the reference phantom method. These images are of *backscatter estimators*, and are obtained for discrete frequencies over the bandwidth of the echo signal.

Equipment was identical to that shown in Figure 6, except a Siemens Sonoline SL-1 scanner and a 3.5 MHz mechanicaly scanned transducer were used for data acquisition. Ten individuals ranging in age from 15 to 54, all assumed to have normal livers, served as subjects. Echo data were acquired from four closely spaced planes that included the right lobe of the liver and the right kidney. Following each liver scan, echo signals were recorded from the B-A phantom using identical instrument settings; the B-A phantom served as the reference. A 7.5 MHz transducer assembly was used to obtain a B-mode image of the body wall and superficial tissues for estimates of fat and muscle layer thicknesses over the scanned region.

Radio frequency echo data were analyzed at frequencies of 2.25, 2.50, 2.75, 3.0, 3.25, 3.5, and 3.75 MHz. The interval between computed data segments along each scan line was 0.64 μsec, corresponding to an axial increment, Δz of 0.5 mm, and a 4-μsec Blackman-Harris window was used. Data from all scans were analyzed to obtain individual $i(\omega_o,\ell,m,n)$ at each frequency, where the integers ℓ, m, and n specify the data segment, acoustic beam line, and scan plane, respectively. Thus, the depth, z of a data segment is given by $z = \Delta z \times \ell$. The filtered, amplitude-squared signals vs. depth for the reference phantom were averaged over all beam lines to form $I'(\omega_o,\ell)$ for each analysis frequency.

$ri(\omega_o,\ell,m,n)$, the ratio of the echo amplitude-squared from segment (ℓ,m,n) obtained from the subject to the average amplitude-squared at the corresponding depth from the reference phantom, was then computed for each data segment for all four scans. These were used to estimate $\Delta\alpha(\omega_o)$ and subsequently, attenuation coefficients for the liver at each analysis frequency. The

TABLE 5
Attenuation Coefficient of Normal Liver *In Vivo*

Attenuation Coefficient in dB/cm
(Standard Deviation in dB/cm)

Subject no.	Frequency in MHz						
	2.25	2.50	2.75	3.00	3.25	3.50	3.75
1	1.13	1.26	1.36	1.43	1.56	1.72	1.93
	(.052)	(.076)	(.048)	(.064)	(.084)	(.112)	(.105)
2	1.13	1.23	1.29	1.50	1.65	1.77	1.95
	(.051)	(.065)	(.070)	(.080)	(.048)	(.065)	(.071)
3	1.48	1.62	1.75	1.81	1.88	2.00	2.11
	(.028)	(.013)	(.074)	(.081)	(.099)	(.139)	(.119)
4	1.36	1.48	1.63	1.77	1.90	2.08	2.24
	(.106)	(.108)	(.102)	(.114)	(.107)	(.120)	(.116)
5	1.07	1.27	1.41	1.52	1.67	1.83	1.91
	(.177)	(.170)	(.194)	(.172)	(.150)	(.154)	(.143)
6	1.02	1.05	1.24	1.33	1.42	1.53	1.66
	(.041)	(.033)	(.069)	(.056)	(.067)	(.071)	(.074)
7	1.39	1.51	1.59	1.62	1.73	1.84	2.04
	(.053)	(.048)	(.068)	(.100)	(.171)	(.215)	(.193)
8	1.26	1.34	1.56	1.65	1.76	1.99	2.17
	(.081)	(.064)	(.053)	(.082)	(.045)	(.016)	(.087)
9	1.44	1.57	1.72	1.86	1.93	1.99	2.01
	(.079)	(.070)	(.070)	(.057)	(.095)	(.129)	(.103)
10	1.37	1.53	1.62	1.69	1.78	1.90	1.94
	(.030)	(.036)	(.051)	(.037)	(.059)	(.077)	(.087)
Avg.	1.26	1.39	1.52	1.62	1.73	1.87	2.00
Avg./freq.	0.560	0.556	0.553	0.540	0.532	0.534	0.533

method for computing the attenuation coefficient has already been outlined following Equation 11.

Results for the attenuation coefficient vs. frequency for all ten subjects are presented in Table 5; average attenuation coefficients are plotted in Figure 14. Values in parentheses in the table are standard deviations at individual analysis frequencies for each subject. The measured attenuation coefficients in the subjects in this series are very close to 0.53 dB/cm/MHz, agreeing with other researchers estimates of attenuation in normal liver.[10,24]

Backscatter estimators are related to the square of the filtered echo signal, corrected for instrumentation and propagation path dependencies. Matrices of backscatter estimators were computed for each image plane and frequency by correcting the $ri(\omega_o,\ell,m,n)$ for the measured attenuation in the liver and the *estimated* attenuation in the body wall, and multiplying by $\eta'(\omega_o)$, the backscatter coefficient of the reference phantom. Similar to Equation 14 for the case of a uniform sample, we have for the backscatter estimators,

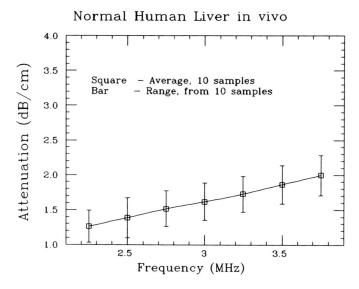

FIGURE 14.   Average attenuation coefficient vs. frequency in the liver for ten normal individuals.

$$BSE(\omega_o, \ell, m, n)_{rp} = \eta'(\omega_o)ri(\omega_o, \ell, m, n)e^{4\int_o^z \Delta\alpha(\omega_o,z',m,n)dz'} \qquad (18)$$

where $\Delta\alpha(\omega_o,z',m,n)$ is the difference between the attenuation coefficients of the reference phantom and of the tissue traversed by beam line m in scanning plane n, at depth z'. A simple-one dimensional model was used in these attenuation corrections (Figure 15). Estimates of thicknesses of overlying fat and muscle layers were made from the 7.5-MHz B-mode image; attenuation coefficients of 0.6 dB cm^{-1} MHz^{-1} and 1.3 dB cm^{-1} MHz^{-1} were used for fat and muscle tissue, respectively.

Images of the backscatter estimators from one scanning plane, obtained from a 46-year-old male subject are presented in Figure 16. The eight images in this set correspond to analysis frequencies ranging from 2.25 MHz (upper left) to 3.75 MHz (second image from the lower right). The lower right panel is of a broad-band image, representing all frequencies in the pulse. Notice that the scattering level in the parenchyma of the kidney is lower than that of the liver, especially at lower frequencies.

These are new and potentially very useful images derived from ultrasound echo data. They depict absolute scattering levels in the liver, kidney, and surrounding tissues. The gray scale has been calibrated such that midgray represents 15 dB above a reference level of 1×10^{-4} cm^{-1} sr^{-1}. Statistical fluctuations in these backscatter estimator images can be reduced by averaging, as shown in backscatter coefficient images generated using the absolute data reduction method. In addition, the frequency dependence of scattering

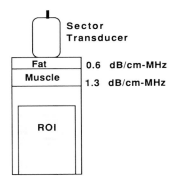

FIGURE 15.   Model for the body wall and region of interest, used to compensate for attenuation along each beam line to form quantitative backscatter estimator images.

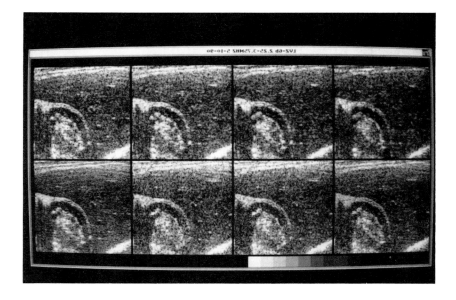

FIGURE 16.   Set of backscatter estimator images for discrete analysis frequencies over the bandwidth of a 3.5-MHz transducer. Frequencies range from 2.25 MHz (upper left) to 3.75 MHz (second from lower right) in 0.25-MHz steps. The lower right image is constructed from the broad-band signal. The center gray bar represents 15 dB above 1×10^{-4} cm^{-1} sr^{-1}, and each step in the gray bar is a 4-dB increment in the backscatter coefficient.

over the bandwidth of the echo spectrum can be appreciated from these images. Comparisons of the images at the higher analysis frequencies with those at the lower analysis frequencies suggests that the frequency dependence of scattering in the liver changes slightly over the bandwidth, with higher scattering at 3.75 MHz than at 2.25 MHz.

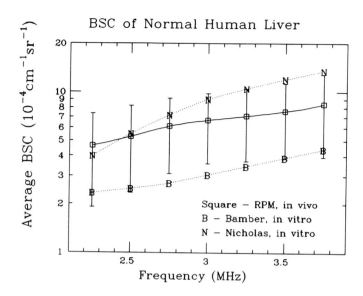

FIGURE 17.   Backscatter coefficients vs. frequency for ten individuals. Also shown are results from *in vitro* studies by Nicholas (N) (Reference 20) and Bamber (B) (Reference 19).

For each image and each subject a region of interest was selected where there were no large vessels or interfaces. The backscatter coefficient from this region was computed for each frequency by averaging backscatter estimators. Figure 17 presents backscatter coefficient vs. frequency for all ten subjects. The average backscatter coefficient at 2.25 MHz (4.6×10^{-4} cm^{-1} sr^{-1} agrees reasonably well with results measured earlier at this frequency by O'Donnell.[24] He found an average backscatter coefficient of 3.5×10^{-4} cm^{-1} sr^{-1} in six normal subjects, though he used a different estimation for body wall losses and processed envelope detected echo signals from the scan converter of a clinical ultrasound imager. The apparent increase in backscatter coefficient with increasing frequency, seen in Figure 16, is evident from the results in Figure 17. Also shown in this figure are backscatter coefficients obtained by Nicholas[20] and Bamber[19] for *in vitro* liver. The measured backscatter coefficients in the present study appear to be intermediate between those of these earlier workers, and the level of agreement with those workers' *in vitro* data is interesting.

The results in this section show that it is possible to construct accurate, quantitative images which depict absolutely the frequency dependent scattering properties of tissues.

IV. STATISTICAL UNCERTAINTIES

There are two categories of uncertainties present in these measurements of backscatter and attenuation coefficients. One is methodological and instrumental due, for example, to electronic noise and instrument inaccuracies, effects of overlying tissues, reverberations, and failure to account for non-uniformities, if present in the medium. Many of these effects, particularly of the overlying tissues, still are under investigation by various investigators. The second category includes errors due to statistical uncertainties introduced by the random nature of the echo signals from media containing spatially randomly distributed scatterers. These are discussed in this section.

Error analysis has been done for measurements of backscatter and atten-uation coefficients done using the reference phantom method.[46] It is assumed that the sample is macroscopically uniform and has a large number of randomly distributed scatterers. Thus, the filtered, amplitude squared echo signal, $i(\omega_o,z)$, is a random variable with expectation value $<i(\omega_o,z)>$ and standard deviation $\sigma_i(\omega_o,z)$. The quantity,

$$SNR_o = \frac{\langle i(\omega_o, z)\rangle}{\sigma(\omega_o, z)} \tag{19}$$

is termed the "signal-to-noise ratio". SNR_o is 1 if Rayleigh statistics apply,[47] although the analysis is not restricted to these conditions.

For any region in the sample, the backscatter and attenuation coefficients are determined from $RI(\omega_o,z)$, the ratio of the mean squared echo signal from the sample to that obtained from the reference phantom. Let σ_{RI} be the standard deviation of this ratio. If N statistically independent echo signal segments are obtained from the region of interest in the sample, and if N' echo signal samples are obtained from the reference phantom, Yao et al.[46] show that

$$\frac{\sigma_{RI}^2}{RI^2} = \frac{N + N'}{NN'} k^2 \tag{20}$$

where $k \equiv 1/SNR_o$. That is, this ratio depends only on the number of inde-pendent samples. The difference between the attenuation coefficient of the sample and the reference phantom is obtained from the slope of $ln[RI(\omega_o,z)]$ vs. z. Using an expression for the standard deviation of the slope due to fluctuations in a linear fitted quantity,[48] σ_α, the standard deviation of the attenuation coefficient of the sample is shown to be[46]

$$\sigma_\alpha = \frac{k\sqrt{3} \sqrt{N + N'}}{2\sqrt{n_z} Z\sqrt{NN'}} \quad \text{(nepers/cm)} \tag{21}$$

where Z is the depth range involved and n_z is the number of independent measurements of RI along each line within Z. In general, n_z is proportional to Z, so the standard deviation is related to the 3/2 power of the depth interval over which the slope is estimated.

A quantity $rb(\omega_o,z)$, an estimator of $RB(\omega_o)$ at depth z, is computed using

$$rb(\omega_o, z) = RI(\omega_o, z)e^{4\Delta\alpha(\omega_o)z} \qquad (22)$$

The actual backscatter ratio $RB(\omega_o)$ is obtained by averaging the $rb(\omega_o,z)$'s over an interval assumed to have the same properties. If there are n_z independent data points within this interval, then straightforward error propagation[46] for the standard deviation of $RB(\omega_o)$ leads to:

$$\left(\frac{\sigma_{RB}}{RB}\right)^2 = \frac{k^2(N + N')}{n_z NN'} + \frac{12k^2\overline{z^2}(N + N')}{n_z Z^2 NN'} \qquad (23)$$

where $\overline{z^2}$ is the mean square depth from which data are acquired for the region.

Equation 23 shows that the statistical uncertainty of $RB(\omega_o)$, and hence, of the backscatter coefficient of the sample, has two sources. The first is related to the statistical uncertainty in the $RI(\omega_o,z)$ ratio itself, which is random and depends on the number of independent acoustic lines as well as the number of samples along each line. The second term in this equation is propagated from the statistical error in the attenuation estimation, and depends also on the depth of the region. If the depth interval, Z, is from $z_o - Z/2$ to $z_o + Z/2$, the mean square depth is estimated using

$$\overline{z^2} = \frac{1}{Z} \int_{z_0 - (Z/2)}^{z_0 + (Z/2)} z^2 \, dz = z_0^2 + \frac{Z^2}{12} \qquad (24)$$

Thus, the statistical uncertainty of the attenuation estimation introduces a dependence on z_o, the center depth of the region (see Figure 8), on the backscatter coefficient uncertainty.

As an example, attenuation coefficients and backscatter coefficients were measured for the H-3 phantom (Table 4) using H-2 as a reference. For each phantom, backscatter signals were recorded from 50 acoustic lines over a range of 4 cm. From these data $\Delta\alpha$, the difference between the attenuation coefficients of the sample and the reference as well as the backscatter coefficient of the sample were estimated. This process was repeated a total of 20 times to calculate experimental standard deviations.

Measurement results of $\Delta\alpha$ are presented in Table 6 for seven different frequencies. Column 2 in this table lists the attenuation coefficient differences obtained from substitution measurements[34] and column 3 lists those computed from scattered echo signals using the methods described in this chapter. The experimentally determined standard deviations are shown in column 4 of this table. The expected standard deviation was computed using Equation 21, with N and N' both 50 and $n_z = 27.2$ (6.8 independent samples per cm, 4-cm range). The result is 0.072 dB/cm, and is in reasonable agreement with the standard deviations reported in column 4. When data from a smaller depth

TABLE 6
Difference Between Attenuation Coefficients of a Test Phantom and a Reference Phantom, Both for Substitution and Reference Phantom Measurements

Frequency (MHz)	Substitution Attenuation difference (dB/cm)	Reference phantom Attenuation difference (dB/cm)	Standard deviation (4 cm range) (dB/cm)	Standard deviation (2 cm range) (dB/cm)
4.25	1.193	1.291	0.071	0.208
4.50	1.257	1.355	0.093	0.208
4.75	1.321	1.431	0.078	0.157
5.00	1.384	1.506	0.069	0.164
5.25	1.446	1.581	0.054	0.197
5.50	1.507	1.656	0.069	0.240
5.75	1.568	1.732	0.069	0.194

interval are used to estimate $\Delta\alpha$, both n_z and Z vary, and this affects the statistical error. Column 5 presents experimental standard deviations for the attenuation coefficient when only a 2 cm depth interval is used. The expected standard deviation in this case is 0.204 dB/cm.

Figure 18 shows predicted and measured standard deviations in the backscatter coefficient of the H-3 sample. When echo signals over the 1 to 5 cm depth interval are used, the predicted percent standard deviation is 11.3%; experimental results for the 20 measurement set bear this out for all seven frequencies involved in this analysis. Halving the depth interval and changing z_o, the center of the region of interest in the sample, has the predicted effect on the uncertainty.

From the above analysis and Equations 21 and 23, it can be seen that increasing the number of statistically independent samples of echo data, by recording signals from multiple acoustic lines and/or at multiple points along each beam line, reduces the statistical uncertainty of backscatter and attenuation coefficient estimates. However, when the volume of the region of interest is restricted, statistical uncertainties cannot be reduced arbitrarily because of correlations in the data from closely positioned scan lines. Similarly, when data are recorded using an ultrasound scanner, the number of statistically independent acoustic lines generally is lower than the actual number of scan lines due to correlations in the data. Thus, it is necessary to account for such effects in the error analysis.

Suppose there are M samples of a signal, $i(j)$, $j = 1, \ldots, M$. Let $\langle i \rangle$ be the expectation value of $i(j)$ and σ_i the standard deviation. The average value of $i(j)$ is

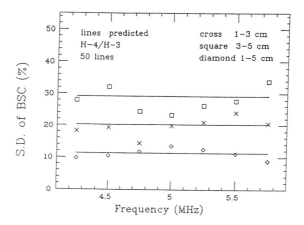

FIGURE 18. Comparison of predicted uncertainties (lines) and measured percent standard deviations in backscatter coefficient determinations as a function of frequency. A 5-MHz center frequency transducer was used. (From Yao et al., *Ultras. Med. Biol.*, 17, 187–194, 1991. With permission.)

$$I = \frac{1}{M} \sum_{j=1}^{M} i(j) \tag{25}$$

The variance of I, σ_I^2, is given by

$$\sigma_I^2 = \frac{\sigma_i^2}{M^2} \sum_{j=1}^{M} \sum_{k=1}^{M} cor_{j,k} \tag{26}$$

where $cor_{j,k}$ is the correlation coefficient between the jth and k^{th} data points. σ_I^2 generally is less than σ_i^2. The number of effective independent samples, *EIS* is equal to the factor by which the variance is reduced due to the averaging, i.e.,

$$EIS = \frac{\sigma_i^2}{\sigma_I^2} \tag{27}$$

$$= \frac{M^2}{\sum_{j=1}^{M} \sum_{k=1}^{M} cor_{j,k}} \tag{28}$$

If the M samples are independent, $cor_{j,k} = 1$ when $j = k$ and zero otherwise, the double summation in Equation 28 will be M and $EIS = M$, as expected. If the samples are partially correlated, *EIS* will be some value between 1 and M.

The number of statistically independent acoustical lines and the number of independent samples along each line can be found by taking into account correlations in the echo data. Correlation lengths can be computed with sufficient knowledge of the transducer geometry, pulse duration and frequency content of the echo signals.[47] In Reference 49 correlations in the data were determined experimentally.

Suppose $I(\omega_o, z)$ in Equation 10 is computed by averaging the filtered echo signal over M_m lines in each of M_n planes. The echo data is used to compute a two-dimensional correlation matrix $cor_{j,k}$, where each entry in this matrix is the correlation coefficient between points j m-intervals and k n-intervals apart, m referring to the acoustic beam line and n the scanning plane. Then Equation 28 can be solved to estimate the number of effective independent acoustic lines in this data. Similarly, a one-dimensional correlation array, $cor(r)$ may be formed using the correlation coefficients between data points r pixels apart in the beam direction. Again, Equation 28 is applied to determine the number of effective independent samples along this direction.

Details of this method, along with results for a limited number of test cases, may be found in References 43 and 49.

V. SUMMARY

In this chapter we have outlined methods for constructing quantitative ultrasound images related to the frequency-dependent backscatter coefficient of tissue. The methods use accurate data reduction techniques originally developed for measuring ultrasonic backscatter coefficients of small samples. An ''absolute'' method, based on a reference reflector and computed beam profiles, and maintaining the temporal nature of the measurement process in the analysis, yielded accurate results in phantoms and was used in initial studies of backscatter coefficient imaging in humans. This method requires precise calibrations to account for the pulser-receiver frequency content, transducer geometry, and time varied gain used in the data acquisition. Such calibrations could be incorporated in advanced, digitally controlled ultrasound imagers.

With present ultrasound equipment a reference phantom method of data reduction was found convenient to implement. The latter method yielded attenuation coefficients for *in vivo* normal liver that were in agreement with results of previous workers. New quantitative images of *in vivo* backscatter estimators at individual frequencies over the bandwidth of the pulse were presented. Backscatter coefficients obtained from data in these images are in the range of those obtained for *in vitro* liver by previous workers. They also agree with *in vivo* measurements at a single frequency by O'Donnell and Reilly.[24] The accuracy of the reference phantom method depends on the accuracy to which the attenuation and backscatter coefficients of the reference phantom are known at the analysis frequencies. Thus, phantoms used in this application must be provided with accurate calibrations by absolute measurement methods.

Future work will explore the potential of these processing methods to differentiate normal from abnormal tissue and thus add to the diagnostic capability of medical ultrasound. Additional work also is necessary to refine methods for accounting for losses in the patient body wall, still a potentially high source of uncertainty of these measurements. Uncertainties ultimately depend on statistical fluctuations in echo signal amplitudes, the effects on derived acoustic parameters decreasing with the number of statistically independent estimates. If diagnostic capabilities are proven useful, the methods described in this chapter could be adopted to real time echo data acquisition and display.

ACKNOWLEDGMENTS

The authors are grateful to Ernst Madsen, Michael Insana, and Timothy Hall for their significant contributions in the development of the methods described in this chapter and to Kathy McSherry for assistance in preparation of the manuscript. This work was supported in part by grants R01CA39224 and R01CA25634 from the National Institutes of Health.

REFERENCES

1. **Holm, H. H. et al.,** *Abdominal Ultrasound Static and Dynamic Scanning,* 2nd ed., Munksgard, Copenhagen, 1980, 96.
2. **Garra, B. S., Insana, M. F., Shawker, T. H., et al.,** Quantitative ultrasonic detection and classification of diffuse liver disease; comparison with human performance, *Invest. Radiol.,* 24, 196, 1989.
3. **Garra, B. S., Insana, M. F., Shawker, T. H., and Russell, M. A.,** Quantitative estimation of liver attenuation and echogenicity: normal state versus diffuse liver disease, *Radiology,* 162, 61, 1987.
4. **Sandford, N. L., Walsh, P., Matis, C., et al.,** Is ultrasonography useful in the assessment of diffuse parenchymal disease?, *Gastroenterology,* 89, 186, 1985.
5. **Lamminen, A., Jaaskelainen, J., Rapola, J., and Suramo, I.,** High frequency ultrasonography of skeletal muscle in children with neuromuscular disease, *J. Ultras. Med.,* 7, 505, 1988.
6. **Helguera, M., Mottley, J. G., Pandya, S., et al.,** Quantitative measures of backscatter from human skeletal muscle: changes with Duchenne's muscular dystrophy, program and abstracts for the 16th Int. Symp. on Ultrasonic Imaging and Tissue Characterization, *Ultras. Imag.,* 13, 1991, 190.
7. **Sommer, F. G., Stern, R., and Chen, H.,** Cirrhosis: US image with narrow band filtering, *Radiology,* 162, 425, 1987.
8. **D'Astous, F. and Foster, F.,** Frequency dependence of ultrasound attenuation and backscatter in breast tissue, *Ultras. Med. Biol.,* 12, 795, 1986.
9. **Landini, L., Mazzarisi, A., Salvadori, M., and Benassi, A.,** On-line evaluation of ultrasonic integrated backscatter, *J. Biomed. Eng.,* 7, 301, 1985.

10. **Kuc, R.**, Clinical application of an ultrasound attenuation coefficient estimation technique for liver pathology characterization, *IEEE Trans. Biomed. Eng.*, BME-27, 312, 1980.
11. **Lizzi, F. L., Greenebaum, M., Feleppa, E. J., and Elbaum, M.**, Theoretical framework for spectrum analysis in ultrasonic tissue characterization, *J. Acoust. Soc. Am.*, 73, 1366, 1983.
12. **Lizzi, F. O., King, D. L., Rorke, M. C., et al.**, Comparison of theoretical scattering results and ultrasonic data from clinical liver examinations, *Ultras. Med. Biol.*, 14, 377, 1988.
13. **Meyer, C., Herron, D., Carson, P., Banjavic, R., Thieme, G., Bookstein, F., and Johnson, M.**, Estimation of ultrasonic attenuation and mean backscatter size via digital signal processing, *Ultras. Imag.*, 6, 13, 1984.
14. **Insana, M., Wagner, R., and Hall, T.**, Describing small scale inhomogeneities from ultrasonic backscatter results, *J. Acoust. Soc. Am.*, 87, 179, 1990.
15. **Sehgal, C. and Greenleaf, J.**, Scattering of ultrasound by tissues, *Ultras. Imag.*, 6, 60, 1984.
16. **Campbell, J. A. and Waag, R. C.**, Measurements of calf liver ultrasonic differential and total scattering cross-sections, *J. Acoust. Soc. Am.*, 75, 603, 1984.
17. **Shung, K. K. and Reid, J. M.**, Ultrasonic scattering from tissues, *Proc. IEEE Ultrasonics Symp.*, IEEE, Piscataway, NJ, 1977.
18. **Nicholas, D.**, Orientation and frequency dependence of backscattered energy and its clinical application, in *Recent Advances in Ultrasound in Medicine*, White, D. N., Ed., Research Studies Press, New York, 1977.
19. **Bamber, J. C. and Hill, C. R.**, Acoustic properties of normal and cancerous liver. I. Dependence of pathological conditions, *Ultras. Med. Biol.*, 7, 121, 1981.
20. **Nicholas, D., Hill, C. R., and Nassiri, D. K.**, Evaluation of backscattering coefficients for excised human tissues: principles and techniques, *Ultras. Med. Biol.*, 8, 7, 1982.
21. **Campbell, J. A. and Waag, R. C.**, Normalization of ultrasonic scattering measurements to obtain average differential scattering cross-sections for tissues, *J. Acoust. Soc. Am.*, 74, 393, 1983.
22. **Fei, D. and Shung, K.**, Ultrasonic backscatter from mammalian tissues, *J. Acoust. Soc. Am.*, 78, 871, 1985.
23. **O'Donnell, M.**, Quantitative volume backscatter imaging, *IEEE Trans. Sonics Ultras.*, SU-30, 1983, 26.
24. **O'Donnell, M. and Reilly, H. F.**, Clinical evaluation of the B'-scan, *IEEE Trans. Sonics Ultras.*, SU-32, 1985, 450.
25. **Boote, E.**, Quantitative Ultrasound Imaging Using Acoustic Backscatter Coefficients, Ph.D. thesis, University of Wisconsin, Madison, 1988.
26. **Boote, E., Zagzebski, J., Madsen, E., and Hall, T.**, Instrument independent backscatter coefficient imaging, *Ultras. Imag.*, 10, 121, 1988.
27. **Boote, E., Hall, T., Madsen, E., and Zagzebski, J.**, Improved resolution backscatter coefficient imaging, *Ultras. Imag.*, 13, 347, 1991.
28. **Boote, E., Zagzebski, J., and Madsen, E.**, Backscatter coefficient imaging using a clinical scanner: tests and preliminary in vivo results, *Med. Phys.*, 19, 1145, 1992.
29. **Madsen, E., Insana, M., and Zagzebski, J.**, Method of data reduction for accurate determination of ultrasonic backscatter coefficients, *J. Acoust. Soc. Am.*, 76, 913, 1984.
30. **Insana, M. F., Madsen, E. L., Hall, T. J., and Zagzebski, J. A.**, Tests of the accuracy of a method of data reduction for determining backscatter coefficients, *J. Acoust. Soc. Am.*, 79, 1230, 1986.
31. **Hall, T., Madsen, E., Zagzebski, J., and Boote, E.**, Accurate depth-independent measurement of acoustic backscatter coefficients with focused transducers, *J. Acoust. Soc. Am.*, 95, 2410, 1989.
32. **Madsen, E. L., Goodsitt, M. M., and Zagzebski, J. A.**, Continuous waves generated by focused radiators, *J. Acoust. Soc. Am.*, 70, 1508, 1981.

33. **Goodsitt, M. M., Madsen, E. L., and Zagzebski, J. A.,** Field patterns of pulsed focused transducers, *J. Acoust. Soc. Am.,* 71, 318, 1982.

34. **Madsen, E. L., Zagzebski, J. A., and Frank, G. A.,** Oil-in-gelatin dispersions for use as ultrasonically tissue-mimicking materials, *Ultras. Med. Biol.,* 8, 277, 1982.

35. **Faran, J. J., Jr.,** Sound scattering by solid cylinders and spheres, *J. Acoust. Soc. Am.,* 23, 405, 1951.

36. **Hall, T.,** Experimental Methods for Accurate Determination of Acoustic Backscatter Coefficients, Ph.D. thesis, University of Wisconsin, Madison, 1988.

37. **Madsen, E. L.,** Method of determination of acoustic backscatter and attenuation coefficients independent of depth and instrumentation, Chapter 7, this volume.

38. **Goss, S. A., Johnston, R. L., and Dunn, F.,** Comprehensive compilation of empirical ultrasonic properties of mammalian tissues, *J. Acoust. Soc. Am.,* 64, 423, 1978.

39. **Chen, C. F., Robinson, D. E., et al.,** Clinical sound speed measurements in liver and spleen in vivo, *Ultras. Imag.,* 9, 221, 1987.

40. **Hall, T. and Insana, M.,** High speed quantitative imaging over extended fields of view, *Proc. 1990 IEEE Ultrasonics Symp.,* IEEE, Piscataway, NJ.

41. **Yao, L., Zagzebski, J., and Madsen, E.,** Backscatter coefficient measurements using a reference phantom to extract depth-dependent instrumentation factors, *Ultras. Imag.,* 12, 58, 1990.

42. **Harris, F.,** On the use of windows for harmonic analysis with the discrete Fourier transform, *Proc. IEEE,* 66, 1978, 51.

43. **Yao, L.,** Reference Phantom Method for Acoustic Backscatter and Attenuation Coefficient Measurements, Ph.D. thesis, University of Wisconsin, Madison, 1990.

44. **Madsen, E. L., Zagzebski, J. A., and Frank, G.,** Ultrasound focal lesion detectability phantoms, *Med. Phys.,* 18, 1171, 1991.

45. **Madsen, E. L., Zagzebski, J. A., Insana, M., Burke, T., and Frank, G.,** Ultrasonically tissue-mimicking liver including the frequency dependence of backscatter, *Med. Phys.,* 9, 703, 1982.

46. **Yao, L., Zagzebski, J., and Madsen, E.,** Statistical uncertainty in ultrasonic backscatter and attenuation coefficients determined with a reference phantom, *Ultras. Med. Biol.,* 17, 197, 1991.

47. **Wagner, R., Smith, S., Sandrik, J., and Lopez, H.,** Statistics of speckle in ultrasound B-scans, *IEEE Trans. Sonics Ultras.,* 30, 1983, 156.

48. **Bevington, P. B.,** *Data Reduction and Error Analysis for the Physical Sciences,* Mc-Graw Hill, New York, 1969, 117.

49. **Zagzebski, J. and Yao, L.,** Statistical uncertainties in acoustic backscatter and attenuation estimates: effects of correlated data, *IEEE Trans. Ultras. Ferroelectr. Frequency Control,* in press.

INDEX